Finding the Secret Space Programme

Removing Truth's Protective Layers

Andrew Johnson

Book details:

Finding the Secret Space Programme: Removing Truth's Protective Layers
by Andrew Johnson
Self-Published – May/June 2018
ISBN-13: 978-1981283705
ISBN-10: 1981283706
BISAC: Technology & Engineering / General

Cover Design by Andrew Johnson (based on a design from www.canva.com)
Rear Photo by Andrew Johnson

No Wikipedia links included!
(Wikipedia censors important information in real-time – use it at your peril!)

The dark ages still reign over all humanity, and the depth and persistence of this domination are only now becoming clear. This Dark Ages prison has no steel bars, chains, or locks. Instead, it is locked by misorientation and built of misinformation. Caught up in a plethora of conditioned reflexes and driven by the human ego, both warden and prisoner attempt meagrely to compete with God. All are intractably sceptical of what they do not understand.

We are powerfully imprisoned in these Dark Ages simply by the terms in which we have been conditioned to think.

-- From Cosmography by R. Buckminster Fuller, 1983

Dedications

I would like firstly, to dedicate this work to Edgar Fouché who passed on in May 2017. Ed made significant personal sacrifices in his revelation of important information. I hope that what I write in this book is inline with his own wishes and actions relating to the disclosures about world-changing technologies that he had knowledge of.

Additionally, I would like, to dedicate this work to Omar Fowler of Derby, UK, who also passed on in 2017. He performed extensive field research, over several decades. He ran a small, local group in Derby for many years and distributed a regular magazine called "OVNI" to many different countries and did so with a considerable investment of time and his own money. Omar also supported my research efforts on a number of occasions and invited me to present them at his Derby group, on a regular basis.

Acknowledgements

Thanks to everyone that has helped in developing this research and to those who make the information, herein, freely available.

Special Acknowledgements

Richard D Hall Edgar Fouché

Important Researchers Quoted:

John Hutchison

Dr Paul A. La Violette

Nick Cook

Michael Schratt

Tom Valone

Ralph Rene

Bill Kaysing

David Percy and Mary Bennett

Jarrah White

Jim Collier

Bart Sibrel

Richard C Hoagland

Mark McCandlish

Other Books by Andrew Johnson

These are available as PDF downloads, as well as other formats:

9/11 Finding the Truth (ISBN: 978-1548827618)

What really happened on 9/11? What can the evidence tell us? Who is covering up the evidence, and why are they covering it up? This book attempts to give some answers to these questions and has been written by someone who has become deeply involved in research into what happened on 9/11. A study of the available evidence will challenge you and much of what you assumed to be true.

9/11 Holding the Truth (ISBN: 978-1979875981)

The truth about what happened to the World Trade Centre on 11 September 2001 was discovered by Dr Judy Wood, through careful research between 2001 and 2008. The author of this book, Andrew Johnson, had a "good view" of how later parts of Dr Wood's research "came together." Not only that, he was also involved in activities, correspondence and research which illustrated that this truth was being deliberately covered up. This book is a companion and follow-up volume to "9/11 Finding the Truth" – and documents ongoing (and successful) efforts to keep the truth out of the reach of most of the population. Evidence in this book, gathered over a period of 12 years, shows that the cover up is "micro-managed," internationally and even globally. The book names people who are involved in the cover up. It illustrates how they often stick to "talking points" and seem to have certain patterns of behaviour. It attempts to illustrate how difficult it is to prevent the truth from being marginalised, attacked and "muddled up."

Climate Change and Global Warming… Exposed: Hidden Evidence, Disguised Plans (ISBN: 978-1976209840)

This book collects together, for the first time anywhere, a range of diverse data which proves that the whole issue of "climate change" is more complicated and challenging than almost all researchers are willing to consider, examine, or entertain. For example, this book contains astronomical data which most climatologists will not discuss in full. Similarly, the book contains climate and weather data that astronomers will not discuss. The book contains some data that neither astronomers nor climatologists will discuss. It contains some data that no scientists will appropriately discuss. It will show the reader why the climate change/global warming scam was invented, and it will illustrate how the scam has been implemented.

Secrets in the Solar System: Gatekeepers on Earth (ISBN: 978-1981117550)

Since 1957, robotic space probes have visited all the planets in the Solar System. Is it the case that they have only found mostly uninteresting collections of gas, rocks, ice and dust? Has any evidence of past or present life in the Solar System ever been discovered?

This book will take you "on a journey" to the Sun and the Moon, Mars, Phobos, Saturn and some of its moons. It will show you some of the numerous anomalies that have been found. Could it be the case that taxpayer-funded space agencies have ignored or even lied about these anomalies, and their significance?

After more than 10 years of ongoing research, collected together here, for the first time anywhere, are over 350 fully-referenced pictures and data from over 50 years of space missions. Anomalous images are presented with some detailed explanations, commentary and analysis, completed by various researchers. The book asks what would happen if NASA or ESA scientists had discovered compelling evidence of past or present extra-terrestrial life in the Solar System? Would they "tell us the truth, the whole truth and nothing but the truth" about such a discovery? Or, would the "scientific technological elite" mentioned in Eisenhower's final address to the USA in 1961, become the "gatekeepers" of "Secrets in the Solar System?"

Author Biography

Andrew Johnson grew up in Yorkshire and graduated from Lancaster University in 1986 with a degree in Computer Science and Physics. He worked in Software Engineering and Software Development, for about 20 years. He has also worked full and part time in lecturing and tutoring and he now works for the Open University (part time) tutoring students, whilst occasionally working on various small software development projects.

He became interested in "alternative knowledge" in 2003, soon after discovering Dr Steven Greer's Disclosure Project. Andrew has given presentations and written and posted many articles on various websites about 9/11, Mars, Chemtrails and Antigravity research, whilst also challenging some of the authorities to address some of the data he and others have collected.

Andrew is married and has two children. You can email Andrew Johnson on ad.johnson@ntlworld.com.

His website is www.checktheevidence.com.

Table of Contents

1. Preface

Why this Book Exists

I sometimes stop to wonder how I can possibly have ended up writing a book like this – and how I can have come to know things that I know. Similarly, I am almost baffled as to how I can have come into contact with the people that I have come into contact with. 20-25 years ago, I was tutoring and lecturing 16-18-year-old pupils in Computing and Maths at a College of Further Education in the East Midlands region of the UK. 10 years ago, I was working from home – still tutoring in a Computing subject, doing some Software Development and also assessing students with disabilities. I had, by that time, started to uncover a different reality – and had realised that there must be a connection between the UFO/ET phenomenon and what has been called "Free Energy" technology.

In 2003, I started to find out more about things that have been hidden from most of the population, using a combination of methods, motives and excuses. When I started this journey of personal discovery, I was investigating a topic which is perhaps more closely related to the subject matter in this book than the others I have compiled. Dr Steven Greer's "Disclosure Project" – which brought forward a compelling body of powerful witness testimony, from many highly qualified and highly trained observers. This body of testimony was also, in many cases, accompanied by important pieces of documentary evidence, detailed in Steven Greer's "Disclosure" book and elsewhere. Since then, I have been immersed in a process of following a number of significant "disclosure threads" and have found they are part of a much larger tapestry of reality than most want to acknowledge. I have found that, surprisingly, even people such as Dr Greer himself do not want to acknowledge that parts of this tapestry are connected to other parts. I have written and spoken, fairly extensively, on the connections between the "free energy" issue and the events of 9/11[1]. This discovery, for me, happened because of the research of Dr Judy Wood into the destruction of the WTC Towers on 9/11[2].

Following my discovery of the Disclosure Project[3] in 2003, I started to research into topics related to Antigravity Technology and theories – and I avidly read Nick Cook's book, The Hunt For Zero Point[4] and watched a number of videos online. Around this time, I began to give presentations to small audiences. I invested in a second-hand LCD projector, which I used in conjunction with a laptop to show the Disclosure Project press conference at a few small venues. In 2004, I compiled some of the additional information I had come across into a PowerPoint presentation which I then narrated, using a desktop microphone, to create a video – called "The Case for Antigravity"[5]. In the video, I included information about current thinking about gravity and also historical research by figures like Thomas Townsend Brown. I uploaded this to Google Video – which had just come into service then. In the years following, I compiled two similar presentations called "Apollo – Removing Truth's Protective Layers," showing the evidence for the Apollo Hoax and "Finding the Secret Space Programme" which showed and discussed further evidence that there has been a secret space programme in operation since, perhaps, the 1960s.

So, I have now expanded re-packaged and updated the information from these presentations into this book.

Interconnectedness

You will likely find that this volume includes much information discussed by other researchers such as Dr Paul A La Violette, whose important work "Secrets of Antigravity" was published in 2008. Similarly, the research of Nick Cook, published in his afore-mentioned excellent book "The Hunt for Zero Point" (year 2000), is also very important.

In this book, I hope to connect together elements of their research with that of others such as Bill Kaysing, Ralph Rene and others – and show the possible reasons for "things being like they are now."

The 9/11 Factor

In two of my other books, "9/11 Finding the Truth" and "9/11 Holding the Truth," I describe how the cover up of 9/11 has been operating since at least 2007, when I realised that a technology or several technologies were being hidden and someone was desperate to prevent any revelations of these things. At this point, I would suggest the reader considers the relationship, in "military terms," between technologies for weapons, energy production and propulsion. Many researchers, I would argue, do not consider these relationships carefully enough. As I mentioned in "9/11 Holding the Truth," there seems to be a direct connection between the UFO/ET cover up and the events of 9/11. That is, some of the same groups that have access to classified information about the UFO/ET issue were involved in the events of 9/11 (or the aftermath of it, at the very least.)

The UFO/ET Dimension

For some people, when researching and discussing the topic of the existence of a Secret Space programme, they are reluctant to "embrace" the UFO/ET topic and related evidence. The UFO research field is indeed highly complex and the use of "normal scientific" and analytical methods in studying data and drawing conclusions are only useful up to a point. Some of the material needing analysis can be considered highly subjective. It is true that an enormous amount of disinformation has been circulated – especially in relation to certain cases like the Travis Walton case and the Roswell case. My own conclusions regarding the Roswell case are as follows.

- A flying saucer, almost certainly of extraterrestrial origin, crashed near Roswell, New Mexico in July 1947. The cause of the crash was either due to a lightning strike or electrical storm, or the craft being targeted with high power radar beams.

- Material from this crash was recovered by the U.S. Army.
- Bodies were recovered and one or more of the occupants survived the crash for a period of time.
- Though many witnesses accounts corroborate, an enormous amount of disinformation has been promulgated and even "fake" witnesses have come forward.

By its very nature, the UFO/ET field is one which is rife with what some have called "high strangeness," and it is often difficult to take accounts and evidence at "face value."

Speaking for myself, I have little doubt that certain well-documented cases strongly suggest or even prove that extraterrestrial beings have been interacting with humans for 1000s of years and they still do so on a daily basis, leaving various kinds of evidence. There are many books about this, of course, a few of which I will reference later in this work. For the moment, I can at least recommend Travis Walton's book "Fire in the Sky"[6] and Dr Steven Greer's Disclosure Project Executive Briefing Document.[7] A number of books from the 1950s such as Frank Scully's "Behind the Flying Saucers" are now available free online[8] and a study of these can reveal important patterns regarding evidence and how it is managed.

One of the goals of this book is to "join up" areas of UFO research in the hope that this will help the overall "disclosure and awakening process," which, it can be argued has been underway for decades.

Style and Structure of This Book

Some parts of this book are technical in nature and so may be difficult to digest for the non-technical reader. I felt it was necessary to include some of the technical details so that technically-minded readers will be able to study the referenced material in a more informed way. There is plenty of non-technical material, however, so I hope the non-technical reader finds the balance to be acceptable.

This book is divided into four parts. In the first part, we will look at what appears to be the main reason for the secrecy of a separate space programme. The reason is related to the reality of working "free" energy (fuel-less) technology, which does not rely on environmental processes (such as the wind or sunlight). Another reason for the secrecy is the inherent and fundamentally transformative nature of working antigravity or gravity control technology. I wanted to organise the material in this book chronologically, but there is quite a bit of overlap in the time line relating to different sections of material and some material could be placed in more than one section of the book.

In the second part, we will look at the Apollo programme – which is still seen as the most advanced manned space programme in history – even though it ended in 1972 – over 45 years before the compilation of this work.

In the third, we will examine the evidence that may have leaked out, which shows that advanced propulsion technologies have indeed been developed. We will consider what specifications these technologies might have, and what principles of operation they may rely on.

In the fourth and final part, we will look at examples of how "truth's protective layers" are kept in place. By doing this, I hope you may be able to "see through" those layers more easily, every time they are applied.

A Note About Internet Forums

If you were to choose a topic mentioned in this book or one of the researchers I mention or refer to, such as Ralph Rene or Richard C Hoagland, you will often be able to find an internet forum, blog or web posting stating that these researchers are mistaken or have been "debunked." These forum postings are often themselves unreliable – especially when they are made by anonymous people who don't have a website or book of their own. Often, posters will not disclose their identity, their professional or academic background - they won't disclose their email or postal address and if you ask them for these things, they will state this is not needed to make their point or they will make some other excuse for keeping their identity a secret. Of course, there are a number of valid reasons for remaining anonymous, but most honest people will allow you to communicate freely with them and won't get aggressive or adversarial when you ask them for more details about themselves. So, if you engage in discussions on any of these forums, "sound the posters out" to see how much they will reveal about themselves, so you can gain a better idea of what their motivation is.

If one studies something called JTRIG, it can be argued that many of these anonymous posters are using tactics similar to those described in a leaked document called "The Art of Deception: Training for a New Generation of Online Covert Operations."[9] Most internet forum users don't use their real name. Instead they use some witty "handle" or pseudonym. Again, some people do this for "fun," whilst others, it seems, do this only to hide their true identity (because, secretly, they are ashamed of what they are doing – or they are making a living protecting secrets, often massaging the truth or just plain lying in the process.)

A whole other set of people do believe the Apollo missions landed on the Moon and claim to have debunked all of the evidence that shows the landings were not possible in the ways claimed by NASA and other agencies. Many of these are just well-meaning level-headed people who don't (at all) match the characteristics I described above. They may be astronomers – professional or amateur – or scientists – with a true passion for their subject. I have encountered a number of these people myself over the years. Indeed, I used to be a committee member of a local Astronomical Society. If I were to try to discuss or present some of the evidence contained in this book with them, they just "can't go there." They may think that if what I was saying in this book were true, it would have such vast implications that, surely, they would already know about it…?

Perception Management

Topics such as 9/11 research and Apollo hoax research seem to have the most obvious ongoing "perception management" activities/operations surrounding them. That is to say, the subject of my earlier book "Secrets in the Solar System" does not seem to attract the same amount of "troll trash" as the two subjects I've just mentioned. I think this is because, as we shall see, knowledge of the reasons for the Apollo Hoax and knowledge of what really happened on 9/11 only lead one to conclude that some group that is running the planet have their own technology and their own science which is "classified," as far as the rest of us are concerned. This book is an attempt to illustrate that what I have written above is true – and everyone who is interested in the future (and the past) should be concerned about this.

Since approximately 2015, the Perception Management strategy has changed slightly and a new (and successful) Psychological Operation was "rolled out," effectively hampering the progress of researchers and thinkers who now know the Apollo craft cannot have landed on the moon. This is/was a heavily resourced promotion of the belief that the earth was, in fact, flat. Prime among the promoters of this observable falsehood was Eric Dubay. I was directly targeted with this operation and therefore wrote about this on my website.[10] It is an operation which relies on people not knowing the value of direct observation and, instead, being influenced by numerous YouTube videos claiming to prove a lie – while ignoring such things as obvious as the presence of day and night and the geometry involved with long distance navigation around our planet. Hapless but well-meaning and concerned "truth seekers" have been suckered-in - doubt is inserted into their minds ("well, I don't know, do I???") and then, they cannot seem to rid themselves of this doubt. This is because they don't know or haven't considered how to make observations and measurements for themselves – observations which prove they have been lied to. Much energy is then wasted on utterly pointless debates – as pointless as a debate about whether you are reading this text – or not.

The reason for this perception management being in operation is to try to make sure that the majority of people regard discussion of topics such as the Apollo Hoax and Antigravity research as being "ridiculous or silly." Revealing the truth about these topics, as I am attempting to do here, could transform the world we live in and neutralise the power that the "secret-keepers" have over us.

Part 1

The Case for Antigravity

2. Gravity and Antigravity

In the first part of this book, the biggest idea that we will concern ourselves with is the idea of antigravity. I will therefore start by giving simple definitions of gravity and antigravity and cover a little of what conventional/mainstream science has to say about gravity. Hence, we will ask:

- What is Gravity?
- What is Antigravity?
- What evidence is there that technology, other than "conventional" aerospace technology exists which could be thought of as being "Antigravity" or something attempting to control Gravity?

Considering Gravity

Newtonian theory is based around this definition:

> *"gravity, Force of attraction that arises between objects by virtue of their masses."*
>
> *(Chambers English Dictionary – CD ROM edition 1996)*

The force of gravity obeys an "inverse square law" – at twice a given distance, the force is only 1 quarter as strong. It is said to be a "long range force," yet is very weak compared to the electromagnetic force. When bodies are in close proximity, the electromagnetic force is typically trillions of trillions times stronger than the gravitational force. A typical demonstration of this involves simply using a small bar magnet to pick up a small object such as a key or a paper clip. It can then be seen that the entire gravitational pull of the whole of the earth can be overcome by the strength of the small bar magnet.

For decades, physicists have attempted to explain what causes gravity and in mainstream thinking, there are three areas of research which are currently said to be "valid."

In the early 1900s, Einstein's work on Relativity is said to have given scientists a new perspective on gravity. Einstein argued that gravity wasn't a real force, but it was something that arose due to the properties of space-time. One might say that "space-time" was the "fabric" from which our 3D universe is made. At the time Einstein did his original research, there was still talk of an "aether" through which electromagnetic waves travel. Talk of this aether gradually declined as many physicists thought that the Michelson Morley experiment[11], performed in 1887, proved there was no aether. Most science and astronomy programmes broadcast on television and elsewhere will repeat the theories of Einstein and perhaps also note that small irregularities in the observed orbit of Mercury represented an early form of proof that Einstein's relativity theories were correct.[12] That is to say, these orbital irregularities could not be explained by Newtonian Physics and equations. Additionally, observations were made that suggested gravity can bend the path of light – as is apparently seen in gravitational lensing.[13]

In Einstein's thinking, gravity isn't a force – attraction between bodies in space happens because the mass that these bodies are made up of bends space-time. Imagine a rubber or elastic sheet – like that on a circular trampoline. If there is a ball at the edge of the trampoline and someone presses the middle of the trampoline down, the ball will travel towards the centre. If the ball is rolling around the circumference of the trampoline when the middle is pressed, the ball will spiral in towards the centre. The argument goes that if there was no friction between the ball and the trampoline, the ball would orbit the centre of the trampoline indefinitely – and hence, this gives an explanation as to why planets orbit stars etc.

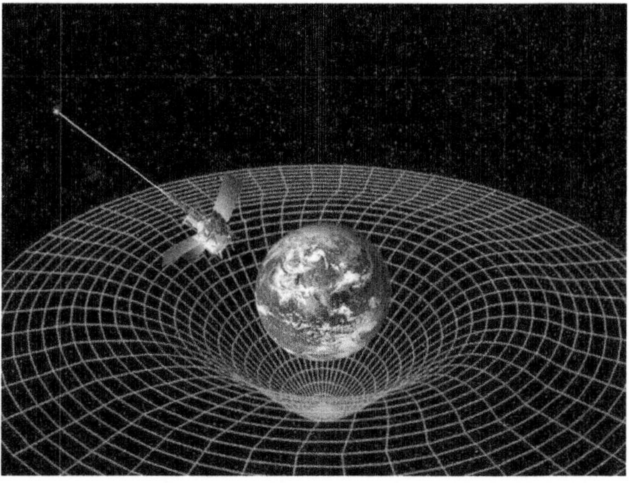

Here, according to Einstein, the mass of the Earth (or any other large body) bends space time around itself. Other masses/matter then take(s) the "path of least resistance" through space time.

Atomic Theory and Quantum Theory

As physicists struggled to understand how matter itself was constructed, theories evolved in the early twentieth century and experiments were used to verify certain parts of the theories. JJ Thomson's "plum pudding" theory[14] of the atom was replaced with Rutherford's "planetary model."[15] This was later replaced with Bohr's theory of the atom.[16]

As physicists did more experiments and discovered more unusual effects, they realised that "strange things" were happening at the level of the atomic nucleus. Attempts to explain these happenings (such as the photoelectric effect) led to the development of quantum theory – often called quantum mechanics[17]. This branch of theoretical physics is mainly concerned with how matter and energy interact at the nuclear and atomic level. Further study and research in this area led to the need for theories about subatomic and sub-nucleic particles to be postulated. Hence, talk of bosons[18] and quarks[19] began to be heard in academia, as the twentieth century passed by. Millions or even billions of pounds have been spent in trying to verify the latest theories about how the various particles interact with each other and create the phenomena we see in the real world. However, only a few practical applications of these theories seem to have been developed.

String Theory

So far in this chapter, we have briefly discussed two branches of physics – the one called "Relativity" is concerned with large scale, long range effects. The second branch, "Quantum Mechanics," is concerned with small scale, short range effects. Physicists realised, after a while, that there were areas where it was difficult to reconcile these two important branches of physics.

Hence, in the 1980s, new theoretical work was developed into something which was called "String Theory."[20] It was an attempt to account for the problems encountered in reconciling quantum mechanics (the physics of the infinitesimally small) with general relativity (the Physics of the cosmologically large). String theory – or another version of the same idea called "m-theory" suggests that all the fundamental particles can be described as being created by "vibrational strings." Different vibration patterns of strings produce different sub-atomic particles. That is, these vibrating strings cause particles to have different charge and mass. String theory contends that the force of gravity is transmitted by gravitons – another class of particle.

There are 5 different formulations of string theory, the latest of which works in 11 dimensions. The mathematics behind this theory are highly complex – and it would seem that only a few people in the world can understand it. Near the end of an article by Professor Brian Green on Britannica.com about string theory[20], the following statement is made:

> *That the pieces fit coherently is impressive, but the larger picture they are filling out—the fundamental principle underlying the theory—remains mysterious. Equally pressing, the theory has yet to be supported by observations and hence remains a totally theoretical construct.*

Hence, at the end of this discussion, we still have little idea of what actually causes the force of gravity. New York University physicist Georgi Dvali, in one discussion of string theory, suggested that if gravitons really were a particle that was transmitting gravity, then it could be the case that a proportion of them "leak" into one or more of the theory's proposed higher dimensions.[21] This would "explain" why gravity was weak in comparison to the other forces of nature. Another suggestion is that gravity may be a force "leaking" from another dimension into our own.

Another theory which attempts to explain the cause of gravity is "loop quantum gravity," which considers more closely the exact nature of Einstein's "space-time" rather than considering the behaviour of forces, or the characteristics of particles.[22]

"It's a Wonderful Theory!"

After this short consideration and summary of decades of research, I can state with confidence that no scientist in any mainstream research establishment has published any peer-reviewed papers which include a method or proposal for using any of the theoretical research done since, say, 1960 to *engineer* gravity in any way, shape or form. Similarly, it seems in the mainstream, there is no demonstrable and practical understanding of what actually causes gravity.

In short, it seems to be "all talk and no action." In a later chapter, we will consider how things were perhaps a little different before 1960. We will also see that someone on this planet knows a whole lot more about "how gravity works" than we do and that this reality is never discussed in the academic community, which is likely because "someone does not want it discussed."

From my understanding of the state of theoretical physics, little of the research has led to tangible, engineerable solutions that are of practical value in daily life. Since the creation of the atomic (fission) bomb and the development of nuclear (fission) power and the laser in the 1960s, there have only been a small number of areas where particle or nuclear physics have been successfully applied to produce new technologies. It could be argued that these include advances in semi-conductor technologies and things like nuclear magnetic resonance imaging (now called MRI). Even these, however, seem to have been developed based on an understanding of the Bohr atom model and closely related research, rather than later theoretical models involving quarks, bosons, hadrons and so on That is to say, the areas of theoretical physics investigated by enormously expensive facilities like the Large Hadron Collider at CERN have contributed little or nothing to creating solutions which help us in our daily lives.

Considering Antigravity

Gravity is one of the main forces which dictates the way we live our lives. It shapes our biology and almost everything related to our way of living. Whilst we rely on it to create certain effects and our bodies are adapted around the effects it creates, it also imposes many restrictions on our freedom of movement. Perhaps we only ever seriously think of how gravity restricts our freedom when we watch birds or other animals that seem to have the ability to "defy" gravity in some way.

As gravity seems to be such a fundamental part of our reality, anyone or thing that could control or overcome gravity has, in our culture, been viewed as being "miraculous" or "god-like."

I often think, now, about how different our world would be if we didn't use ordinary cars/automobiles and how different things would be if there was no environmental destruction and pollution that is part and parcel of the development of overland transportation systems. I now know that this sort of destruction is completely unnecessary – most or all of this destruction is preventable, because working antigravity and free energy technology has been perfected. I have already discussed a number of reasons why this is true in my earlier books, which are listed near the start of this book. Please consider downloading and reading them. As part of the process of connecting the different areas of research, we will revisit some of the same topics and evidence in this volume.

Antigravity Devices

Before we consider anything related to antigravity – such as the creation of devices which seem to defy gravity, I want to give some examples of things which don't fit my definition of an antigravity device. I feel this is necessary, due to past conversations I have had. Consider the following examples:

- A chair is not an antigravity device, simply because it stops you falling down. A chair is simply a rigid structure which resists the force of gravity.

Being even more facetious, neither does your skeleton, nor your muscles form part of an antigravity "system."

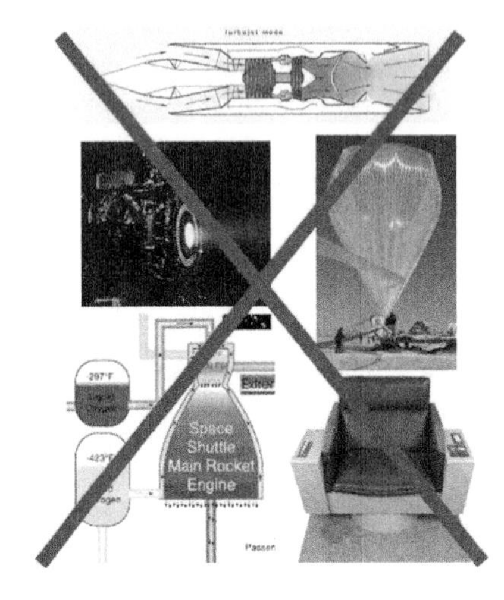

- A balloon is NOT an antigravity device - it floats because of mass displacement (due to Archimedes Principle).
- A propeller/jet/rocket engine is NOT an antigravity device. It works by displacing mass to generate thrust.
- An ion engine is not an antigravity device – it also works by mass displacement but is rather more efficient with propellant mass than chemical rockets.

So, what *is* antigravity or an antigravity device? I would define it as:

- Any device / method capable of changing the local gravitational field strength or having the effect of reducing an object's *weight*, without changing its *mass*. There is no obvious "propellant" or particulate fuel involved.

We should pause for a moment to define the terms "mass" and "weight." Normally, we consider these as being the same thing. However, "mass" in a "physics domain" means "amount of material" and "weight" means the "force that the mass experiences when it is in a gravitational field." Hence, gravity *does not* create mass, but it *does* create weight.

We might also consider that antigravity could be described as "gravity control" or "gravity reduction." Below are photos of what I think are antigravity devices.

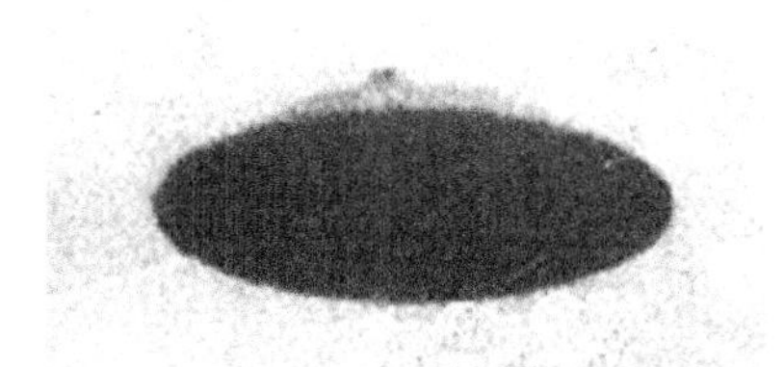

McMinville Saucer (Paul Trent, May 8, 1950)[23]

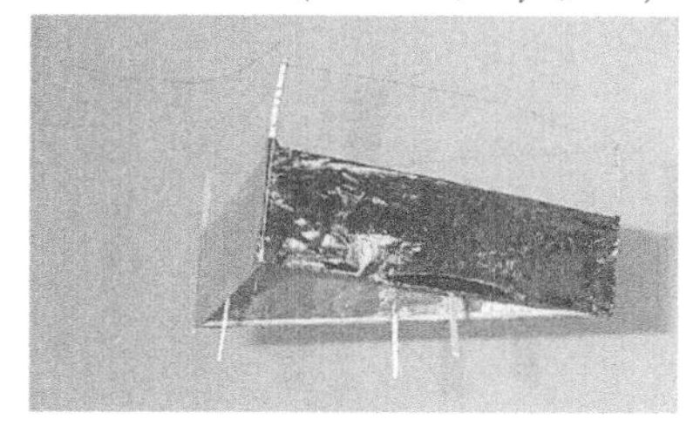

Lifter 1 – JL Naudin Experimental Model[24]

Above: TT Brown's Experimental Gravitator Disk[25]

We will study the "experimental items" later in this book.

Antigravity and "Free Energy"

The consensus view that has now been established tends to say that there is no such thing as "free energy" and there is no way to significantly change the force of gravity with any known technology. As is often the case with knowledge that is world-changing, profound and so forth, the consensus is wrong about said knowledge and this "wrongness" is a big factor in how the knowledge is suppressed.

3. Electrogravitics and Thomas Townsend Brown

Some Useful Sources of Information

There are probably a fair number of researchers that have, over the years, investigated the possibility of creating a device or technique for reducing an object's weight. Most are regarded as cranks, hoaxers or dreamers and those making judgements such as these have not done much research into the history of antigravity research.

Others who *have* concluded that serious and important antigravity research has been conducted in the past – and with success - include Nick Cook (Defence Journalist)[26], Dr Paul A La Violette (physicist and energy consultant)[27] and Joseph P Farrell (author and doctor of Patristics)[28]. Unfortunately, for reasons described in "9/11 Holding the Truth," I find I cannot trust the work of Joseph P Farrell, due to him helping the cover up of what really happened on 9/11. Additionally, his books, I would argue, contain a mixture of information and disinformation. I would also argue that Dr Farrell should, in a number of cases, be able to better distinguish between the two. His books include, among others, "Roswell and the Reich" and "SS Brotherhood of the Bell: The Nazis' Incredible Secret Technology."[29] They may provide some useful information, but I would recommend checking the references he uses more carefully. There are other researchers not included in my list, so I apologise to any of those researchers, if they feel I have wrongfully omitted them here!

In 2000, Nick Cook's very important book, called "The Hunt for Zero Point: Inside the Classified World of Antigravity Technology,"[30] was published. The title speaks for itself and is especially "brave," considering that Cook was a Defence Industry Journalist. I will be continually referring to his book at appropriate points in this work. Also, of significance is the documentary that Cook wrote and developed in 1999 called "Billion Dollar Secret,"[31] where Cook speaks to several important people about Black Programmes (which we will discuss later, too.) Nick Cook became intrigued when someone sent him an article about antigravity research. He tried to follow up a contact, George S. Trimble, whom, he had read, had worked in the field of antigravity research in the 1950s. Trimble initially agreed to be interviewed, but then withdrew suddenly – reportedly "scared". Cook continued his research "quietly" over 10 years and journeyed some way into the "black world" of World War II defence research. His book "The Hunt for Zero Point" is the story of that journey.

In 2008, Dr Paul A La Violette (PLV) published another important book called "The Secrets of Antigravity," which we will also be referring to. I recommend the reader studies both these books, to flesh out some of the material in this book. PLV has a science background and therefore, his insights are very valuable. He has also written several other interesting books, often ignored by other researchers.

Finally, another knowledgeable researcher who has written on the subject of antigravity is Dr Tom Valone[32] – he has compiled two books just on Electrogravitics research, for example.

Characteristics of Antigravity Devices and Experiments

Before we discuss more details about what research has been done, I want to make some general observations about what becomes clear when working through various types of antigravity research. One tends to read about:

- Devices spinning very quickly or having components spinning at high speed.
- Use of very high voltage – electrogravitics.
- Use of very strong magnetic fields – magnetogravitics.
- Use of high frequency oscillations of electric current or magnetic field.

It seems to me that what is basically happening (and I am *not* the first to say this) is that the underlying aether – from which matter and energy seem to appear – can be "stressed." Stressing the aether in particular ways can produce specific effects – which can include an enormous release of energy, or a change in the force of gravity – and other forces too. This is something that Nick Cook spoke about on several occasions, when he was making radio appearances in the early 2000s. We will discuss all of this in more detail in later chapters.

"Mad Science"?

Of course, tales of "mad scientists" doing "weird experiments" and producing all kinds of effects have been reported since the early 1900s, and it is again assumed in the mainstream that if anything significant and real had been discovered, it would, by now have been developed into something advanced. And, with the ability to freely share so much information via the web, the fact that this "has not happened" means there is no reality to any of these stories…

In this chapter, we will look at some of the pioneering work done in the early 1900s.

Nikola Tesla (1856-1943)

Tesla was born in Serbia and emigrated to the US in 1884. Tesla was mainly a pioneer in the fields of Electrostatics and Electricity Generation (with Thomas Edison) and invented the Alternating Current-based electricity distribution system we use today. The unit of Magnetic Field Flux Density is named after Tesla. Though Tesla is a "giant" in the sense of his scientific and particularly his engineering achievements, we will only devote a very small amount of text to his work here, because he is not someone who appeared to have done specific experiments in antigravity.

Nikola Tesla, c. 1890

After his death, most of his notes and papers disappeared. Many stories about him have also been snowballed in myth and legend, in much the same way as those in the UFO field. A good source of information on Tesla is Marc Seifer's book "Wizard: the Life and Times of Nikola Tesla,"[33] which even includes a discussion of Tesla's alleged "death ray" development and the possible link to the Tunguska devastation which occurred in June 1908[34]. The book also mentions the claim that Tesla built and put a "free energy motor" into a Pierce Arrow car, though the book concludes this story is probably untrue.

It has, to my knowledge, never been claimed that Tesla worked on or developed an antigravity device or technique, but his work with electrostatic fields, power generation and transmission overlaps with areas of antigravity research, so it can be said that some of his work paved the way for similar pioneering work in the field of Electrogravitics.

PLV discusses Tesla in more depth in his "Secrets of Antigravity" Book. In chapter 6, he mentions how Tesla had noticed certain "force" effects when he was working on his high voltage transmitters. In a 1980s video of experiments performed by Eric P Dollard[35] (who has been harassed and had his equipment stolen[36]), Dollard seems to describe and demonstrate something similar.[37] This sounds quite similar to effects that another researcher/experimenter, Eugene Podkletnov, investigated in the 1990s.[38]

George Piggot

Again, here, I will only mention Piggot, as he seemed to be a forerunner of the figure we will devote more of the rest of this chapter to.

Piggot, apparently, performed levitation experiments in the early 1900s – and an article by him, about his own experiments, was published in the July 1920 issue of "Electrical Experimenter."[39]

> *A strong electric field was produced by means of a special form of generator and when the metallic object was held within its influence, it drew up to approximately a distance of 1 mm from the Centre of the field, then was repelled backward toward an earthed contact, going within 10 cm of the same when it was again attracted toward the field's Centre but this time getting no nearer than 5 cm from the polar nucleus. This backward and forward movement contained for some time until the metallic object at last came to a comparatively stable position, about 25 cm from the field's Centre where it remained until the power was shut off.*

The article, referenced above, contains further descriptions and illustrations (one of which is reproduced below) of the apparatus he used. It is stated that he was using about 500,000 volts (500 kv) and ¼ kw, which is 250 watts (which could power two and a half electric light bulbs).

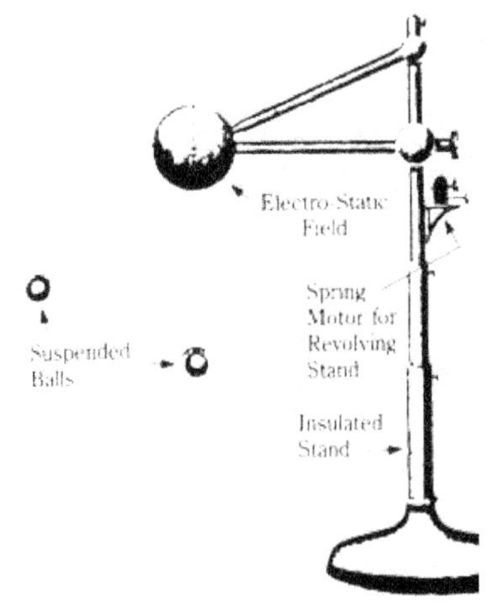

From Piggot's article - "Figure 3 --- A close-up view of the charged metal sphere mounted on a pedestal together with a spring driving motor, whereby the electrode or charged ball could be rotated. The two smaller silver balls are shown as suspended in mid-air, the earth's gravitational pull having been nullified."

Thomas Townsend Brown (1905 – 1985)

Brown is a very significant figure in the field of antigravity research and not surprisingly, PLV's book contains a detailed account of his activities which started in the 1920s. I include a summary of Brown's Electrogravitics work here, but also recommend that the interested reader completes additional research.

Thomas Townsend Brown c. 1950

Thomas Townsend Brown was born in Ohio and though he became an experimental scientist, he never obtained a degree. He spent time studying at several schools including Caltech, Kenyon College and Denison University. He has been called a Physicist but was probably an Electrical Engineer more than anything else. He served in the US Navy in the 1930s at the Naval Research Laboratory. He died in California in 1985. Much of his work still seems to be kept "under wraps" and his family keep a low public profile.

He worked with Dr Paul **Biefeld** (often mis-pronounced or mis-spelled Bifield**)** who was a friend of Albert Einstein and was the Director of Swazey Observatory in Denison, Ohio.

The Biefeld-Brown Effect

This effect was originally discovered when Biefeld and Brown were working with high voltage X ray tubes. It was discovered that they moved excessively when turned on and off.

In 1923, Brown started to experiment with "gravitators" and found an interaction between electric charge and the force of gravity. He investigated the properties and behaviour of (electrical) capacitors and completed research and experiments with the dielectrics used in capacitors (see below).

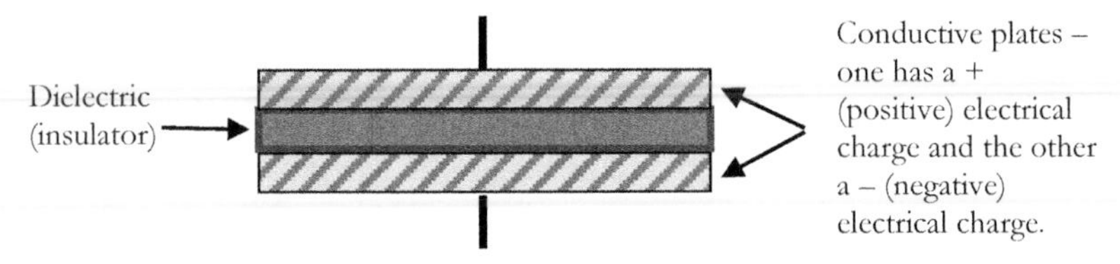

A simple view of a capacitor.

Brown's work was in the field of electrogravitics and essentially involved experimenting with different forms of capacitors – or objects which were acting as capacitors. He would suspend them in an arrangement so that they were free to spin and move around when charged.

This was really the start of a whole new field of research – electrogravitics, a term we mentioned earlier in this chapter. In various experiments, Brown discovered that if a capacitor/gravitator unit was charged, there would be motion in the direction of the charge. There is plenty of additional information online, including a fair collection on the Townsend Brown family website.[40] Another useful resource is a paper by Tom Valone published in the Journal of Geosciences in 2015.[41]

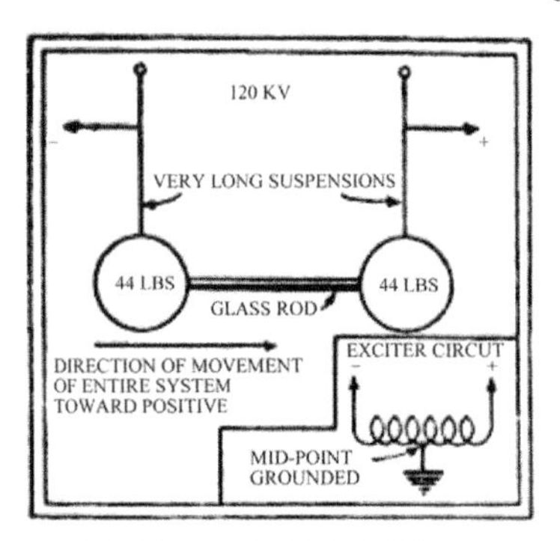

A SIMPLE TYPE OF GRAVITATOR IS SHOWN IN THE ABOVE ILLUSTRATION

PLV, in his "Secrets of Antigravity" book, describes the experiments in considerable detail and even includes some equations and calculus, for those that understand these things!

I will only give an overview of his research here – enough to show you that what Brown did was real, and it was an important effect that he discovered. Others have also reproduced parts of his research, as can also be found online.

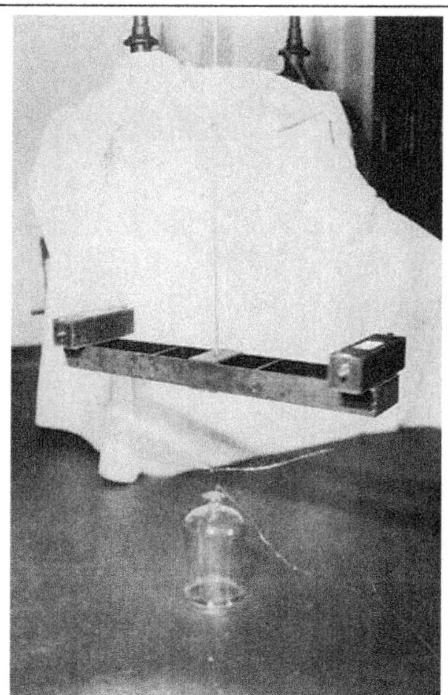

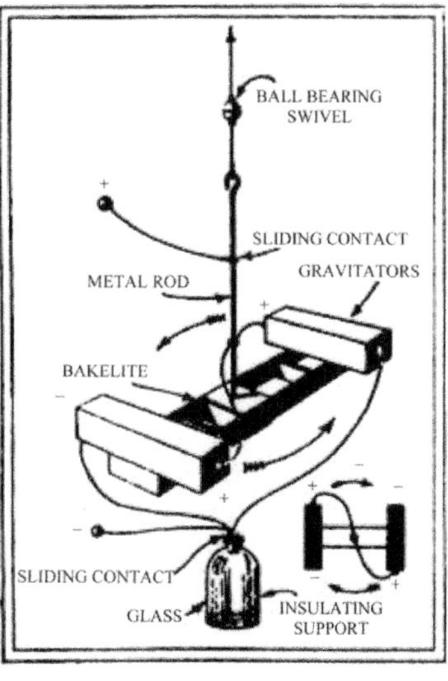

A GRAVITATOR ROTOR IS SIMPLY AN ASSEMBLY OF UNITS SO MADE THAT ROTATION RESULTS UNTIL THE IMPULSE IS EXHAUSTED

One of Brown's gravitator experiments.

PLV also notes how Brown's idea was patented in 1928[42]. Of course, there was resistance from the scientific community, partly because Einstein's relatively new (pun intended) theory of gravity could not explain the effects Brown had created. PLV describes the situation thus:

> *Brown entered the California Institute of Technology in 1922. He spent a good part of his freshman year attempting to win the friendship of his professors and to convince them of his abilities as a first-class "lab man." However, when he began mentioning his ideas about electrogravity, no one would listen. At the end of the year, he had his laboratory equipment shipped from Ohio, set it up in his quarters, and sent invitations to several of his professors, including the renowned Dr Robert Millikan, to witness a demonstration of the new force he had discovered. No one came. Some time later, one of Brown's friends tested Millikan by asking him whether he knew of anyone who had ever found a way of modifying or influencing the force of gravity. Millikan is said to have replied brusquely, "Of course not; such a thing is impossible and out of the question."*

Millikan became a well known name in the Physics world because of his "oil drop" experiment, first performed in 1909, to determine the charge of a single electron.[43]

Brown continued to experiment with various types and forms of "gravitator" – and used, at various times, disks and even a pendulum. From 1931 to 1933, he also worked at the Naval Research Laboratory where he was still investigating effects related to electrogravitics – this time in fluids. PLV also explains how Brown documented changes in the electrogravitic effects he observed in his many experiments "...in a way that related to solar and lunar tides and a sidereal correlation of unknown origin."

It proved difficult to explain the effects, though and according to PLV, Brown's explanations were only written up in his laboratory notes – he did not publish scientific papers, as such.

Subquantum Kinetics

In chapter 1 of his "Secrets" book, PLV writes about how his own theory, called "Subquantum Kinetics," can explain some of the effects that Brown was creating. PLV argues that gravity can attract or repel matter, depending on the arrangement of positive and negative charge on a body. He argues that when bodies on the earth have no net surface charge, they normally attract each other, because the gravitational force generated by "proton matter" is slightly stronger than that generated by "electron matter." Hence, if a body gains enough electrons, if can begin to have a repulsive force of gravity, as well as a negative electrical charge.

The Philadelphia Experiment

Another topic that PLV deals with his Thomas Townsend Brown's alleged involvement in the Philadelphia Experiment. This is something that is alleged to have taken place in 1943 and involved making a Navy Ship – the USS Eldridge – invisible to radar and, allegedly, optically invisible. (Several books discuss this alleged project, but it is extremely difficult to know what actually happened. I would guess something important and interesting happened, which needed to be covered up by circulating disinformation.) In his "Secrets" book, PLV writes:

> *Later in his life, Brown was privately asked by family friend and business associate Josh Reynolds about his involvement in the Philadelphia Experiment. Brown answered that he "was not permitted to talk about that part of his work"; however, he did comment that "much of what has been written about the project is grossly exaggerated." Here, he was probably referring to claims some have made that the ship had been made to travel through time or that it had teleported itself to Norfolk Harbor, where it was alleged to have reappeared for a few minutes before disappearing and reappearing once again in the Philadelphia Navy Yard.*
>
> *Yet the fact that he did not flatly deny his involvement in the project leads one to suspect that the rumors of his involvement are true. Moore, co-author of the book "The Philadelphia Experiment," once asked Brown to edit a rough draft of an article he was writing on Brown's life. Moore had planted a paragraph describing a series of experiments, sponsored by the Navy, that were based on the effects and equipment later associated with the Philadelphia Experiment. He had done this intentionally to see Brown's reaction. Although Brown made other corrections and notes for changes to the manuscript, he allowed the entire test paragraph on the Philadelphia Experiment to remain intact. Thus, we are left to conjecture that tales of the existence of this project may be true and that Brown had somehow been involved in this project, although what his involvement was is open to speculation.*

Also, Brown was working in the Navy on a project related to mine sweeping, so it was not a "big jump" to be working on something like the Philadelphia Experiment. However, information on the Townsend Brown family website[44] is a little more "reserved," it criticises the Moore book thus:

> *Various significant errors in research procedures and reliance on unreliable sources has come to light in more recent years.*

PLV discusses Brown's ongoing work and demonstrations in the 1940s with electrically charged discs. Also mentioned is the possible parallel classified work being done by the Office of Naval Research (ONR) and/or the US Air Force, at the Wright Air Development Centre.

Following his work for the Navy, Brown went to work for the Lockheed Vega Aircraft corporation[45] – funnily enough, a forerunner of the enormously powerful defence contractor and owner of "Skunkworks," Lockheed Martin. We will mention this company again later.

Townsend Brown and NICAP

Separate to his study of Physics, Townsend Brown founded NICAP (National Investigations Committee on Aerial Phenomenon)[46] in 1956, to serve as a clearing-house and "listening post" for information regarding Saucer sightings and evidence. He served for some time as Director of NICAP (as did Major Donald Keyhoe). So, it is clear, then, that Brown had a serious interest in the UFO/Flying Saucer issue – and he probably thought that studying it could yield information that was useful to his own areas of research – i.e. electrogravitics and related effects. We will re-visit Brown's interest in this topic later in this chapter.

Project Winterhaven

It seems that Brown realised the potential of the technology and techniques he was developing and so, by January 1953 he had put together a 66-page proposal which was codenamed Project Winterhaven.[47] This was a wide-ranging proposal – which included research in roughly 30 different areas (some theoretical, some practical), as shown on page 7 of the report.

The proposal even named companies and organisations that would be involved. These included the US Department of Defence (DoD), Lear Inc (presumably owned by Lear Jet), University of Chicago, The Franklin Institute in Pennsylvania and even… the Stanford Research Institute (SRI – which later became a host for research into things like Low Energy Nuclear Reactions – LENR – and Remote Viewing. I also mentioned SRI in my "Secrets in the Solar System" book). The organisational chart shows the Townsend Brown Foundation at the same level of control as the DoD. It appears that the project was, indeed, a serious proposal to develop usable electrogravitic technology. On page 1 we can read:

> *PURPOSES: For the last several years, accumulating evidence along both theoretical and experimental lines has tended to confirm the suspicion that a fundamental interlocking relationship exists between the electrodynamic field and the gravitational field.*
>
> *It Is the purpose of Project WINTERHAVEN to compile and study this evidence and to perform certain critical or definitive experiments which will serve to confirm or deny the relationship, If the results confirm the evidence, it*

> *is the further purpose of Project WINTERHAVEN to examine the physical nature of the basic "electro-gravitic couple" and to foresee and develop possible long-range practical applications.*
>
> *The proposed experiments are to be limited at first to force measurements and wave propagation. They are to be expanded, depending upon results, to include applications in propulsion or motive power, communications and remote control, with emphasis on military applications of recognized priority.*

On page 5, we read:

> *RESEARCH ON THE CONTROL OF GRAVITATION: In further confirmation of the existing hypothesis, experimental demonstrations actually completed in July 1950, together with subsequent confirmations with improved materials, tend to indicate that a new motive force, useful as a prime mover, has in reality been discovered.*

Another area Brown had done research in was communication, which was described on page 10 of the Project Winterhaven proposal thus:

> *ELECTROGRAVITATIONAL COMMUNICATION SYSTEM*
>
> *(Electrogravitic induction between systems of capacitors involving propagation and reception of gravitational waves) Project started at Pearl Harbor in 1950 - Theoretical background examined and preliminary demonstrations witnessed by Electronics Officer and Chief Electronics Engineer at Pearl Harbor Navy Tard, Receiver already constructed - detects cosmic noise which, according to supporting evidence, appears to emanate from that portion of the sky near the constellation Hercules (16h RA, 40° N Decl,), Transmitter designed and now partly completed. Radiation is more penetrating than radio (has been observed to pass readily through steel shielding and more than 15 feet of concrete).*
>
> *In 1952 a short-range transmitting and receiving system was completed and demonstrated in Los Angeles. Transmission of an actual message was obtained between two rooms - a distance of approximately 35 feet. Transmission was easily obtained through what was believed to be adequate electromagnetic shielding, but this test must bear repeating under more rigorous control. See definitive experiments (Group B) hereinafter proposed.*

This sounds quite similar to communications techniques that have been developed based on Tesla's work. Some people erroneously use the term "scalar communications."

On page 23, following earlier discussion of propulsion systems for ocean liners, we find thoughts about developing propulsion for use in space:

> *The second type of electrogravitic reactor now demonstrated in disc airfoils may find its principal field of usefulness in the propulsion of spaceships in various forms. For the moment, at least, the disc form appears to have the greatest promise, largely because there is reason to believe it can be self-levitating and, therefore, made to possess the ability to move vertically (as well as horizontally) and to hover motionless, in complete control of the Earth's gravitational field.*

Nick Cook writes about Townsend Brown's "Project Winterhaven" in his "Hunt for Zero Point," but points out that there isn't any evidence that leads to an obvious conclusion that Project Winterhaven was converted into a "Black Programme" to develop electrogravitic technology. Whilst I agree, in principal with this conclusion, there is other evidence we have yet to cover that leads us on a journey to the Black World quite quickly.

Townsend Brown's Later Work

PLV's "Secrets of Antigravity" book also documents work completed by Brown and Bahnson in the late 1950s and early 1960s. Brown secured several more patents related to his electrogravitics research. A summary of his career is posted on the Townsend Brown family website[48] (some formatting issues seem to be present in the text), along with a number of letters that Brown received and responded to.[49]

Dipping in to the letters reveals some areas that Brown was working in. A letter written in November 1955[50] is particularly interesting. It describes certain events which took place after Brown's work had featured in the newsletter distributed by a London-based Aerospace consultancy group called "Aviation Studies." Below, I have quoted from the letter Brown wrote to an associate named Ed Hull:

> *The same report, incidentally, was read by Douglas, and undoubtedly by many other aircraft manufacturers in the United States, Martin, as you know, is starting a new project and I believe Lockheed will too. According to the grapevine, the studies in gravitation are already under way at both Northrop and North American. According to the New York Herald Tribune, Convair is now interested and so is Bell. Then, also, we must give credit to my old friend Bill Lear for continuing interest in the spreading of the good word.*

Interestingly, Brown then openly discusses the flying saucer phenomenon:

> *Several new flying saucer books are being released here in the United States, and I understand there is one in England too. Donald Keyhoe's new book (The Flying Saucer Conspiracy) has already gone to press and copies should be available soon. I notice also in the last issue of Life magazine, dated December 5th, 1955 what amounts to an admission that the unidentified flying objects, as observed by the Air Force and other observers, are not all explained away by the recent announcement of the Air Force and it gives some drawings in color of the various shapes of U.F.O.'s that have been reported.*
>
> *Of course, to me, this indicates that something very real is going on behind the scenes, and that where there is so much smoke, there certainly must be some fire. I feel sure that these things are electrically driven. It is true, of course, that the halo effects surrounding these craft in flight can come from nuclear radiations, but more than likely they would come from simply electrical corona and this of course means electrical potentials of the order of several million volts.*

Brown then goes on to discuss some of his own experiments and the ones he would like to do, using higher voltage supplies.

The Aviation Studies group published a summary of the ongoing Electrogravitics research in February 1956[51] and this, too, makes for interesting reading, indicating in the final section, on the last page:

> *Aviation Report 15 November 1955*

ELECTRO-GRAVITICS EFFORT WIDENING

Companies studying the implications of gravitics are said, in a new statement, to include Glenn Martin, Convair, Sperry-Rand, Sikorsky, Bell, Lear Inc. and Clark Electronics. Other companies who have previously evinced interest include Lockheed, Douglas and Hiller. The remainder are not disinterested but have not given public support to the new science - which is widening all the time.

Wright Patterson AFB and Electrogravitics

In chapter/section 2.4 of his "Secrets of Antigravity" book, PLV describes how he discovered a document with a title "Report CRG-013/56—Electrogravitics Systems" and a subtitle of "An examination of electrostatic motion, dynamic counterbary and barycentric control." Although PLV found a reference to this 1956 document when he was in the Library of Congress in Washington, DC, it was quite difficult to locate a copy, as the only available one was at… Wright Patterson AFB. He was able to get hold of a copy (but not the original). It has since been scanned and posted on the web.[52]

Dr Tom Valone also wrote about PLV's discovery and summarised the relevance of this document thus:[53]

- It validates T.T. Brown's experiments;
- It lists the major corporations that were collaborating on electrogravitics;
- It includes the requirements for supersonic speed;
- It shows the continuity from Project Winterhaven in 1952;
- The report includes a list of electrostatic patents;
- It had been classified by the Air Force for an undetermined amount of time which underscores its importance.

Other Publications that Mentioned Antigravity Research

It wasn't only industry newsletters that were talking about the research. A number of articles were published in special interest magazines in the 1950s and 1960s. A few examples are shown below, thanks to them being uploaded onto the Scribd website by people interested in the topic.

From the November 1956 issue of "Young Men - The Magazine For Tomorrow's Technicians And Engineers"[54]

The G-Engines Are Coming! By MICHAEL GLADYCH

By far the most potent source of energy is gravity. Using it as power future aircraft will attain the speed of light.

Nuclear-powered aircraft are yet to be built, but there are research projects already under way that will make the super-planes obsolete before they are test-flown. For in the United States and Canada research Centres, scientists, designers and engineers are perfecting a way to control gravity—a force infinitely more powerful than the mighty atom. The result of their labors will be anti-gravity engines working without fuel— weightless airliners and space ships able to travel at 170,000 miles per second. If this seems too fantastic to be true, here is something to consider—the gravity research has been supported by Glenn L. Martin Aircraft Co., Convair, Bell Aircraft, Lear, Inc., Sperry Gyroscope and several other American aircraft manufacturers who would not spend millions of dollars on science fiction. Lawrence D. Bell, the famous builder of the rocket research planes, says, "We're already working with nuclear fuels and equipment to cancel out gravity." And William Lear, the autopilot wizard, is already figuring out "gravity control" for the weightless craft to come. Gravitation—the mutual attraction of all matter, be it grains of sand or planets—has been the most mysterious phenomenon of nature.

Perhaps even more significant is an article in the June 1957 issue of "Mechanix Illustrated,"[55]

ANTIGRAVITY: Power of the Future

By G. Harry Stine

Chief, Navy Range Operations, White Sands Proving Grounds

Rumors have been circulating that scientists have built disc airfoils two feet in diameter incorporating a variation of the simple two-plate electrical condenser which, charged to a potential of 50,000 volts, has achieved a speed of 17 feet per second with a total energy input of 50 watts. A three-foot diameter disc airfoil charged to 150 kilovolts turned out such an amazing performance that the whole thing was immediately classified. Flame-jet generators, making use of the electrostatic charge discovered in rocket exhausts, have been developed which will supply charges up to 15 million volts. Several important things have been discovered with regard to gravity propulsion. For one, the propulsive force doesn't act on only one part of the ship it is pushing; it acts on all parts within the gravity field created by the gravitic drive. It probably is not limited by the speed of light. Gravity-powered Vehicles have apparently changed direction, accelerated rapidly at very high g's and stopped abruptly without any heavy stresses being experienced by the measuring devices aboard the vehicle and within the gravity-propulsion field. This control is done by changing the direction, intensity and polarity of the charge on the condenser plates of the drive unit, a fairly simple task for scientists.

From Practical Mechanics, Vol. XXIX October 1961 No. 330[56] we can read:

ANTI GRAVITY The Science of Electro Gravitics by I E van As

New Type of Lift

Electro-gravitics is a vertical lift which is not accomplished by means of rocket thrust or propulsion by airscrews and airfoil and is not dependent on any type of atmosphere, air or a vacuum as would be found in outer space. All these methods seek to overcome gravity, but we approach a new era where man no longer seeks to leave our Earth by overcoming gravity, but instead by utilising it to his own advantage. An antigravity machine is not impossible and many countries including Russia are at present investigating this new approach to aviation. Canada has its "Project Magnet" which is the production of an antigravity machine using the electro-gravitic principle. Many American aircraft manufacturers are spending millions of dollars on the use of gravity as applied to their industry. A number of universities are also going into the problem, which, incidentally, is not a new one. An actual flying model using this principle was made in England before the war.

Referring to this quote above, is "I E van As" a pseudonym? The "Project Magnet" quote is something we will return to in a later chapter, but for now, we will just mention it is not quite accurate.

Both Nick Cook and PLV observe in their books how, not long after these writings – i.e. the late 1950s and early 1960s, talk of the development of antigravity drives or electrogravitic propulsion went into a sharp decline. Little was heard about it in the "white world."

1980s – Secrecy Still Prevails

Moving forward 25 years, for example, we see in a response to a letter written to an interested person in February 1982, Brown wrote[57]:

> *I regret to advise you that electrogravitic research has been taken over in its entirety by a Californian corporation which has imposed secrecy - at least until their investigations are completed. No further publication or release of information is permitted, possibly until next year.*

He wrote other similar letters, too, around the same time. Some months later, he received another letter from "William S," dated 17 September 1982[58], enquiring about Brown's knowledge of the alleged Saucer crash at Aztec, New Mexico in 1948[59]. "William S" also mentioned Brown's involvement in NICAP. It seems that Brown felt rather differently about the UFO/Flying Saucer issue in 1982, as he replied:

> *September 29, 1982*
>
> *Dear Mr. S,*
>
> *This will acknowledge your letter of Sept. 17.*
>
> *Your interest in UFO's is commendable but I am wondering if it is a bit futile. I myself have no evidence, even in recent years, that valid encounters have ever actually taken place. The thought, however, persists that there may be something to the stories. Only time will tell. For my part, I have never personally ever seen anything to support the rumors.*
>
> *You ask if I ever had any part in the alleged UFO investigations near Aztec, N.M. The answer is definitely NO, your Dr. Gee must be someone else, for I don't even know of such investigations and, even now, question their authenticity.*
>
> *Sorry I can be of so little help to you.*
>
> *Sincerely,*
>
> *T. Townsend Brown*

I can only guess what had caused Brown to be so much more reserved about the UFO issue than he appeared to be in the 1950s. It seems reasonable to suggest, however, that military interest in Brown's work was ongoing and reading through the available information seems to give a strong indication that Brown was "never too far away" from classified projects.

The B2 Bomber

With regard to military black programmes, chapter 5 of PLV's book, covers the Northrop B2 bomber which, it is surmised, employs some kind of electrogravitic system to enhance its performance. A French webpage by Jean Pierre Petit highlights interesting features of the B2[60]. It shows a very short sequence from a Northrop B2 "promo" video and suggests that a glow seen in a vapour cloud created by the transonic flight of the B2 over the sea might be the result of some kind of electrostatic field being in use[61] in the plane's air frame. Having looked at this myself, I do wonder if it is just illumination from sunlight, at a lowish angle, however.

As we shall see later on, there is some evidence to suggest that the US military was already working on antigravity technology in the 1950s, and this was not without results...

Simple Electrogravitics Experiments?

When you read about the Biefeld-Brown Effect online, it is frequently dismissed as simply being caused by an "ion wind." That is, when using high voltages to charge up disks or other models, charged air molecules (ions) are created and these generate mechanical thrust, due to their motion in an electric field.

Again, in PLV's "Secrets" book, this is explored in some detail. Brown himself was quite aware of this effect and that it was sometimes used to "explain away" the anomalies he was studying. It is also discussed in an Army Research Laboratory 2003 report called "Force on an Asymmetric Capacitor" by Thomas B. Bahder and Chris Fazi[62]. Calculations included in this document essentially prove that the ion-wind effect alone cannot explain the observations (and Brown was well aware of this). Similarly, it is easy to demonstrate that this cannot account for all the "force" effects observed. For example, experiments shown on JL Naudin's (very interesting) website[24] show this to be the case. Perhaps the simplest of these is an experiment called "Frolov's Hat." Here, a simple, home made capacitor made from aluminium foil and expanded polystyrene is rigged up to a chemical balance. The "hat" is counter-balanced with a small weight, to be in equilibrium. Then, a voltage – upwards of 10,000 volts (10 kv+) is applied from an external power supply. The "hat" then becomes lighter and the balance tips towards the weight!

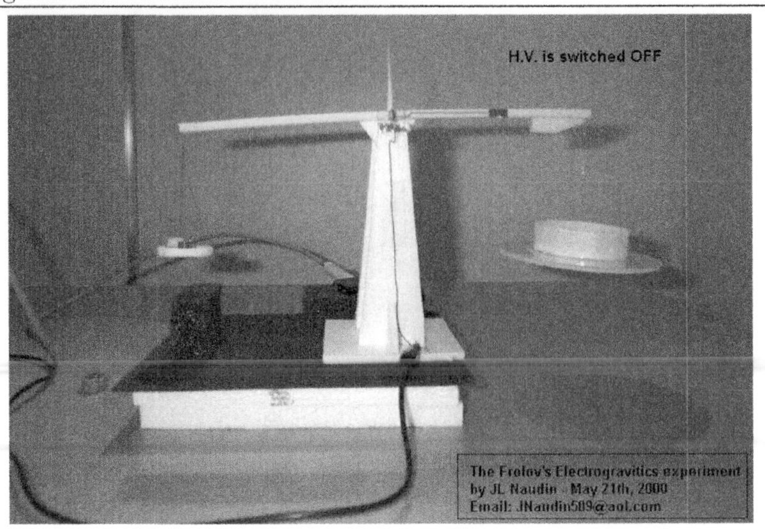

Frolov's Hat Experiment reproduced by JL Naudin[63]

The voltage is removed and, after a few seconds, the balance returns to equilibrium again. This, to me, demonstrates that the change in weight is something to do with the gain in electrical charge of the "hat" itself, as the weight slowly changes – the chemical balance does not level itself immediately when the voltage is removed.

To try and further remove doubt that ion wind effects are involved, the experiment can be repeated with the "hat" contained inside a plastic bag.

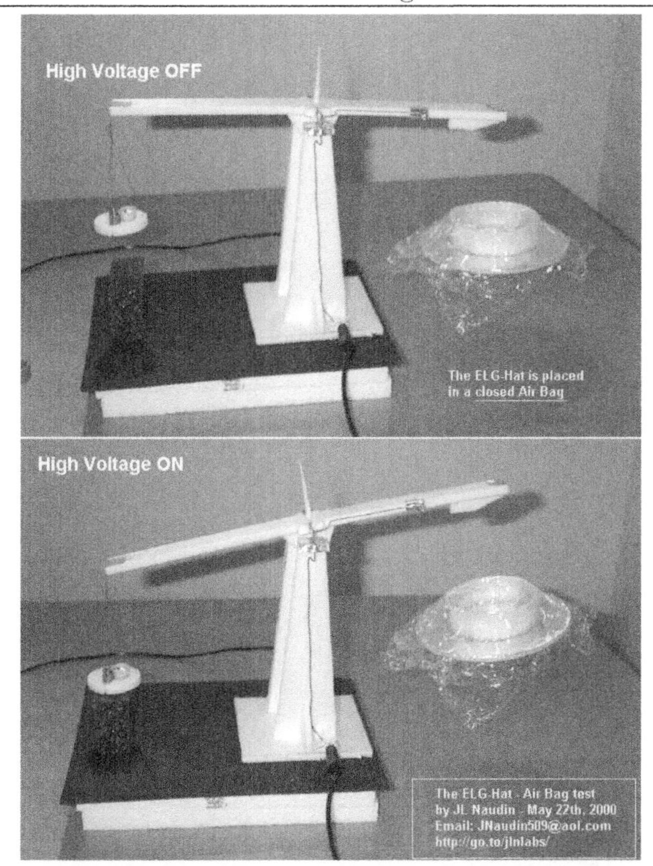

Experiment is repeated with "hat" inside a plastic bag, to reduce any "ion wind" effect.

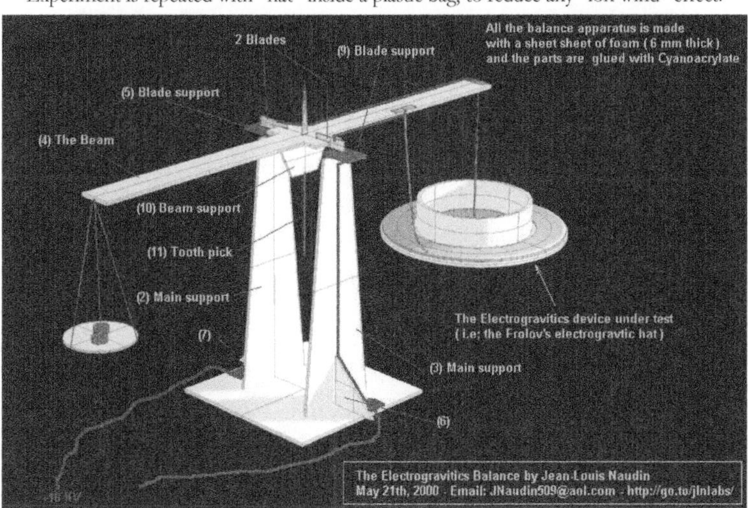

Schematic of Frolov's Hat Experiment reproduced by JL Naudin[63]

In the early 2000s, there seemed to be a wave of renewed interest in the topic of electrogravitics and I, myself, came across J L Naudin's fabulous website[64] in

2003. Some months later, I followed the instructions he had published[24] and I made my own "Lifter 1," which *just* flew (I was a bit ham-fisted with the construction!) To provide the required voltage, I used a high tension/high voltage power supply from an old PC monitor I had.

Naudin has previously posted links to many reproductions of his "Lifter" experiments – in various forms – from around the world[65]. Up until 2006, he reports that there were 351 reproductions of the experiments.

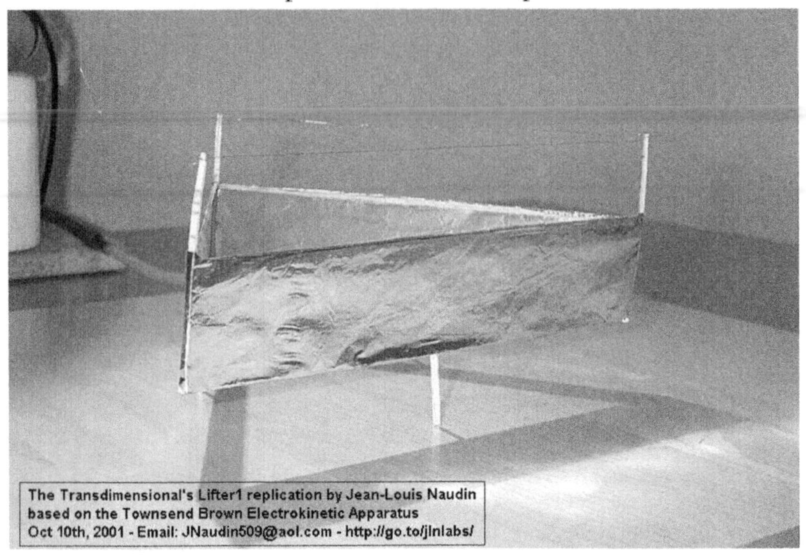

The Transdimensional's Lifter1 replication by Jean-Louis Naudin
based on the Townsend Brown Electrokinetic Apparatus
Oct 10th, 2001 - Email: JNaudin509@aol.com - http://go.to/jlnlabs/

Made of Balsa Wood and Aluminium Foil, I have included JL Naudin's description of the Lifter 1 below.

The Lifter1 specifications

The Lifter1 is an asymmetrical capacitor with one electrode made with a thin corona wire placed at 30 mm from the main rectangular electrode constructed "ala" Townsend Brown.

Weight: 2.3 g

Triangle size: Equilateral with each side 200 mm wide and 40 mm high, made with a thin aluminum sheet.

Mounting legs: 30 mm length.

Main frame: balsa wood 15/10 mm thick and 2 mm wide.

Power required to compensate the weight: 18 Watts (40 KV @ 450 µA)

Power required for a stable flight above the ground with a payload of 1 g: 23.9 Watts (41.9 KV @ 570µA)

A schematic by JL Naudin is shown below.

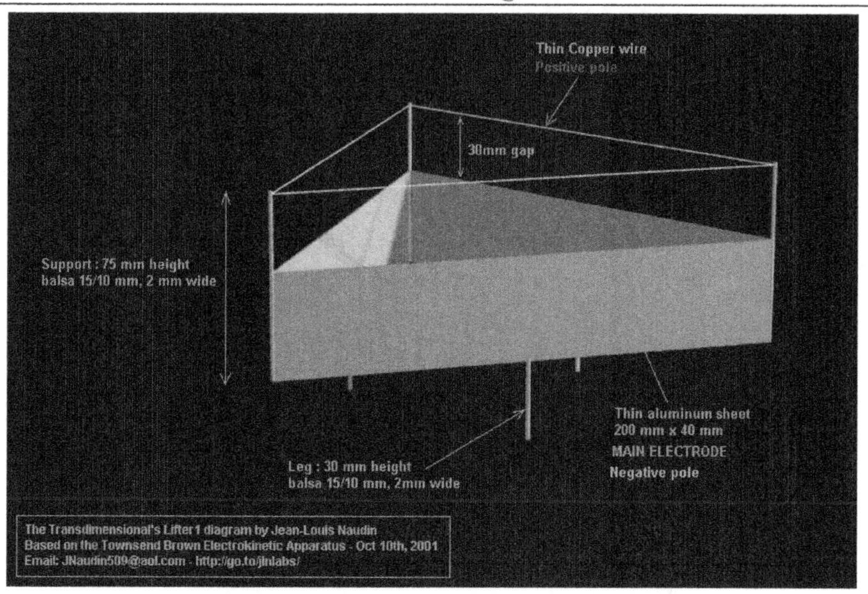

Schematic Diagram of Naudin's Simple "Lifter 1" model

A number of derivative designs can produce greater lifting effect, by combining several triangles into a "hexalifter."

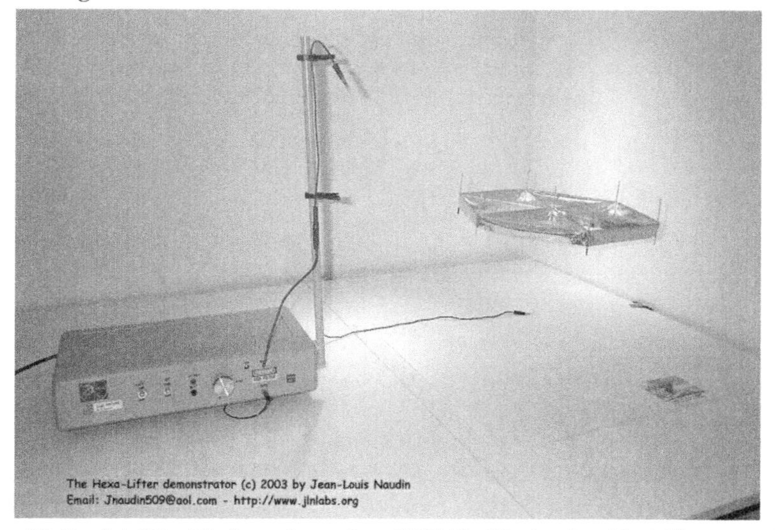

One of JL Naudin's "Hexalifter" experiments from 2003[66]. The lifter is tethered by 2 thin copper wires which provide the electrical power.

Naudin also carried out tests to see what effect the ion wind was having[67] – and he concludes from his own calculations with a test model:

> *For ion wind to cause levitation, the air flow below LIFTER3 should be about 0.913 meter per second or 2.04 miles per hour. If some air also flows outside the borders of the device, then the "virtual area" would be larger and the air flow somewhat slower than 2 mph. If the actual wind is *far* smaller than 2 mph. (e.g. 10x smaller,) then I would conclude that most of the lifting force comes from electrogravity. (Well, this is true if my calculation has no flaws!)*

Chapter 12 of PLV's "Secrets of Antigravity" contains many further technical details of these experiments, including tables, figures and equations related to J.L. Naudin's experiments and similar ones performed by other researchers.

Lifters and "Unusual Effects."

Another website which became popular in the mid-2000s, perhaps partly because it included videos of "Lifter" experiments, was Tim Ventura's American Antigravity[68]. This site (as it is organised currently, in 2018) has many interesting interviews "hidden" in the "News" section. Most of these, however, focus on white-world research – such as the electrogravitics material we have already covered.

Among many interesting interviews, Ventura recounts some of his experience in experimenting with his own lifters. In an interview from April 2005[69], with famed Canadian Researcher and experimenter John Hutchison (whom I have written about in both my 9/11 books and in chapter 12), Ventura describes how he and others experienced strange effects when running the lifters for long periods. Ventura states:

> *I didn't notice that the lifters do seem to have side effects. And it's one of those things that very few people like to report. We did testing with a shielded magnetometer and we were unable to pick up the Earth's magnetic field while the lifter was operating and we also had some unexplained signals in the background. And it's one of those things where, when you mention it online, reports start pouring in from people all over the world saying "you know, I had this happen and I can't really explain it. It's not really something that happens all the time..."*
>
> *Another thing that I had happen, interestingly, was I had sparks that appear out of nowhere. And in a way, they're almost not really bound in time at all. When I operate the lifter consistently for several days on end, I can turn it off, but there is a period of a week or so where I'll get unexplained electromagnetic effects. They are uncommon enough and random enough that there is no real way to pin them down.*

Ventura then repeats that he observed "residual energy effects" after his equipment was turned off and the lifter and associated equipment has been grounded and fully discharged. Hutchison confirmed that he had observed similar residual effects following his own energy-related experiments. In other interviews, Hutchison himself had reported that he detected changes in the radiation background count when running his experiments.[70] Typically, it would drop, then return to normal, some time after he had switched off his equipment.

In the next chapter, we will again travel back to the 1950s and review some gravity related experiments that were performed in Canada, with an entirely different approach to the electrogravitics methods used by Thomas Townsend Brown.

4. Wilbert B Smith and Gravity Control

Wilbert Brockhouse Smith is not a name which normally comes to mind for most people when they are discussing the evidence pertaining to antigravity. Indeed, most people tend to confuse Wilbert Smith with Wilbur Smith - the author of numerous historical fiction novels.

Wilbert Brockhouse Smith was himself a writer, though his main work "A New Science" was never finished nor widely published.

Wilbert Brockhouse Smith (1910-62) Canadian Radio Engineer

Wilbert Smith was born in Lethbridge, Alberta, Canada in 1910 and he graduated from the University of British Columbia (UBC) in 1933 with a B.Sc. in Electrical Engineering. Following this, he obtained his MSc (Master's Degree) in 1934 also at UBC. In 1934/5, he became chief engineer for radio station - CJOR - in Vancouver. In the years following, at certain times, he acted as a consultant to the Government and in 1939 he joined the federal Department of Transport.

Smith was engaged in the engineering of Canada's war-time radio monitoring service and following this, in 1947, he was put in charge of establishing a network of ionospheric measurement stations, several of which were in isolated parts of Northern Canada. He eventually became Superintendent of Radio Regulations Engineering with the Department of Transport [DOT]. His technical research was in Radio Wave Propagation, where he came across such topics as auroras, cosmic radiation, atmospheric radio-activity and geo-magnetism (the Earth's Magnetic Field). It was likely this area of science which made him more interested in the Flying Saucer Phenomenon (he rarely called them UFO's – as that term was not coined until 1952, by Edward J Ruppelt).

WB Smith's interest in "flying saucers" probably was triggered by a magazine article. Initially an extreme sceptic, he began to investigate "saucer" cases himself, developing questionnaires for witnesses and contactees and carrying out methodical study to gather and analyse data about the phenomenon. Unusually, for someone like Smith, through the 1940s and 1950s, he rose to a high position in the Canadian Government and eventually became the Superintendent of Radio Regulations.

He continued to work for the Department of Transport until his death in 1962 (he died of Cancer of the Lower Bowel). At the time of his death, he held 37 patents. He was married, to Murl and they had 3 children.

Wilbert Smith was a rare individual indeed – naturally inquisitive, kind-hearted, methodical, analytical, thorough, resourceful - yet open-minded. Some of his writings read like those of a spiritual leader, whilst remaining grounded, straightforward and accessible.

I discovered WB Smith, some time in 2003, thanks to the wonderful website of the extremely knowledgeable Grant Cameron. I strongly identified with Smith's writings and conclusions – perhaps because we have both worked in engineering disciplines – which are all about solving problems. That website is www.presidentialufo.com – and it contains a great deal of useful and compelling information and evidence pertaining to the US Government's and US Presidents' knowledge of the UFO/ET cover up. Cameron, who became interested in this subject following his personal UFO/ET related-experiences in the 1970s, had been collecting information on Wilbert Smith for about 25 years and he had posted a small selection of it on his website. Other researchers, such as Arthur Bray have also collected and preserved Smith materials and prevented their untimely removal by government employees. Back in 2003, I obtained a CD copy of Grant Cameron's collection of Smith related files. When I "opened" this treasure trove of information, it became clear to me that a lot more was known about UFO's/Flying Saucers in the 1950s than I ever realised.

The "Top Secret Memo" and "Project Magnet"

If Wilbert Smith's name comes up in UFO research, it is usually in relation to the "Top Secret Memo" which he wrote to the Canadian Dept. of Transport in 1950.[71] This is widely regarded as one of the most important documents that we have in relation to the UFO cover up. A draft copy of this vital document was declassified by the Canadian government in 1979 and discovered by Nick Balaskas in the University of Ottawa Archives. In the memo, Smith discloses the secrecy classification on the study of the "saucer" phenomenon. Smith wrote:

> *"a. The matter is the most highly classified subject in the United States Government, rating higher even than the H-bomb.*
>
> *b. Flying saucers exist.*
>
> *c. Their modus operandi is unknown but concentrated effort is being made by a small group headed by Doctor Vannevar Bush.*
>
> *d. The entire matter is considered by the United States authorities to be of tremendous significance."*

Dr Robert Sarbacher in 1950

When you really "get down to it," and understand that this was a real document, written by a gifted radio engineer, you have to consider the question "Who gave Smith this information?" Further research, in 1983, by Nuclear Physicist and veteran UFO investigator Stanton Friedman, revealed that the information about the classification of the UFO subject appears to have been given to Smith by Dr Robert Sarbacher - a Physicist. Sarbacher, who worked as a consultant to the DoD (Dept. of Defence), would later go on to create the Washington Institute of Technology.

Of equal or even greater interest, however, is what Smith wrote in the next paragraph:

> *"I was further informed that the United States authorities are investigating along quite a number of lines which might possibly be related to the saucers, such as mental phenomena and I gather that they are not doing too well since they indicated that if Canada is doing anything at all in geo-magnetics, they would welcome a discussion with suitably accredited Canadians."*

As I understand things, it is not clear who gave Smith the information about "mental phenomena" – it wasn't Sarbacher. It is rare to see these sorts of concepts brought up in de-classified government documents. Smith, like any good engineer, was curious to find out more – in the hope that he could use any findings to solve problems. So, Smith proposed setting up a project to investigate Saucer Phenomena:

> *"It is therefore recommended that a PROJECT be set up within the frame work of this Section to study this problem and that the work be carried on a part time basis until such time as sufficient tangible results can be seen to warrant more definitive action. Cost of the program in its initial stages are expected to be less than a few hundred dollars and can be carried by our Radio Standards Lab appropriation."*

The "Project" became known as Project Magnet.

Project Magnet

Though officially, Smith's Flying Saucer interest was private, in reality he used his knowledge and government position to instigate a semi-official study of the Saucer Phenomena. His proposal to set up a "station" at Shirley Bay, about 20 miles outside Ottawa (where the Department of Transport's Communication's monitoring facility was situated) was accepted and Project Magnet ran for about 4 years, with the aim of gathering information about "magnetic phenomena." It wasn't officially called a "flying saucer monitoring station," but Smith strongly implied it was meant to have that sort of capability. It was meant to be a classified project in case the results yielded a new insight into magnetic phenomena which might be exploited. Its existence was only ever reluctantly acknowledged during Smith's lifetime. However, word of this "got out" and made headline news, as Smith later discussed in a lecture he gave in Vancouver in 1958[72]. On one hand, he seemed to play down the project's significance, but on the other, he seemed to make some incredible statements. In the lecture, he said:

> *I might point out that the Project Magnet that I was associated with, which received a great deal of publicity, was not an official government project. It was a project that I talked the Deputy Minister into letting me carry out making use of the extensive field organisation of the Department of Transport. No funds were spent on it and we merely have access to the very large field organisations and opened a number of files. Unfortunately, the gentlemen of the press climbed on this and made a big deal out of it. As a matter of fact, some of the headlines concerning Project Magnet were set in larger type than the declaration of war.*
>
> *However, we carried the project through officially for about four years and the last year of the four, we had a little shift setup in which we had a number of instruments - the idea being to try and coordinate sightings with scientific measurements. We had equipment for detecting any radio noise that might emanate from these objects and any gravitational disturbances which might result, any radioactivity which might be connected with it and any magnetic disturbances. These four instruments were of a recording type and they produced graphical lines on a recording tape, using a four-pen recorder. This ran 24 hours a day with alarms attached so that in case any of the pins moved beyond a prescribed limit, the horns would blow, and bells would ring and the boys in the nearby ionosphere station we're supposed to come running to see if they could see anything.*
>
> *Towards the end of the year we hadn't received anything of any great importance on the instruments, so we had been following up a number of contacts with these people from outside, so we arranged a special circuit and one Sunday afternoon we asked them if they would bring a craft down, close enough, so we could see if we got any indication. Unfortunately, the afternoon we chose was very heavily overcast - we were not able to get a visual coordination, although we did get a number of squiggles on our instruments. The craft that they sent down they said was a about 80 feet long about 14 feet in diameter and was a cigar shaped effort and was primarily for scientific observation, but for some reason or other they didn't get it any closer to the*

> *earth than about 50 miles, so it wouldn't have been visible with the naked eye anyway.*
>
> *However, the gentlemen of the press apparently found out about our long-distance telephone circuit anyway they were out there [inaudible] shortly after this experiment and they made us so thoroughly annoyed that we closed the whole business down. It went underground from there on and that's where it still is as far as the gentlemen of the press are concerned.*

To add to what Smith stated, something was detected on August 8th, 1954, at 3:01 P.M, but then Project Magnet was officially shut down on August 10, 1954. Smith later stated[73]:

> *Project Magnet was officially dropped from the Department of Transport in 1954, although the Department indicated its willingness to permit the continued use of laboratory facilities, provided that this could be done at no cost to the public treasury. The project continued under these conditions, and to this extent may be said to have gone underground. The government of Canada was therefore not participant in the continuation of the project and not in any way responsible for its conclusions.*

After Smith's death, the strength of government denials about the nature of Project Magnet seemed to increase.[74] A statement prepared for the Minister of Transport to be presented in the House of Commons in the 1950s confirmed that the project was Smith's and the government had approved it. But, a few months after Smith died, in late 1962, Mr. Dupuis, Minister of the Department of Transport, stated something rather different in the Canadian House of Commons:

> *"Between December 1950 and August 1954, a small program of investigation in the field of geomagnetics was carried out by the then communications division of the Department of Transport with a view to obtaining, if possible, some physical information or facts which might help to explain the phenomena which was generally referred to as unidentified flying objects. Mr. W.B. Smith was the engineer in charge of this program."*

However, in 1964 an Ottawa UFO researcher Arthur Bray, wrote to request information about Project Magnet. Bray received a letter that started the denial,

> *". . . at no time has this Department carried out research in the field of unidentified flying objects . . .The department did not take part in any of his (Smith's) research work nor did Mr. Smith provide the Department with any useful information arising out of his work."*

Arthur Bray later became a custodian of many of Smith's files and kept them safe for about 20 years. The written statement that Arthur Bray received conflicted, to a degree with statements made in 1968 by Dr. Peter Millman who knew Smith. He worked at the Dominion Observatory and it was he who wrote the statement that would become the "official line" in relation to what happened with Smith and Project Magnet. Millman wrote,

> *"The project was a personal one carried out by Mr. Smith with the knowledge of his department, but without any official sponsorship."*

Smith Built Unusual Hardware

As part of Project Magnet, Smith conducted some experiments in which he showed it was possible to extract energy from the Earth's magnetic field. He thought this is what the saucers used for both propulsion and as an energy source. He described this device as a magnetic "sink" (which energy would "flow into").

After a time, Smith seemed to have established a line of communication with "The Boys Topside," as he often called them. He actually built several items of simple experimental hardware according to instructions given to him by "Topside People". In the 1958 lecture mentioned above, he said he thought it was "absolutely regrettable" that his group (in Ottawa) was the only one turning the information into real hardware. He went on to describe the hardware they built in some detail. The first item was a "binding meter[75]" – it was said to measure "binding forces" in the environment.

Smith said he was told that materials were subject to "binding forces" and, for example, objects coming close to the saucers could be affected because the binding forces were altered by a flying saucer's operational field. He stated he thought that the now well-known incident when Thomas Mantell's plane broke up - when Mantell was following a flying saucer - happened because Mantell got too close to the craft. Smith stated Mantell's plane had got inside the craft's operating field and this altered binding force in the region. This caused the plane to "fall apart". Current physics does not acknowledge this type of force. However, I can certainly see, in the research of Canadian Experimental Researcher, John Hutchison (see chapter 12) that what we witnessed in the destruction of the World Trade Centre on 9/11 (as analysed by Dr Judy Wood in her book "Where Did the Towers Go"[76]), could very well be the result of an alteration of the "binding forces" in the materials that turned to dust.

Another item, which Smith made and was quite proud of, was a special type of coil (as used in AM radios, transformers, car engines and other electrical apparatus). It is prepared and wound in a very particular way. Smith found that when he tested it with his radio equipment, it would completely absorb radio wave energy of a certain frequency. This appears to violate the accepted laws of electro-magnetism.

Smith Saw Unusual Hardware

Over the years, it has become fairly clear that flying saucers have crashed and been recovered. We will revisit this issue later in this book, but it seems that because Smith was very well qualified, knowledgeable and was well-regarded in the Canadian Government and he had studied the flying saucer phenomenon in great detail, it is hardly surprising to read accounts of him being involved with analysing pieces of hardware from saucers. In an article on Grant Cameron's website[77], Smith is quoted from a 1962 interview by Bob Groves:

> *'In 1952 we had a noteworthy or a notorious sight over Washington DC. During this time an Air Force jet shot a piece right off a UFO. It was found two*

> *hours later. It had a glow to it - a white glow to it - after two weeks it had diminished to a brown texture. The part that was shot off was about as big as could be held in a couple of hands. It had a very distinct edge. It was curved. It had tapering sides so that it appeared that it had been shot off the edge of a double saucer shape. The typical shape.'*

Further information from Grant Cameron (for example, an audio interview with Art Bridge, one of WB Smith's associates) suggests that Smith and his associates received various pieces of hardware which they were asked to study and report on.[78] We will return to the subject of "recovered saucer hardware" later in this book.

Gravity Control

Another area that Smith spent a long time investigating involved experiments with a rotating disk and with rotating magnets. Due to his investigation of the Saucer Phenomenon, Smith postulated (in simple terms) that the force of gravity was actually the result of interacting magnetic fields that objects themselves generated, rather than it being a separate force, as traditional physics suggested. He determined that the speed of rotation of an object can be used to influence gravity and he and his team designed experiments to test this. (It seems very unlikely that Smith would lie about his experimental results or misinterpret them.)

In a lecture given in 1958, Smith stated[79]:

> *...we figured out that we were able to go into the laboratory and conduct a series of experiments which proved beyond doubt that this is true.*
>
> *Our laboratory experiments have allowed us to make about a 1% change in the weight of objects --- we can make them about 1% heavier or 1% lighter. Now that is a long way from holding a spacecraft up, because we have to go over 100% to do that. But the fact that we can do it --- the fact that the principles which these people from outside gave us and guided us to finding out for ourselves are valid --- certainly indicate that, first, these people are what they say they are, and, secondly, that their technology is what they say it is, that it is superior to ours and that ours is inadequate in many respects.*

It seems he was referring to experiments conducted with apparatus that he designed. It is very likely that some of this apparatus is the one shown in photos from William Treurniet's Website, reproduced below[80][81][82][83][84]:

An enlarged version of this photo shows an overlapping structure of the magnets around the circumference of the rotor. We know that the rotor is six inches in diameter, so we can estimate the size of the magnet wafers. Unfortunately, we do not have information on how the magnet poles are oriented.

This photo shows the connection to the motor driving the rotor. The red disk-shaped object on the ground might have been used to cover the rotor, perhaps for safety and/or to hold an object to be levitated.

This photo shows the precision balance scale supporting an object suspended above the rotor. Note that the indicator is calibrated to its Centre position. The implication is that the rotor is not spinning.

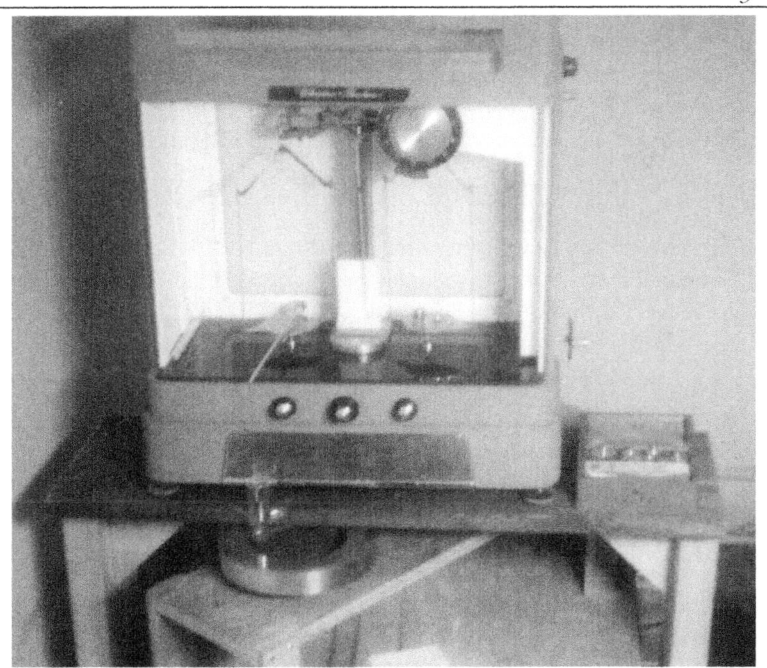

Here we see the same scale and suspended object, but with the indicator deflected to one side. The scale deviation implies that the rotor is spinning, thus causing a weight change. Note that the suspended mass appears to be inside a glass tube to prevent air currents from affecting the measurement.

In a short paper that Smith wrote (after August 1958, but exact date is unknown), called "Suggestions on Gravity Control Through Field Manipulation"[85] we read:

> *The Fisher Hooper experiment, which we have successfully repeated several times suggests that there is a relationship between gravity and the motional magnetic field, and that they might even be the one and same thing (2). Our experience in electrical engineering indicates that the motional magnetic field is electric in nature in that it can move electric charges. We know that we can shield out the effects of displaced electric charges but we can't shield out the effects of gravity or of motional magnetic fields.*

Smith then tries to discuss and analyse the relationship between the electric and magnetic fields and considers how non-conducting matter might experience a kind of "stress" induced by the electric field.

> *We have good evidence (in the Betatron) that the induced electric field is still there and if the region is occupied by material which is non—conducting, such material will obviously be subjected to exactly the same electrical stress as was the conductor. Now, since we believe that matter is fundamentally electrical in nature such stress must result in some kind of "polarization" which will not be offset by the movement of charges. From the foregoing we may conclude that the electric field induced by the motional magnetic field could and probably does have very much the same properties as gravity, and in fact might be the same thing.*

Again, he is talking about "polarisation" in a way which, to me, sounds like what PLV discusses in his theory of subquantum kinetics. (This idea has also been considered by a fellow researcher I know called Frederik Nygaard, following our discussion about the reason for the force of gravity being lower on the earth during the time of the dinosaurs.[86])

Later in the paper, Smith considers that if there is a polarisation effect which causes gravity, then logically, controlling this polarisation effect (electrically or magnetically) may allow "gravity to be controlled."

In a later article that Smith appears to have written in 1961[87] for his special interest "Topside" newsletter, he includes more detailed speculations on gravity – which are based on theories by P.G.A.H. Voigt. (I have not been able to find out much about this person). Smith quotes Voigt quite a lot in this article and one interesting quote is given below:

> *Imagine a primitive gas. A gas with particles much smaller than electrons. Particles which I shall call 'microls,' so fine that they can pass or bounce their way freely through the atoms. These microls move at an incredible speed and this sub-atomic gas occupies all space, from the vast distance between the galaxies down to the small ones between parts of each atom.*

Again, this sounds very similar to PLV's subquantum kinetics ideas and reading the rest of the 1961 Smith article, referenced above, only seems to make this comparison stronger.

Smith on Relativity

Of course, Smith knew about Einstein's theory of relativity, which had been around for almost 50 years by the time Smith was conducting his own research. In the same 1958 lecture, referenced above,[79] Smith elegantly and simply highlighted a basic flaw in the theory of relativity.

> *Now that is a result, apparently, of time dilation in the theory of relativity, in that the spacecraft in which I travel was moving relative to the earth at a velocity nearly equal to the velocity of light. The paradox arises when you consider that relative to the spacecraft, the earth was traveling away at exactly the same velocity. Therefore, to the people on the spacecraft, who are relatively stationary, 10 years should have passed, and by the time the earth comes back to them, it should have only been away a year. So, you can see right away that the very premise upon which the theory of relativity is predicated --- namely, that if A is relative to B, then B must be relative to A --- leads you to an impossible paradox. This paradox is resolved completely if you recognize the variable nature of time, and as you move round from one part of the Universe to another you'll encounter all sorts of values of time in certain given intervals.*

I seem to remember, when I covered relativity theory in my University Physics course, there was talk of "non-inertial frames of reference[88]" and also we would discuss how we would make measurements "relative to distant parts of the universe." At this point, however, I can't fault Smith's observation above and it is clear that Relativity theory has not, in the white world at least, been used to engineer gravity.

I would guess that Smith felt quite strongly that relativity would not help anyone solve the gravity mystery.

Smith and Gravity Day

Smith faced the same sort of intellectual intransigence in the academic community as is evident today. He expressed this, to some degree, in an article he wrote for his special interest group magazine "Topside,"[89] following an event he attended in New Boston, New Hampshire on 27 August 1960. He described that, whilst he found some of the presentations as being interesting, his article concludes thus:

> *It was noteworthy the omission of any papers, even speculative, on the possibility of gravity control, or a recognition that such control might be possible. It seemed that the entire thinking was dominated by the philosophy of modern science, with a great inertia to deviate from the well-trodden path of orthodoxy.*

On Grant Cameron's site, an article[90] suggests that near the end of his life Smith had become quite disillusioned – "He had taken his antigravity experiment apart telling his wife that the world was not ready for it."

New Science

Considering what we have discussed in this chapter, it is perhaps not surprising that Smith turned his attention to the task of trying to rewrite the whole idea of scientific thinking from the ground up, based on what he had learned from his experience and his experiments and from "The Boys Topside". At the beginning of his unfinished book The New Science[91], Smith writes:

> *Assembled by W. B. Smith from data obtained from beings more advanced than we are.*

The work is available on the Internet, and deals more in "Metaphysics" initially, rather than Science, describing concepts of "nothingness", "awareness", "reality", "space" - which he shows are necessary to define properly so that one can develop a new understanding of the Cosmos around us - and our place within it.

In 10 years of research, Smith's understanding developed and he came to see "the bigger picture". He realised the key role that awareness and consciousness played in the "flying saucer"/UFO phenomena that he was investigating. Over 50 years later, there is much more general discussion in the Alternative Knowledge Research Community of the topic of consciousness in relation to ET's and UFO's. Smith had a more analytical approach than most and his writings are much more concise and focused than most of the wide range of material that is now available. Perhaps I will revisit these topics in a separate volume.

5. Douglas Aircraft and BITBR

In the late 1960s, much of the earlier "chatter" within the aerospace industry about antigravity drives and gravity control had died down, but it had not fallen completely silent.

The previous chapter documented a small portion of the Wilbert Smith story – which, really, deserves a book in its own right. As I mentioned, this story was catalogued in considerable depth by Grant Cameron – with the earlier help of other researchers such as Nik Balaskas and Arthur Bray. I corresponded with Grant Cameron fairly regularly in 2004 and 2005 and, in March 2005, someone made contact with Grant regarding some documents they had acquired which formerly belonged to Douglas Aircraft Ltd. This company later became McDonnell Douglas, which was bought by Boeing in 1996[92]. As Grant's website was a serious effort to uncover some of the truth regarding the US (and other) government's knowledge of the UFO/ET phenomenon, he was a good person to contact to assess the quality of any official-looking documents relating to UFO research. Such was the case here. So, Grant passed my contact details onto a chap called Louis and I gave him an FTP login for my computer. In the next couple of hours, he uploaded about 170 pages on 5 and 6 of March 2005. Another 100 pages were given to us in August 2005. By this time, I had set up my website (http://www.checktheevidence.com/) and all these pages have been hosted there ever since[93].

According to what Louis told us, the documents were left in a barn which may have belonged to a Douglas Aircraft employee who died or moved. The new owner of the place found the files and sold them on Ebay for $31.00!

It seems that the project that generated the documents came about as a result of a similar set of circumstances to Project Magnet, but the "project" was done inside a private company, rather than within a government department. That is, there was clearly someone who was well-qualified and could also see there was something in the UFO/ET phenomenon which was real, and worthy of serious scientific investigation. They managed to persuade their superiors in the company that this was the case and they were then given some freedom to investigate. Hence, the BITBR project or "Boys in The Back Room" project came into being…

The documents illustrate some of the research conducted by a small team of people working at the Douglas Aircraft corporation in the period 1967 - 1969. The "prime movers" in this secret research appeared to be J M Brown, Bob Wood, W.P. Wilson and D.B. Harmon. The most familiar of these names, perhaps, is Robert M Wood, Ph. D, who has authored and co-authored books and papers on the UFO/ET subject[94].

A New Theory of Gravity

As it says in one of these documents, the goal of the study was, ultimately, to develop a new method of propulsion based on a new theory of physics. This research was, of course, secret.

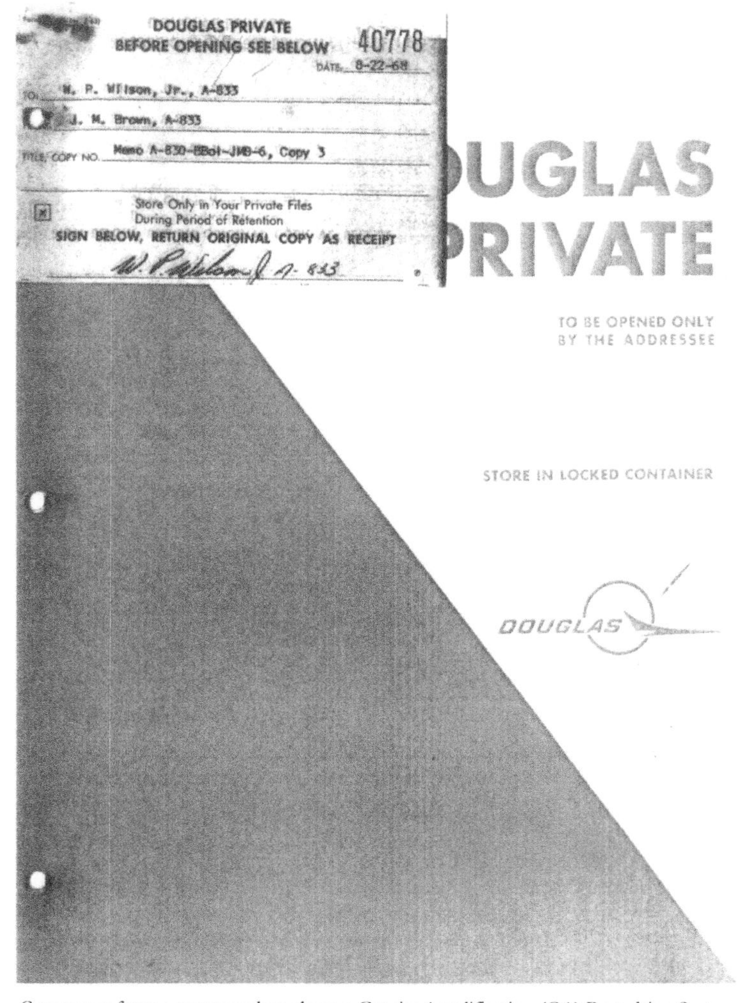

DOUGLAS PRIVATE
BEFORE OPENING SEE BELOW 40778
DATE 8-22-68
TO: W. P. Wilson, Jr., A-833
J. M. Brown, A-833
TITLE, COPY NO. Memo A-830-BBol-JMB-6, Copy 3
[x] Store Only in Your Private Files During Period of Retention
SIGN BELOW, RETURN ORIGINAL COPY AS RECEIPT
W. P. Wilson Jr A-833

DOUGLAS PRIVATE
TO BE OPENED ONLY BY THE ADDRESSEE
STORE IN LOCKED CONTAINER
DOUGLAS

Cover page from a memorandum about a Gravity Amplification (GA) Propulsion System

As may be expected from specialist researchers in the Douglas company, there is quite a bit of technical content in the documents, which would not be particularly interesting for the average reader. Similarly, there are a number of equations and a considerable level of theoretical discussion. However, it is worth mentioning that this research team, too, considered the idea of moving particles as a way of explaining the force of gravity. For example, in a document called "Proposed Vehicle R&D Program (Project BITBR)"[95] authored by J.M. Brown on page 8, we read:

> ***Gravitation is believed to be due to the gradual collection of basic particles from the background by all matter and then a pulse emission of a group of the basic particles in the form of a non-interacting particle (graviton or neutrino).***

It appears that J.M. Brown was developing his theory as the research progressed and, in another document entitled "A Unifying Kinetic Particle Theory of Physics", he describes some of his fundamental ideas. Here is part of page 2:

THE BRUTINO

The basic particle which makes up the universe is named the brutino. The mass of an item is defined as the number of brutinos which comprose the item. The brutinos move in a straight line except when they collide. The collision interaction time for brutinos is instantaneous. Collisions are such that for a reference frame in which the normal component of velocity just prior to impact are equal and opposite, then the normal components are reversed while the tangential components are not affected. This collision mechanism provides the definitions of "elastic" and "smooth."

The set of postulates given above rigorously results in the following six universal laws of physics:

1. Everything in the universe is made up of one type of particle, the brutino.
2. Everything always moves with constant velocity unless it collides with something else.
3. Mass can neither be created nor destroyed.
4. Linear momentum can neither be created nor destroyed.
5. Angular momentum can neither be created nor destroyed.
6. Energy can neither be created nor destroyed.

Another document – one of the later or even the last in the sequence, written in 1969, has a title "Electron Gas Analogy Experiment"[96] which again indicates they seemed to be thinking along similar lines to both Wilbert Smith and PLV – in terms of very small moving particles, or a sort of "gaseous aether".

Not all of the documents are theoretical concepts, however. A few documents have proposed experiments. One experiment described is a "gravity amplifier," and is shown on page 3 of a short memo about the topic[97]:

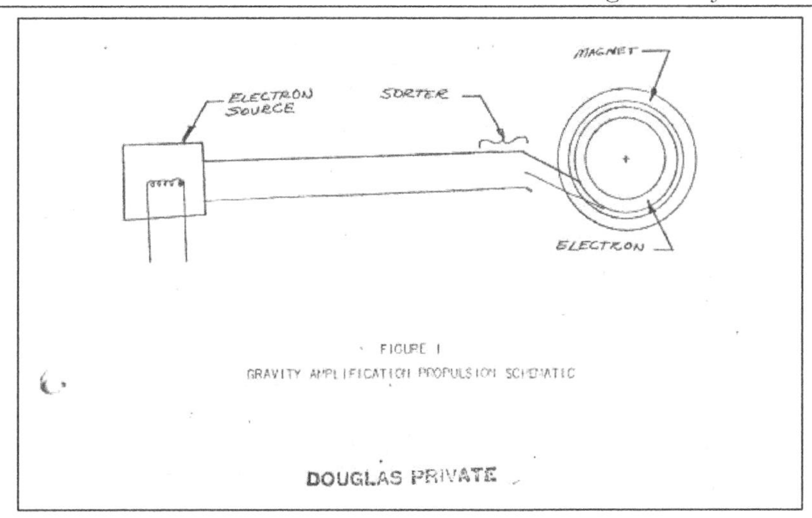

Another memo with a subject line "Light/Magnetic Field Interaction Experiment" describes an attempt to measure the effect of a magnetic field on a beam of light.

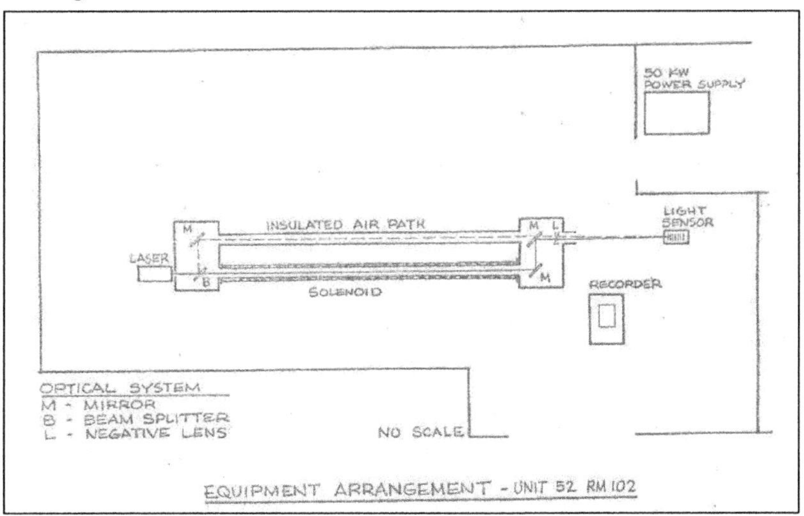

As can be seen in the diagram, a laser light beam is split and then passed through a tube surrounded by a solenoid. A variable current is passed through the coil in an attempt to affect the light beam. If this had worked, it would have caused the beam to be deflected slightly, lengthening its path. This would be detected by an interference pattern at the light sensor. The diagram above shows a 50 KW power supply (about the power needed by 20 electric kettles to boil water), so this would have been a good, strong magnetic field!

The 12-page memo describes that they did not detect any effects, even though their theory had predicted them. However, they suggest that the effects may actually be present, but they were too small to be detectable with their experimental set-up. Suggestions on how the experiment could be improved are also made.

"A New Communications Mode"

A 20-page document with this title[98] discusses ideas about how communications might work when vehicles have to travel at high speed. On page 3 we read:

> *With the advent of longer-range supersonic flight vehicles, and the approach W of possible interplanetary travel, communications and navigation needs play an ever-Increasing role in the system performance of any vehicle-ground system complex. This paper examines the possibility of a new communications-navigation concept utilizing a mode of Information transfer adapted from natural phenomena other than electromagnetic.*

There follows a list of 29 requirements of such a communications system or method and then, on page 4:

> *These requirements eliminate systems utilizing the following modes:*
>
> 1. *Electricity*
> 2. *Magnetism*
> 3. *Light (Optics)*
> 4. *Heat*
> 5. *Hydraulics*
> 6. *Electromagnetics*
> 7. *Nuclear Energy*
> 8. *Solar Energy*
> 9. *Combustion*
> 10. *Sound*
> 11. *Mechanics*
>
> *It leaves little with which to work. The two modes which possibly could satisfy most if not all of the requirements are:*
> 1. *Gravity*
> 2. *Magnetohydrodynamics*

There is discussion of "instantaneous communication" by something like gravity waves and later in the document, it seems like someone makes a "pun" with a heading shown below:

> <u>*A Possible Example: ESP (Equilibrium System Perception)*</u>
>
> *If a communication mode already exists in nature utilizing a gravity system, we possibly could find it in that portion of mammalian physiology which is dependent upon gravity for its continuous and successful operation.*

The text then discusses how animals detect gravity – as a way of keeping their balance or to use for navigation. However, on page 11, we read of tests that were done a couple of years or more earlier, up to 1965:

> *It was decided that inasmuch as a reception was made, a transmission had been made, and that a controlled test should be performed. First, however, requirements should be established for being a good receiver, and for a good transmitter.*

> *From the small experience up to April 1965, we had learned that the best Description of a good receiver is a person who has found his inner peace. One who has learned that concern and worry are not the same; who has found minute-to-minute, hour-by-hour, day-by-day, month-by-month, year by-year ways of meeting everything, ranging from happiness and joy to boring nomalcy to extraordinary adversity with calm acceptance, gaining victory when necessary; accepting failure when unavoidable, and turning it to success whenever possible. It takes a relaxed mind and body to be a good receiver.*

It appears they are talking about some kind of telepathic communication. This is very similar, again, to what Wilbert Smith talked about – and seemingly was doing – with his "Topside" contacts…

Acquiring Data about UFOs

A document entitled "Field Data Acquisition Requirements," authored by Robert M Wood,[99] goes into quite a bit of detail about what they need to acquire data about UFOs. On page 2, we read:

> *UFO TARGETS*
>
> *A basic analysis of UFO reporting strongly indicates that their presence and operation may be associated with any one or a combination of several 'observable physical phenomena. They may produce steady state and cyclic changing, magnetic, electric, electromagnetic (photon) and gravitational fields. They may emit nuclear particles, generate steady state or acoustical atmosphere pressure fields and leave pronounced residual effects. the targets may produce weak or strong signals with respect to the ambient background and may be within range of the sensors for long periods to - short time intervals. The shortest interval would most probably be associated with a close-range flyby. For this reason, it may be seen that the shorter times might produce the strongest signals. For example, a very close fly—by at 10,000 feet per second could be within the range of practically all sensors for a period of several seconds. A data system that would not saturate and could record all possible signals for these conditions would provide significant information. Therefore, sensor system capabilities which will respond in the magnitude range of ambient to a high level, to give Spectral content (and polarization, where applicable), and to be activated over the full time of event, would be the ideal system for these extremes.*

The document then goes into quite some detail about detection parameters – such as timings, wavelengths of radiation that might be detected by a given sensor etc. It also considers whether a detector would be in motion, or in a fixed position.

Even the gathering of information from contactees is described as the second main avenue of deriving useful data on UFO propulsion.

Their rationale for the use of contactee data can best be derived from the so-called "Advanced Vehicle Concepts Research" presentation dated May 2nd, 1968, where it is shown on a slide - "Philosophy: Hear everything and use some of it to our advantage" and on page 25 "contactee data could be useful, experiencing high accelerations - without forces, visual internal appearance of vehicles."

They even considered paying a contactee – Barbara J Hickox:

> *"We discussed the possibility of employing Mrs. Hickox as a consultant. We told her that if she accepted employment as a consultant the company would expect to own any Ideas divulged by her. She would, in return, receive the agreed upon hourly compensation. We agreed to proceed with background information gathering to prepare a recommendation to our management that she be employed as a consultant at a rate of slightly over $3.00/hour. If our checks on her resulted in our recommendation to management and if management concurred, then a few exploratory hours of her time would be utilized. Further time might then be warranted to go into various areas in great depth. Primarily in her description of the vehicle and its propulsion system."*

The documents even contain some satirical cartoons:

M—43081
UNITED STATES
AIR FORCE
UFO
INFORMATION
CENTER
"I assure you, Madam, if any such creatures as you describe really existed, we would be the first to know about it."

As far as I could tell, there is no mention of Thomas Townsend Brown's research in any of the documents, nor of the term "electrogravitics," although I could easily have missed some of these terms, as some of the pages are handwritten.

Postscript – Robert M Wood and the Condon Committee

Around the same time that the Douglas Aircraft BITBR project was progressing, an official report was being put together for the US Air Force on the subject of UFO's – it was entitled "The Scientific Study of Unidentified Flying Objects."[100] The study was completed by a committee of researchers[101], headed up by Edward Condon, a physics professor at the University of Colorado. Much of the report was contracted through the same university. Among evidence-based UFO researchers, the Condon report is notorious because Condon himself more or less stated what the report would conclude quite some time before the report was even finished.[102] It seems that in the 1960s, there were a greater number of scientists who took the UFO phenomenon seriously, and when the report was published, quite a number of them, such as Dr James McDonald, expressed their dissatisfaction with the report.

Before the report was published, Robert M Wood, one of the BITBR team, mentioned above, had a meeting with some members of the Condon Committee. Wood's article makes for interesting reading and he describes how he explained to some of Condon's committee members what they were working on, but Condon only wanted to discuss Wood's lack of concrete results and talk about UFO hoaxes. Wood was "stunned" by the committee's apparent lack of knowledge of some of the key UFO evidence and the best cases. He was also shocked at their lack of curiosity.

Louis, the person who released the BITBR documents to us, also provided a copy of an article the Dr Robert Wood wrote which was published in the 1993 Jul/Aug issue of International UFO Reporter (IUR) journal.[103] Louis also provided a summary of the article, thus:

> *Wood and Co. briefed the Condon Committee which was ongoing at the same time. They explained that it was possible to make a superconductive ring of current float in a magnetic field with only 10 times the current capacity that they had available at that time. Condon wasn't interested. Wood then wrote and suggested that the committee work in 2 groups, believers and nonbelievers. Condon retaliated by writing to the head of the company and trying to have Wood fired. On page 437 he mentions that they spent around $500,000 between 1968 and 1970. This figure is close to what the Condon Committee spent around the same time according to David Saunders' book "UFOs... Yes!" (page 182).*

Wood's 1993 article also mentions the AAAS (American Association for the Advancement of Science) meeting about UFO's that was meant to have taken place in 1968 (I discussed this event in "Secrets in the Solar System"[104]). Also included is a copy of Wood's letter to Edward Condon. In this letter, Wood mentions two contractors that are involved in the development of the Condon Report – Stanford Research Institute (SRI) and Raytheon. SRI appears six times in the Condon report's Appendix X[101]. (SRI got involved with Richard Hoagland's Mars/Cydonia research, they were involved in Operation Stargate – the remote viewing programme, they were involved in LENR research...)

Post Postscript – "The Mechanical Theory of Everything"

Some 10 years after I first posted the Douglas Aircraft BITBR documents, a new book was published by J.M. Brown[105] (in 2015) – one of the researchers in the original team. The description for the book reads:

> *The Mechanical Theory of Everything is a comprehensive and unifying look at how the universe works. Through fresh insights and rigorous derivations, readers will learn*
>
> - *where energy comes from*
> - *how a photon dissipates in ten billion years*
> - *what electrons and protons are made of*
> - *the solution to Einstein's Unified Field theory*
> - *how language is made*
> - *and why we age.*
>
> *The evidence presented is compelling and spectacular: the universe we live in is mechanical.*

I think it is correct to consider that studying the nature of flying saucers/UFOs etc is one way to unlock some of the fundamental secrets of the Universe. Indeed, someone has already done this, in modern times and then managed to keep most of their research and results hidden from the rest of us…

6. Coral Castle, A Lost Love and the Case for Antigravity

For hundreds of years – millennia even, there has been an enormous amount of speculation, debate and argument about how the pyramids in Egypt were built. Experiments have been performed, models have been built and theories have been put forward. Considering the incredible engineering in some of the Egyptian pyramids, and similar ancient structures like the mostly destroyed temples in Puma Punku[106], one of the issues that has to be considered is how enormous blocks were quarried, lifted and moved to their final positions. The precision of movement and details of the construction of these monuments is a separate issue. In this book, I merely wanted to introduce a place called Coral Castle to the reader who is not familiar with it. There are many interesting books, videos and websites which go into a lot more detail than I do here. Part of this chapter is based on an article I wrote in 2005.

Coral Castle

Perhaps more accessible for tourists than either the Giza Pyramids and the Tiwanaku site of Puma Punku in South America is "Rock Gate Park" or, as it is now known, "Coral Castle" in Homestead, Florida, about 40 miles south of Miami.[107]

As with the Pyramids and similar structures, one asks "how was this made?" Can it be correct that these things were constructed using only simple stone tools, rollers, piles of earth, huge numbers of men and brute force? Theories that this was not the case are never taken very seriously. But could a place like Coral Castle ever be taken as proof that such theories about "brute force" methods of construction are not necessarily correct in all cases?

Aerial view of Coral Castle, Homestead, Florida

The aerial view may be a little deceptive – and it might not seem that surprising that this structure was built by a diminutive man called Edward Leedskalnin. However, visiting the site, examining it close up and considering and understanding the circumstances and nature of its construction will leave most people scratching their heads.

A Lost Love?

Leedskalnin was born on August 10, 1887 to a farming family at Stramereens Pogosta, a small village near Riga, Latvia, but he emigrated to North America in about 1913. According to one account, the reason why Ed emigrated seems to be related to the fact that, when he was 26 years old, he was engaged to be married to his one true Love, Agnes Scuffs. Agnes was ten years younger than Ed; he affectionately referred to Agnes as his "Sweet Sixteen". Agnes cancelled the wedding just one day before the ceremony – apparently saying Ed was too poor and too old! Ed was heartbroken.

After emigrating, Ed moved to Canada and, while working in a Canadian lumber camp, he contracted tuberculosis. So, in 1918, he moved to the warmer climate of Florida. There, he purchased an acre of land near Florida City, for $12. He had no formal education.

This story is the central theme of a book by Joe Bullard[108] called "Waiting for Agnes".[109] The story goes that after settling in Florida, some 5 years after this great disappointment, he began a "project" which is said to be the result of his will to create a monument to his lost love – to show her what he was capable of.

According to the "Sweet 16" story Coral Castle, like the Taj Mahal, was built as a testament to his loved one.

Ed started to build Coral Castle where he originally lived, in Florida City, in about 1924. It is called Coral Castle, because Coral is the material from which it is made. Ed had acquired some skills working in lumber camps and some stone mason's skills whilst he was still in Latvia. His blocks were carved at the original Florida City site, as he constructed the place.

The construction work continued until about 1936, when he found out that someone planned to start building housing (or at least some kind of building) next to him. Being intensely private and reclusive, Ed decided to "up bricks" and leave, and he transported the nearly-complete Coral Castle to a new site, which was 10 acres of land he had purchased at Homestead (still in Florida), 10 miles away from the original site. This move took him 3 years to complete.

This story doesn't sound so unusual, until you realise that Ed built the whole of Coral Castle himself, using only tools and equipment that he made. Ed was 5 ft. tall and weighed about 100 lbs – a diminutive 7 stone! This tale still doesn't sound very unusual until you realise that Ed built his "humble abode" using blocks that weighed up to 30 tons!

On the following pages, I have included some photos I took in August 2017, when I was finally able to visit this incredible place.

There are quite a number of intriguing structures at Coral Castle. One such structure called "The Great Obelisk" is over 25 feet high and weighs over 28 tons - taller than the Great Upright at Stonehenge. Carved on its surface is the year of completion, the year it was moved and the year and country of Ed's birth. The hole near the top is carved in the shape of the Latvian star.

Among its other oddities is a scattering of oversized chairs also made of coral, each one weighing half a ton. Although they look extremely uncomfortable, I can say that I have sat in 2 of them and they were quite comfortable. Ed actually sculpted and placed 3 reading chairs at right angles to one another, so that he could move between them to make best use of direct sunlight.

Entrance

The Obelisk

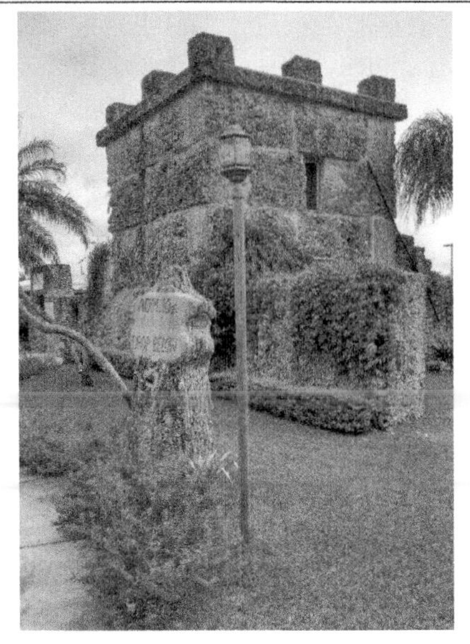

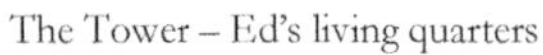

The Tower – Ed's living quarters

Features from left to right: Polaris Telescope (aligned to Pole Star), Crown (on North Wall) Mars, Saturn, Crescent of the East, The Well (behind railings)

Tool Selection (in Base of Tower)

Wheel holding opposing magnets (see below).

Closer view of Polaris Telescope

Looking down into the well.

North Wall and Crown

9-ton rock gate – originally span freely

"Feast of Love" table.

Part of one of the Rocking Chairs (top) and Ed's sophisticated Sundial, which showed the time of year (on the vertical) as well as the time of day (on the horizontal).

He also made several rocking chairs – which were perfectly balanced. Only one of these can now be "rocked". Another unusual construction is a heart-shaped "Feast of Love" table – which is estimated to weigh about 5 tons. This had an Ixora bush placed in the centre, where it remained for about 50 years – it was lost during Hurricane Andrew.

When Ed moved all the blocks he had used, many people saw the coral carvings being transported along the Dixie Highway, but no one had actually ever seen Ed loading or unloading the trailer. In 1940, after the carvings were in place, Ed finished erecting the walls. The coral walls weigh about 125 pounds per cubic foot. Each section of wall is 8 feet tall, 4 feet wide, 3 feet thick, and weighs more than 58 tons!

In December 1951 Ed fell ill. He put a sign on the door of his Castle saying, "Going to the Hospital" and bussed himself to Jackson Memorial Hospital in Miami. Three days later he died in his sleep, of malnutrition and kidney failure, at the age of 64. After his death, his only surviving relative, Harry who lived in Michigan, inherited the Castle. In 1953, shortly before Harry's own death, he sold the Castle to a family from Chicago, who gave it its present name. During the take-over, a box of Ed's personal effects was found. It contained a set of instructions that led to the discovery of 35- $100 bills, Ed's life savings - which he made by giving tours of the castle for 10¢ or 25¢, selling pamphlets and from the sale of the land where U.S. Highway 1 passes the Castle.

For those wanting to learn more about Ed's life and the circumstances which led him to build Rock Gate Park, I recommend that they read Carol A Lake's "Coral Castle Book".[110]

How Did He Do It?

It has been estimated that 1,000 tons of coral rock were used in construction of the walls and towers, and an additional 100 tons of it were carved into furniture and art objects, which adorn the site. These facts, then, present us with several puzzles. Firstly, how did Ed single-handedly carve out these blocks?

Scott Russell's "Made by One Man"

One answer to how the rock cutting may have been achieved is contained in a video which was posted on YouTube in 2015, made by Scott Russell[111].

Russell has built replicas of a number of Ed's tools – including the unusual "magnet wheel" pictured above. Russell contends that this wheel was part of a "voltage pulsing system" which Ed used to trigger relays which would move cutting blades up and down or from side to side. Russell claims that Ed built a number of these wheels, changing the design from having 16 spokes to 24.

Some of Russell's arguments seem quite powerful. For example, he points out that the iron door at Coral Castle has the same length as one side of some of the blocks and he therefore states that the door was used to cut the blocks. I don't find this statement convincing.

Russell does make another logical suggestion, stating that Ed worked at night because it was cooler. Russell suggests that Ed set up a number of old car headlights, powered from about 20 car batteries, wired together in parallel. He points out there appears to be an old diesel generator dynamo at the museum and suggests this was used to recharge these batteries during the day. Russell states that Ed's machine, then, could provide both a cutting action and light. Russell also demonstrates the use of two or three tripods which he claims Ed used to move the blocks. He illustrates his idea quite well but does not show any similar sized blocks actually being lifted. Neither does Russell run his replica cutting machine long enough to cut any sizeable rock. However, it seems clear that Ed used wedges to break a slab of rock off, as is shown in Russell's video.

The photo of the tripod, above, includes a black box at the top, which Russell seems to suggest would contain a car battery. Why this would be needed for lifting a block up just with an ordinary block and tackle, is not made clear by Russell. One can imagine that a car battery might be used to power a rocking or forward and backward moving cutting blade as he suggests, but why would it then be so far away from the surface to cut?

There has been some conjecture, therefore, over the function of this black box.

Centre of Gravity

One of the things which is not mentioned in Russell's video, and is often overlooked in other explanations of Ed's construction, is how Ed balanced perfectly such heavy blocks. A prime example is the 9-ton stone door. Not only did it have to be raised and lifted into place, its centre of gravity had to be known accurately, so that it could be placed on a bearing that could then be turned. How did Ed do this? He achieved a similar feat with the 3-ton block near the entrance. It can still be turned by a child (I did this myself on my visit).

It is still hard to imagine how Ed lifted these blocks precisely into place without large cranes or other heavy-duty equipment. Even if you decide that he must have used some kind of crane, as a comparison, consider this account from Christopher Dunn about a modern construction project:[112]

> *"My company recently installed a hydraulic press that weighed 65 tons. In order to lift it and drop it through the roof, they had to bring in a special crane. The crane was brought to the site in pieces and was transported from 80 miles away over a period of five days. After 15 semi-trailer loads, the crane was finally assembled and ready for use.*
>
> *As the press was lowered into its specially prepared pit, I asked one of the riggers about the heaviest weight he had lifted. He claimed that it was a 110-ton nuclear power plant vessel. When I related to him the 70 and 200 ton weights of the blocks of stone used inside the Great Pyramid and the Valley Temple, he expressed amazement and disbelief at the primitive methods that are promoted by Egyptologists."*

We can ask, how can Ed possibly have moved these blocks a distance of 10 miles? It is reported that Ed had the chassis of an old truck on which he laid two rails. Ed would load the trailer himself. He had a friend with a tractor move the loaded trailer from Florida City to Homestead. Ed lived a very simple life and did not own a car. Instead, Ed would ride his bike 3½ miles into town, to do his shopping! So somehow, Ed managed to construct coral Castle *and move it* a distance of 10 miles, apparently with no special equipment *at all!*

The details of how he did it are not clear. If it was done by some kind of trickery or as a "magician's stunt", then it must surely rank as the greatest (and longest) ever performed. It certainly does not seem to be possible that the blocks are hollowed out – unless someone has carefully replaced them one by one, as they must have crumbled over the years.

Levitation?

As mentioned above, Ed Leedskalnin worked in secrecy, after sundown - by lantern light, when he was certain no one was watching him. It was reported that some curious neighbours did see Ed move the stones. They say he placed his hands on the stone to be lifted... but what was in his hands? Somehow this levitated the blocks. According to an article in *Fate* magazine, "some teenagers spying on him one evening claimed they saw him 'float coral blocks through the air like hydrogen balloons,' but no one took them seriously. When he was personally asked how he managed the feat, Leedskalnin replied only that he understood the laws of weight and leverage. He is also quoted as saying,

> *I have discovered the secrets of the pyramids. I have found out how the Egyptians and the ancient builders in Peru, Yucatan, and Asia, with only primitive tools, raised and set in place blocks of stone weighing many tons.*

The very stones of Coral Castle support his story - at an average of six tons, they are twice the weight of the blocks in Egypt's Great Pyramid at Giza.

What other evidence is there that he did not use cranes or other heavy equipment to build his masterpiece?

The Case for Antigravity

At the castle, you can see the tool room where some of Ed's equipment is on show. One item (pictured above) is said to be an AC generator he built, though how he used this is not really known. At one time Ed erected a massive grid of copper wire poised above his quarried stones – there is quite a lot of copper wire in his workshop. Pictures also show Ed working with tripods, though none of these tripods are on the site now.

Engineer Christopher Dunn suggests[112] that the device shown above is not an AC ("alternating current") generator, but an "alternating magnetism" generator. He proved this by holding a bar magnet near the device and setting the "generator" in motion and found that the North end of the magnet was alternately repelled and attracted. Dunn goes on to suggest that gravity may not be a "real" force – it arises due to an object's own magnetic field interacting with the Earth's magnetic field. He also suggests that Ed had devised a method for making the magnetic fields within the coral blocks line up in opposition to the Earth's magnetic field.

As a guess, Ed wrapped the blocks with a copper wire grid and then connected this wire to his magnetic field generator. Starting up the generator then had the effect of changing the magnetic field in or around the block in such a way that it was levitated. It is therefore suggested that the tripods were used merely as a way of supporting the wires and chains which were used on the copper grid – or maybe they were used in the photographs as a "decoy", to throw people off the scent of how he actually did it (i.e. they may not have been used in the lifting procedure at all). Carrol A. Lake, a colonel in the U.S. Army Corps of Engineers, whose book I mentioned earlier in this chapter stated "Leedskalnin proved for all the world to see today that he knew the construction secrets of the ancients." Vincent H. Gaddis, who wrote a number of articles for *Amazing Stories* magazine, said of the mysterious Latvian immigrant, "There is no doubt that he applied some principle in weight lifting that remains a secret today."

According to a Ray Stoner's 1983 book "The Enigma of Coral Castle",[113] Ed disagreed with modern science, and claimed that the scientists were wrong, 'that nature is simple.' This idea may also perhaps be considered in terms of what Ed wrote in his 1945 book "Magnetic Current"[114]. Near the end of this short, but difficult to grasp book, he wrote:

> *As I said in the beginning, the North and South Pole magnets they are the cosmic force. They hold together this earth and everything on it, and they hold together the moon, too. The moon's North end holds South Pole magnets the same as the earth's North end. The moon's South end holds North Pole magnets the same as the earth's South end.*

Aquifer and Alignment

One thing I learned when I visited Coral Castle was that it was built on the Biscayne Aquifer[115]. This reminded me that when considering "unconventional forces," water can be an important factor. For example, Wiltshire in the South of England has been the site for the highest concentration of crop formations in the world. Wiltshire is also on top of the biggest chalk Aquifer in Europe.[116] Could this be significant? Earlier in this chapter, I mentioned how Ed moved the site because of "new neighbours." However, an article by an unknown author on the Labyrinthina website states[117]:

> *These telluric grid dynamics played a vital role in the construction of the Castle according to author Ray Stoner in his book, "The Enigma of Coral Castle." Stoner speculates that the complex was originally moved from Florida City to Homestead not because of privacy issues (as most historians suggest), but because Ed realized he had made a mathematical error in his original positioning and moved the entire structure to take advantage of an area with greater telluric force.*

It is Still There!

One would have thought that because Coral Castle was built and still exists – a testament to "lost love" - people might take the theories of someone like Leedskalnin much more seriously. He was either onto something, or he pulled off perhaps the greatest civil engineering hoax in history.

Footnote - A Link Between Gravity and Magnetism?

Now that we have mentioned "Magnetic Current" by quoting a document written by Ed Leedskalnin in 1945, I will now include another quote, made 59 years later:[118]

> *Gravitomagnetism is produced by stars and planets when they spin. "It's similar in form to the magnetic field produced by a spinning ball of charge," explains physicist Clifford Will of Washington University (St. Louis). Replace charge with mass, and magnetism becomes gravitomagnetism.*

The source of this quote is a NASA Science Story (20 April 2004). It is headlined thus:

> *In Search of Gravitomagnetism*
>
> *Gravity Probe B has left Earth to measure a subtle yet long-sought force of Nature.*

In the next chapter, like Wilbert B Smith, we will consider the effects that spinning objects have on gravity.

7. Is Antigravity "in a Spin"?

Readers with less technical backgrounds and knowledge may find parts of this chapter hard to understand. I will therefore give a simple summary at the end.

A recurring theme which appears when researching what is known about antigravity devices and techniques, as we mentioned briefly in chapter 3, is that high-speed rotation can seemingly have an influence on the force of gravity that an object experiences. Of course, this is a well-known effect, in certain cases. For example, when you turn corners (change direction) at high speed, such as when riding a spinning fairground ride, you experience a force which may seem to push you down into your seat. This is centrifugal force (although by some definitions, this isn't a real force). It has been postulated that in a space station, you could have a circular living section which, when spun at a certain speed, would seem to create its own gravity, but this would really be another example of centrifugal force.[119]

Similarly, there is centripetal force – which is the force that keeps an object moving in a circular path[120]. What we are discussing in this chapter might be *related* to one or both of centrifugal and centripetal forces, but it is clearly not the same thing. Anomalies in the experienced force of gravity can be demonstrated fairly easily, as we can see, but the effects are very small, so sensitive equipment and careful experimentation is required to make the required measurements.

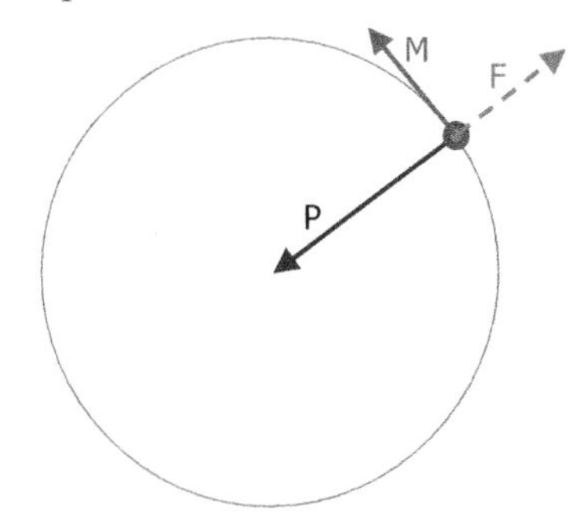

Imagine a ball on a string being spun in a circle, M would be the direction of motion. 'F' would be centrifugal force and 'P' would be centripetal force.

The space station in the film 2001 was shown as rotating at a constant speed, so that occupants could walk around on the inside of the outside rings. Though this was based on accepted physics, it has never been tested on this scale in space.

Bruce De Palma, Gravity and Free Energy

Bruce Eldridge De Palma (1935-1997), was the son of an orthopaedic surgeon and he was the elder brother of film director Brian De Palma[121]. He studied electrical engineering at Harvard (1958) and taught physics at the Massachusetts Institute of Technology (MIT) for 15 years. He was also employed for a period of time by Edwin H. Land of the Polaroid company (the one that made the pictures which would develop themselves just after you took them!)

De Palma is a very important figure in the realm of what might be called "Alternative Science." He did some very basic experiments which led him to question the conclusions about the "way gravity worked." He also developed a device called the N-machine, which was based on an experiment performed by Michael Faraday[122]. The N-machine[123] was a free energy device and could generate a lot more energy than was fed into it. However, the energy which came out was in the form of a very high current (hundreds of amps) and low voltage (typically less than 10 volts) – this is the "opposite way round" to most electrical generators, where the voltage output is typically a lot higher than the current.

It was seemingly development of the N-machine which, ultimately, cost De Palma his life.

Here, however, we will cover De Palma's other experiments with a spinning ball and gyroscopes which seemed to be superficially similar to the experiments performed by Wilbert Smith, that we covered in chapter 4. I would bet that the effects De Palma observed have some relationship to what Boyd Bushman described to Nick Cook in 1999, as we shall see in chapter 10.

DePalma's scientific career was deeply affected when he was a lecturer at M.I.T., in the late 1960s - when he started to look at the behaviour of gyroscopes, which exhibit certain unusual effects.

Spinning Ball and "Force Machine" Experiments

De Palma wished to determine if the force of gravity experienced by a spinning object was different to a rotating object (just like WB Smith!)[124] Working with a student, initially, they could not find an experiment that had already been performed to test this, so they designed their own. Below, I have included the details of the experiment he performed:

> *He designed an experiment using two 1-inch diameter ball bearings, one not rotating and one rotating 18,000 rpm produced by a hand router. The assembly then was given a precisely measured thrust and photographed in the dark with a 60 cycle strobe light. Repeating this numerous times and analysing the parallel trajectories of the ball bearings as documented photographically, did indeed reveal a variation in the gravitational behaviour of the rotating ball bearing verses the non-rotating ball bearing. The rotating ball given the same thrust, went to a higher point in its trajectory, fell faster, and hit the bottom of its trajectory before the non-rotating ball bearing.*
>
> *A second test repeatedly demonstrated a small but significant and clearly perceptible effect with a stationary mechanism designed to drop the ball bearings from a height of only six feet.*

The results of the Spinning Ball Experiment were published in the British Scientific Research Association Journal in 1976[125].

De Palma found in his experiments that a rotating object experienced a *slightly stronger* force of gravity and fell more quickly than when the exact same object wasn't rotating.

Also, in the 1970s DePalma developed a device or apparatus he called a "Force Machine," which consisted of suspended, spinning gyroscopes which were rotated, at high speed, in opposite directions. The "machine" was suspended from a spring scale and the gyroscopes were spun at 7600 rpm. The gyroscopes themselves were contained in a cylinder. A diagram of a version of this "force machine" is shown below.

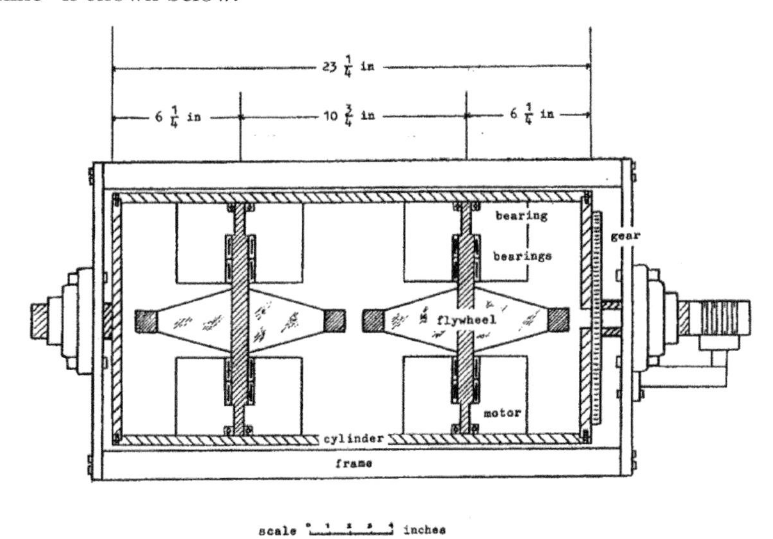

Schematic of Bruce De Palma's 1970s "Force Machine"

In his experiment, De Palma rotated the support cylinder at 4 revolutions per minute (rpm). Accurate measurements consistently showed the unit experienced 4-6 pounds of weight loss!

De Palma also noted that when the suspended force machine was gently swung – like a pendulum, the way in which it moved while the gyroscopes were spinning was different to how it moved when the gyroscopes were stationary. He observed that the "peaks" of the swinging motion would be slightly flattened. Normally, swinging pendula undergo something called "Simple Harmonic Motion"[126] and this means the motion of the pendulum itself follows a path which can be represented as a Sine Wave. De Palma found that the motion of his "Force Machine" pendulum no longer followed a Sine Wave – the pendulum's rate of swing would be slower.

In his article "The Secret of the Force Machine," Bruce DePalma writes[127]:

> *When a real mechanical object, a flywheel, is rotated, forces appear, the centripetal forces of rotation within the material of the flywheel. These forces are the counterpoise to the spatial distortion created by the centripetal acceleration applied to the mass elements of the rotating wheel. Although these forces are not available for explicit measurement, their presence is evidenced when the wheel is rotated at a high enough speed such that the forces exceed the tensile strength of the flywheel material and an explosion*

> *results. The interesting phenomenon is that no work is required to maintain these forces at arbitrarily high values.*

De Palma then tries to draw an analogy between "gravitational power" and "electrical power," though I must admit that I find his thinking rather difficult to follow here. In the article, he argues that

> *Gravitational energy is a flow - not a force, which distinguishes it from Newtonian forces arising from the acceleration of masses.*

Interestingly, De Palma states, later in the article

> *Actually, what is connoted as gravitational power flow and mechanical power output derived from Free Energy anti-gravitational apparatus is Time-Energy.*

This sounds similar to what Wilbert Smith discussed in his "New Science" book – where he talked about "a tempic field."[91] Smith said he was told by the boys "topside" that it was possible to extract energy from the tempic field.

De Palma argued the Force Machine was generating a "force against space itself" or a "space drive." He said that he had made a similar device which could propel someone across the room on a small cart or trolley. (Professor Eric Laithwaite, whom we will discuss later in this chapter, performed similar demonstrations.)

Accutron Watch Experiment

In further experiments, De Palma even noted effects on time keeping devices in the presence of a spinning flywheel. The Accutron Watch[128] predated the wide availability of "quartz watches" (i.e. those watches which used an oscillating quartz crystal as their time-base). It used a "tuning fork" mechanism which oscillated 360 times per second. It is not clear why, exactly, De Palma's experiment would affect the Accutron Watch specifically – particularly in view of the observations he makes in writing about the experiments he did.

Here, I reproduce part of a paper he wrote in 1993 called "On the Nature of Electrical Induction"[129]

> *The Experiment: A good way to detect a field whose effect is a spatial inertial anisotropy is to use a time measurement based on an inertial property of space and compare it to a remote reference. With reference to figure (1) we have a situation where the timekeeping rate of an Accutron tuning fork regulated wrist watch is compared to that of an ordinary electric clock with a synchronous sweep second hand.*
>
> *The Accutron timepiece is specified to be accurate to one minute a month. Examination of the relative time drift of the Accutron - electric clock combination shows a cumulative drift of .25 second Accutron ahead for 4 hours of steady state operation. This is within the specification of the watch.*
>
> *With the flywheel spinning at 7600 r.p.m. and run steadily for 1000 seconds (17 minutes), the Accutron loses .9 second relative to the electric clock.*
>
> *Much experimentation has shown that the effect is greatest with the position of the tuning fork as shown. Magnetic effects from leakage fields from the gyro drive motors are almost entirely absent; any remaining leakage is removed by*

> *co-netic magnetic shielding. The Accutron is also in a "non-magnetic" envelope.*
>
> *The purpose of the experiment is a simple demonstration of one of the effects of the OD field of a rotating object. The demonstration may easily be repeated using any one of a variety of rotating objects, motor flywheels, old gyrocompasses, etc. The rotating mass of the flywheels used in these experiments is 29½ pounds. The rotational speed of 7600 r.p.m. is easily accessible. The effect is roughly proportional to the radius and mass of the rotating object and to the square of the rotational speed.*

At the end of this paper he states:

> *Conclusions and Observations: The proper conclusions and evaluations of the above experiment will affect present conceptions of Cosmology. Before this can happen, simple tests must be performed to show the existence of a new phenomenon. It is hoped the apparatus for the performance of these tests is widely enough available to lead to quick verification.*

Others have tried to reproduce this experiment, but apparently didn't use an Accutron watch. The effects on the watch are, seemingly then, related to the particular way in which the watch works, rather than the effect being on the rate at which time passes. This is essentially what is explained by De Palma in the quote from the paper, referenced above.

Apparently, this experiment was explained to Dr. Edward Purcell, one of the most eminent experimental physicists from Harvard at that time. According to DePalma, Purcell, after considering the findings of the experiment for several minutes, remarked "This will change everything."

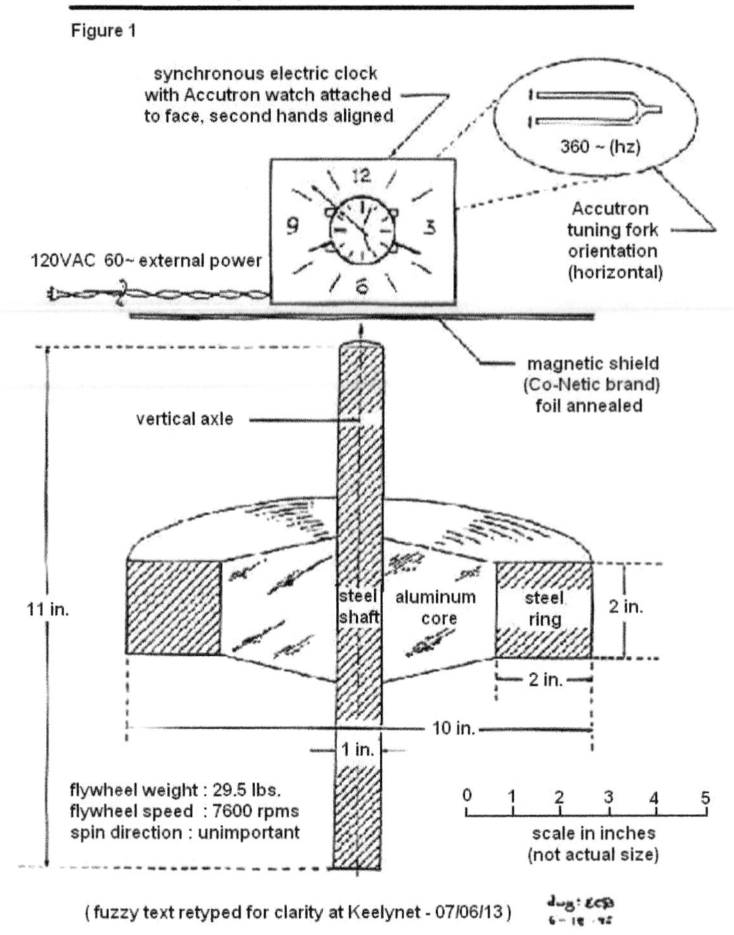

De Palma's Conclusions

By 1977, De Palma had come to the following conclusions

1. *Rotating objects falling in a gravitational field are* ***accelerated at a rate greater*** *than g, the commonly accepted rate* ***for non-rotating objects falling in a vacuum****.*
2. *Pendula utilizing bob weights which are rotating,* ***swing non-sinusoidally*** *with time periods increased over pendula with non-rotating bobs.*
3. *A precessing gyroscope has a measurable anomalous inertial mass,* ***greater than its stationary mass.***
4. ***An anomalous field phenomenon*** *has been discovered, the OD field, which confers inertia on objects immersed within it. This field is generated by the constrained forced precession of a rotating gyroscope.*

In his 1977 "Spinning Ball" paper[121], De Palma tried to explain what appears to be going on. Under the heading "Form of the Gravitational Interaction," he writes

> *We infer that the behavior of the falling non-spinning object, falling in accord with Newton's Laws, is a special case of the motion of objects in general. The more general case, involving rotation, is obscured by the gravitational interaction.*
>
> *We would expect, if we could increase the inertia of an object (through rotation of by some other means), that the object would fall more slowly in a gravitational field. Let us consider however that while a conferred inertial property, od, would reduce the acceleration of a given body acted on by a given force in outer space, in the presence of a gravitational field, the conferred inertial property would be an additional mechanical "dimension" for the extraction of energy from the gravitational field in falling. Conversely, enough energy could be delivered from this "dimension" to cancel, or overcome, the mechanical energy extracted from an object raised in a gravitational field.*

He then attempts to quantify the effects of this "inertial field" on gravity by suggesting modified versions of the equations which quantify an object's kinetic energy due to it falling in normal gravity. He then suggests:

> *In a gravitational field, the possibility of extraction of greater energy by a new mechanical dimension opens the possibility of an anti-gravitational interaction.*

He freely admits, however, that "he doesn't know everything…"

> *What is not determined at this point is the necessary increment of energy required to neutralize the weight of a given object, viz., it might take 1.1 foot pounds of work to lift a one pound object one foot.*

De Palma then considers the wider implications of his discovery and states that care is needed to consider experimental error and accuracy of measurements when trying to determine the reality of results. He then also introduces the concept of "simularity" (yes, spelled with "u") as a way of considering that our understanding of the way gravity works is approximate and we need to realise that there is a "defect" in the way we measure certain effects. On this theme, De Palma writes:

> *What is considered is the real properties of the level of causes and effects. What this represents physically as a form of inertia and a connection between rotation and gravitation. The "connectivity" of defect and the other real properties of inertia fields is better left to discussions to begin with the data presented herein. The theory is more properly left to the serious students of these ideas. As apprehension of the theory of Simularity necessarily entails the dropping of certain restrictions on the mind of the experimenter.*

Though the concepts he is trying to communicate seem quite subtle, like others who have studied anomalous gravitational effects, some of which we discuss elsewhere in this book, De Palma then states.

> *In the search for the gravitational interaction, we have long been hampered by the erroneous equation of inert and gravitational masses.*

He also notes that one must be careful in comparing experiments which may seem to be the same, but actually are not the same. An example might be repeating the Accutron Watch experiment without using an Accutron watch – it gives a different result!

Dr Eric Laithwaite and the "Abominable Knowmen"

Eric Laithwaite was born in 1921 in Atherton, Lancashire – not too far from Manchester. He was the son of a farmer. He joined the Royal Air Force in 1941, and from 1943 he worked on automatic pilot systems at Farnborough, UK. It was perhaps this area of work that stimulated his interest in gyroscopes and his desire to understand more about them.

In 1946 he went to read Electrical Engineering at Manchester University, graduating with a BSc in 1949 and an MSc in 1950, when he joined the staff as an Assistant Lecturer. He worked at first under Professor F.C. Williams on the Ferranti Mk I computer, but his real interest was in power engineering.

Though he became a Professor of Engineering, Laithwaite was a polymath – an expert on many diverse subjects - and he was very good at designing experiments and studying how physical systems worked. He was an expert on butterflies[130] and I remember a later documentary he featured in in the 1990s where he was, for example, asking questions and making observations about simple things – like how generations of weeds grow.

One of his main areas of research (with some success) was with Linear Motors (which are the power system in Maglev Trains, for example)[131] and he demonstrated these at the Royal Institution too.

Dr Eric Laithwaite and a working model of a Maglev Train.

From 1964 to 1986, Laithwaite was a Professor at Imperial College London, where he continued his research. In some videos of his presentations[132], he can be heard to talk about "a magnetic river" and "magnetic flow" – the similar terminology to that which Ed Leedskalnin had used in the 1940s[114].

Gyroscopes Galore!

Laithwaite created "a small stir" with a lecture he gave at the Royal Institution in 1974 – about gyroscopes[133]. I remember hearing this lecture when I was just 10-years old, as it was aired on BBC television as part of the Royal Institution Christmas Lectures.

What is particularly fascinating about this lecture is the demonstration of the apparent weight loss shown by gyroscopes. Laithwaite does not claim he has invented an antigravity machine, he merely points out, as Bruce De Palma did, that the laws of Newtonian Gravitation do not seem to be applicable to objects when they are moving like a gyroscope does. Again, I am reminded of Wilbert Smith's own antigravity experiment [134]– with a rotating disk. Laithwaite and Smith were both engineers and clearly held a similar fascination for observation, analysis and experimentation.

About 10 minutes into the lecture (Part 2 of a set of YouTube videos[135]), Laithwaite demonstrates some very interesting results of attaching weights to a gyroscope, by hanging them on a rod which is attached at the end of the gyro's axis of rotation. Laithwaite makes an interesting observation about the way the gyroscope's speed of precession changes suddenly as a weight is attached or removed. The acceleration seems instantaneous! This would mean that there is an "infinite force" at work!

In parts 4 and 5 Laithwaite shows a number of demonstrations which do seem to show that the gyroscope, when spinning, loses weight.

However, most people would explain this as simply the result of the gyroscope's angular momentum vector – which acts, along with the torque of the gyroscope's spin, to cancel out the force of gravity. Additionally, the moment of inertia must be taken into account, at some point.

Laithwaite was teaching Physics, however, and his explanations seem to imply that he understands all of the foregoing arguments clearly. He goes to some pains to state that if one changes how the gyroscope can move against the force of gravity, the torque produced does not seem to act as expected.

Demonstration Similar to Laithwaite's[136]

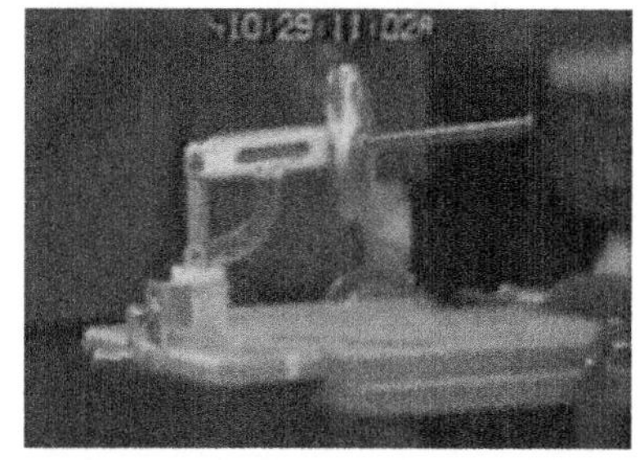

Laithwaite Shows how a heavy precessing gyroscope does now cause the hinged metal plate on which it sits to tip over.

Laithwaite uses a similar model to the image shown above, left, but he has an additional pivot in the rod, between the gyroscope and the centre pivot – so that the gyroscope, when in motion, could pivot at either point. He illustrates how its expected behaviour differs from the gyroscope on the hinged plate and implies this difference is inexplicable with current Newtonian physics.

An Analogy

In the last part of the lecture, he draws an analogy with electromagnetic theory, using another demonstration. He compares the situation with alternating and direct current flowing through the same coil of copper wire. His simple demonstration shows the coil has a *resistance* of 2 Ohms, producing a current of 4 amps with 8 volts applied. However, when the same coil has 4 amps AC running through it, the voltage is 32 volts – so the "resistance" (actually impedance) is 8 Ohms rather than 2 Ohms. He points out that Ohm's Law was not then said to be wrong, it just became understood that Ohm's Law only worked for DC current.

Laithwaite makes this argument to illustrate that, with the copper wire coil, the only thing that is different is that the current is alternating, but the measured figures are different. He therefore argues that the forces on the gyroscope, when it is spinning ("alternating," one might say), are also different. That is, "Newton's laws are restricted to objects that move in straight lines" (which is the same as what De Palma said). His demonstrations, like De Palma's, also imply that these effects are more clearly seen with higher spin rates – for example greater than 5000 rpm. (That is, he doesn't explicitly state this in his lecture, but various aspects of the demonstrations he does suggest that this is what he understands.)

Acknowledging Anomalies

Laithwaite was very careful in dealing with the way established figures dealt with the sorts of things he was highlighting, but he was willing to speak out. People accused Laithwaite of claiming he had created "momentum out of nothing." He stated:

> *But when journalistic editor got at the headline was something like Newton was wrong says professor. "Professor said nothing of the kind." But from there, it's a short step to perpetual motion energy creation from nowhere and half the crackpots in the kingdom are writing to me, solving the world's energy crisis.*

Near the end of the lecture[137], he states more research into particular gyroscope experiments is needed and then he says:

> *Gyroscopes do not exhibit a new force they show a lack of a force where there should have been one. That's why it was so hard to see a lack of centrifugal (force).*

I will just switch away from Laithwaite here for a moment to include a couple of still frames from a more recent lecture, which covers similar ground, but in a University Physics class. A lecture by Walter Lewin at MIT (date unknown, but probably recorded in the 1990s and posted in 2015) included some similar demonstrations to Laithwaite's[138]. One particular demonstration stands out. Having done several demonstrations with gyroscopes in different configurations, Lewin asks a student to help him with a suitcase that comes out from the back of the theatre. He asks the student to pick up the case and make some turns.

Walter Lewin's "suitcase Gyroscope" demonstration.

As the student turns, the case twists towards and away from his body. Again, physicists will just say that this is the result of what we discussed earlier – the interplay of forces of torque, angular momentum and inertia.

Back to Eric Laithwaite – whilst he seemingly did not wish to get embroiled in discussions about "mystical forces," he was nevertheless very critical of what he called "The Abominable Knowmen" or "Abominable No-men" – who "say 'No' before they have even watched…" In the lecture he states:

> *Are these discoveries new? Well not entirely and I've had letters from all kinds of people who've done similar experiments before and for one reason or another have given up - for (1) lack of know-how and scientific background (2) lack of money and workshop facilities (3) lack of courage (4) lack of anyone to listen*
>
> *Quite the most amazing letter I had was from a man who wrote quite unemotionally that he performed some of the experiments that I'd done some 10 years ago and he knew that they worked but he added:*
>
> *"I've not told any of my colleagues or friends here, because even though I occupy quite a senior post in the scientific civil service, I still have a few years to go to retirement and I still cherish hopes of further promotion."*
>
> *That is a hell of a thing to say in the scientific world isn't it? This man dare not tell his colleagues! We talk about black birds pecking out the white feathers of their colleagues or killing them if they can't because they didn't conform. This man is afraid of being put away, because he wouldn't conform!*

So Laithwaite, seemingly, started to get wind of how cover ups work! Sadly, he died in 1997 and an obituary in the UK Independent Newspaper said[130]:

> *In 1986 Laithwaite was delighted to receive the Tesla Award of the (American) Institute of Electrical and Electronics Engineers "for contributions to the development and understanding of electric machines and especially of the linear induction motor".*
>
> *In his later years at Imperial College Laithwaite pursued his interest in gyroscopes; he felt their behaviour had never been adequately explained and they had properties which might be exploited in space travel. He sought to demonstrate his ideas and raise some of his questions in a Friday Evening Discourse at the Royal Institution. The attempt brought him much criticism and some personal hurt when colleagues dissociated themselves from him. But he*

> *persisted and, as a colleague observed recently, some of his questions remain unanswered.*
>
> *Laithwaite worked to the end and died without regaining consciousness after collapsing in the laboratory.*

As we shall see, Laithwaite was right – there are aspects of gravity which physics cannot explain, but it isn't just related to the force of gravity that is experienced by an object being affected by whether or not that object is spinning. In any case, De Palma was spinning small ball bearings, not flywheels and he observed anomalous results too (at sufficiently high spin rates.)

Summary

In this chapter, we have had insights into decades of research by what I would call two *real* scientists – who designed simple experiments, performed these experiments, observed the results and then designed additional or improved experiments to find out more about why things worked in the way they did.

They demonstrated that the observed effects of gravity on objects that are spinning at a sufficiently high speed (greater than about 5,000 revolutions per minute) are not the same as those on stationary objects. The observations cannot be explained by Newtonian physics and neither can they be explained by relativity or quantum mechanics. My guess is that this high speed rotation affects the normal electric and/or magnetic field structure within atoms of the material, such that small effects on gravity can be observed.

8. Antigravity in 1940s Germany – For Whom The "Bell" Tolls

Much has been written about alleged secret Nazi technology and weapons programmes in the WWII and what is presented in this chapter covers just a small selection of information that seems to have become available.

In the 1990s and early 2000s, Nick Cook, a journalist and a former editor for Jane's Defence Weekly, made a significant contribution to the field of antigravity research. We have already referenced his book "Hunt for Zero Point," (which I will refer to as "HZP" from now on)

Cook's HZP book also covers some of the WWII period. It is a period where it becomes more difficult to establish the truth about most things, as Cook himself states in his book. For example, he talks about a 1950s book by Major Rudolf Lusar called "German Secret Weapons of World War II." It is this book that mentions German Flying Discs. Cook further discusses another book called "Intercept—But Don't Shoot" by an Italian named Renato Vesco which also discusses some elements of secret German research into aerospace and weapons technology. However, he finds that much of the information is unverifiable. It is easy to find many strange photos of alleged Nazi saucers or similar vehicles on the web, but finding more information about what they actually show seems to be rather difficult – one quickly seems to get drawn into stories of channelled information and the Vril Society[139]. Again, the reader can research such topics for themselves.

From what I have read about alleged Nazi flying disc programmes, I too have found nothing much that is concrete and agree with Cook's assessment of the available information, some of which we will discuss below.

Books by Joseph P. Farrell, such as "Reich of the Black Sun" and "S.S. Brotherhood of the Bell" also discuss some of this research, but for reasons outlined in my 9/11 books (see the relevant page near the start of this book for details), I cannot unreservedly recommend Farrell's writings, as it is clear that Farrell himself undertook some type of role in attempting to manage information about the events of 9/11[140] which was written up in Dr Judy Wood's book, "Where Did the Towers Go?"[76] Hence, Farrell cannot be trusted in relation to accurately reporting any important elements of secret German research in the 1940s. There are several other books and online videos regarding alleged German saucer development projects, but this is not a focus of this book, so I leave the reader to investigate those for themselves.

Schauberger and the Repulsine

Another device that is usually spoken of in relation to alleged Nazi flying Disk projects is the Repulsine. This was an invention of an Austrian, Viktor Schauberger (1885 – 1958). Schauberger developed the device because he had studied the behaviour of vortices in flowing water. He had observed these behaviours while working in the natural forest environment. Schauberger was clearly a gifted and highly observant man. One can even visit a museum where reproductions of some of his experiments[141] can be studied (though apparently, they though don't have a reproduction of the repulsine device).

In chapter 21 of his HZP book, Nick Cook does an excellent job of covering Schauberger's story – and how it links in to the picture of German / Nazi black programmes. I will only cover a summary here and refer the reader to that book for the rest of the details.

An article from 1997 by Callum Coats[142] summarises Schauberger's work thus:

> *Through his study of the vortices occurring naturally in flowing water and in the air in the form of cyclones and tornadoes, Viktor Schauberger developed his theories of Implosion. It was through the research and development of these theories that he was able to produce spring-quality water and generate considerable energies in and with water and air.*

From my understanding, Schauberger essentially claimed that with manipulation of vortex effects, unusual effects could be achieved. He even patented some of his ideas[143]. One patent was for a Jet Turbine[144]:

> *The object of the invention is a hydro-electric device, which exploits the kinetic energy of a water jet for the purposes of generating electricity.*

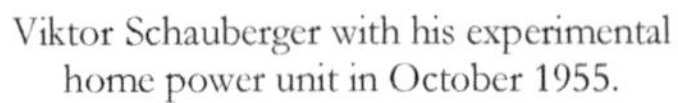

Viktor Schauberger with his experimental home power unit in October 1955.

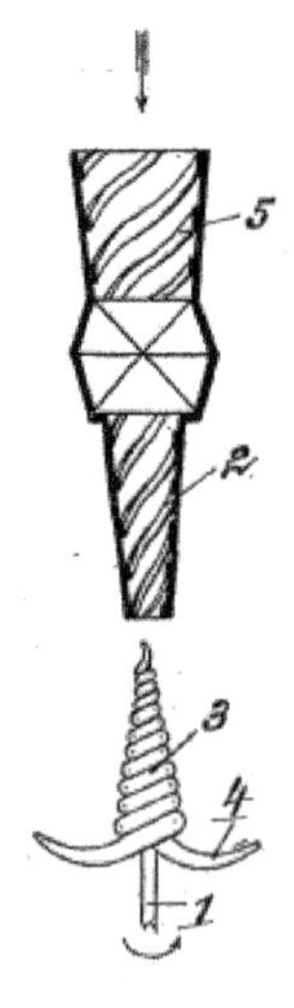

A diagram of a Jet Turbine device that Schauberger Patented in 1930 in Austria

In the 1940s, Schauberger designed a number of devices – which were given the name "Repulsine" – but this name seemed to have been re-used for several devices, according to what is in Schauberger's notes. The most well-known of these are pictured below:

According to one source[145], these were Schauberger's experimental models of flying saucers made from copper. Built in 1940 by Kertl co. In Vienna.

Jean Louis Naudin describes these units as the Type A[146] and Type B[147] Repulsine. On his website, Naudin attempts to illustrate how the units might work (which is not really understood):

> *The Repulsine device is an Electro-Aero-Dynamic device (E.A.D.) and uses two effects:*
>
> - *The Coanda Effect, a pure Aerodynamic effect based on the Bernouill's principle (see the Coanda Saucer Test)*
> - *The ElectroDynamic effect: The high speed vortex in the vortex chamber produces an electric charges separation effect, called "the diamagnetic effect " by Schauberger.*
>
> *These two effects combined create the so-called "implosion effect".*
>
> *When the main electric engine is started, the Coanda effect begins to create a differential aerodynamic pressure between the outer and the inner surface of the primary hull. At a higher speed, the vortex chamber becomes a kind of high electrostatic generator due to the air particles in high speed motion acting as an electrical charges transporter. The "Repulsin" will begin to glow due to the strong ionization effect of the air. Now, we have all the ingredients for a continuous and strong Aether Flow along the main axis from the top to the bottom of the craft...*

Naudin includes some diagrams to illustrate these explanations. He describes the Type B Repulsine, including the fate of one of the test models, thus:

> *In the Repulsin "type B" the vortex turbine has been improved for increasing the "Implosion Effect" and thus the lifting force, so you will find below a possible explanation about the working principle of "the Repulsin" type B flying saucer from Viktor Schauberger.*
>
> *The enhanced vortex turbine increases significantly the "implosion" effect in the vortex chamber. This contributes to generate a stronger thrust than the centrifugal turbine used in the Repulsin type A. By means of a suction screw-impeller, which revolved from the outside towards the inside along a cycloid spiral space-curve, the same force is generated which creates twisters, cyclones, typhoons through the effect of suction.*
>
> *In 1941 some models had been built, one of them had a diameter of 2.4 meters with a small and very fast electric motor. It climbed straight up into the air so suddenly that unfortunately it hit the workshop ceiling and crashed to the ground in pieces...*

Again, the Repulsine was not operating on conventional principles of aerodynamics and its behaviour could not be explained in this way. What is mentioned in the HZP book is the high speeds of rotation that are involved in the implosion process. The Repulsine was essentially a kind of Turbine which would rotate at between 15,000 and 20,000 rpm (once a starter motor had been employed initially). It apparently became self-sustaining. On page 213 of the HZP book, Cook writes:

> *This capacity to fly Schauberger partly attributed to the creation of the vacuum in the rarefied region immediately above the plates. But the primary levitating force, he claimed, was due to some other process altogether—a reaction between the air molecules in their newly excited state and the body of the machine itself.*

Cook lays out the story of how Schauberger was forced to work for the SS and diaries at the Schauberger Museum in Austria give quite a lot of detail about his movements during and after the war. The Nazis did fund Schauberger's research and he even went to work in a place called Jablonc, in what is now the Czech Republic – which was very close, apparently, to where Hans Kammler (covered later in this chapter) was working on other secret programmes. Also, there is some anecdotal evidence that the German aircraft designer, Heinkel, had copied (or tried to copy) some of Schauberger's ideas when building his own engines. Schauberger had apparently designed several devices – even some type of energy generator. Regarding what happened at the end of the war, Cook writes, summarising Schauberger's diary entries:

> *The diary had made it clear that Viktor Schauberger had built a machine that had flown earlier in the war at Kertl (and almost certainly during Schauberger's secret period of research in Czechoslovakia). It was also quite clear that the device's modus operandi was wholly unconventional—that is to say, the method by which it generated lift was insufficiently explained by current scientific knowledge.*

After the war, it is documented that Schauberger was apprehended by US Intelligence:

> *For the next nine months, until March 1946, Schauberger was held under close house arrest and extensively debriefed by U.S. technical intelligence about his activities during the war.*

Later, in 1957, a German American named Karl Gerchsheimer, who seemed to be connected to US Intelligence, went to see Schauberger and persuaded him to travel to the USA to set up a facility to manufacture Schauberger's devices. Schauberger agreed, but only if this was being done for "the common good." Schauberger's diaries, however, relate the story of how he was tricked into signing a contract, which resulted in the confiscation of some of his property and many of his ideas. Shortly after this, Schauberger returned to Austria, where he died.

Vortex Effects and Antigravity

It is worth mentioning here how Schauberger's technology appeared to be based on the high-speed flow of fluids (that's gases and liquids) in a chamber. Here again we are back to "spin" – which we encountered in the gyroscope experiments by Eric Laithwaite, the antigravity experiment by Wilbert Smith and the spinning ball experiments of Bruce De Palma.

Others have written about the apparent antigravity effects seen in tornadoes. Certain effects seen cannot be due to "high speed gusts of wind" alone. For example, changes in material properties, as seems to be shown in photos like this:

Wood pierces a relatively thin piece of stone – and somehow, the pieces of stone land on top…[148] This is a photo from Joplin, Missouri, USA.

In the HZP book, Nick Cook writes, as many others have, how something like "torsion fields" appear to be relevant in explaining the phenomena we see. Again, explaining these effects is not the focus of this book – but a focus of this book is to try and show that these effects are *real* and *someone has engineered with them.*

Hans Kammler

Both Cook and Farrell write about a German Officer who, by all accounts, became very powerful in Hitler's regime. In chapter 15 of HZP, Cook describes Kammler's rise in the SS and his involvement in the construction of a fair number of underground factories for the German "war machine".

Hans Kammler – Was he in charge of "Black Programmes?"

These underground installations were built using slave labour from the concentration camps and many died in the process of their construction. Cook makes the argument that someone like Kammler was in a prime position to know about secret technology and weapons programmes within the German Military. Speaking to Barbara Bogave on an American Radio show called "Fresh Air" (Aug 14, 2002), Cook stated:

> *Kammler was this mercurial, incredibly unpleasant bureaucrat who had his hands in the architecture of the death camps, for example, but it's significant that at the end of the war, he disappeared off the face of the earth.*
>
> *So little is known about him and yet he was one of the most senior individuals outside of the German cabinet - Hitler's inner circle during the Second World War. His mandate towards the end of the war was to amass all the high technology that Germany produced ... For instance, the V1 and V2 missiles and rockets that were developed and fired on the Low Countries and on England ... were under his control at the end of the war...*

In the interview, and in the HZP book Cook reports that Kammler:

> *...built the largest underground facility in the world... [in the Harz Mountains of central Germany]. The Mittelwerk is a huge underground facility - it's still there you can go and see it - was built to construct the V2 rockets and other weapon systems and it was put underground so that US and UK the British strategic bombing during the war couldn't touch.*

He also gives further details about another plant that Kammler set up.

> *Kammler set up a secret research cell in Czechoslovakia which was ring-fenced even from other high-ranking Germans. Within this research cell he authorized research into weapons that went beyond the V1 and the V2, which in themselves were highly advanced...*
>
> *The plant that Kammler oversaw was actually a sectioned-off site of something called the Skoda Works ... It was in the Russians designated zone of influence at the end of the war. What is interesting, though, is that Patton's army, right at the end of the war, in early May, swept in and overtook the Skoda works for a brief period of time before the Russians got there. Now, what evidence there is suggests that Kammler got a lot of his research out. We don't know where that research went - some of it may well have gone to Russia. Some of it may well have gone to the United States.*
>
> *The intriguing aspect is we don't know what happened to Kammler. There are four conflicting accounts of his death. None of them stand up to scrutiny. This was the most influential and most important guy, at the end of the Second World War, who had all Germany's secret weapons research at his disposal - and somebody like that, of course, would have had a great bargaining chip to deal with any of the victorious Allies that he cared to have chosen to deal with - to bargain for his life.*

In "Reich of the Black Sun," J.P. Farrell recounts that General Patton was killed in a mysterious car accident not long after this "swoop" into Germany.[149]

One can surmise, then, that there were some secret aspects of research that did end up either in Russia or in the USA.

The Bell and the "Henge"

Perhaps the most well-known (now) of these secret research projects involved something called Die Glocke – "The Bell". Much of this information came via a researcher that Nick Cook references many times in HZP – Igor Witkowski[150]. Witkowski is also a military journalist/historian. He accompanied Cook on a visit to an underground facility in what is now Poland. The Wenceslas Mine in the Sudeten Mountains was allegedly where the Bell unit ended up. It seems that most of the information that Witkowski obtained about the Bell came from testimony of Nazi SS Officer Jakob Sporrenberg. This testimony, according to Witkowksi, was included in a classified document, shown to him by a trusted source. Witkowski has written his research up in his own book called "The Truth About the Wunderwaffe," a revised edition of which was published (in English) in 2013.[151] Though some people state there is no proof that the Bell unit ever existed, Witkowski relates some interesting details about the unit, and where it was supposedly taken.

"Fly Trap" or "Henge" as shown on Google Maps (Satellite)[152] near the Wenceslas Mine. Was this a test rig?

Nazi SS Officer Jakob Sporrenberg

Igor Witkowksi and the site of an alleged test rig near the Wenceslas Mine. The rig is about 10 metres high and 30 metres in diameter. Photo by Tim Ventura, 2003[153]

In HZP, Cook quotes Witkowski

> *In this case, he said, a series of experiments had taken place in a mine in a valley close to the Czech frontier. They had begun in 1944 and carried on into the April of the following year, under the nose of the advancing Russians. The experiments required large doses of electricity fed via thick cabling into a chamber hundreds of meters belowground. In this chamber, a bell-shaped device comprising two contra-rotating cylinders filled with mercury, or something like it, had emitted a strange pale blue light. A number of scientists who had been exposed to the device during these experiments suffered terrible side effects; five were said to have died as a result. Word had it that the tests sought to investigate some kind of anti-gravitational effect, Witkowski said.*

He then includes the information about the Bell which is alleged to have come from Sporrenberg (who was a prominent SS Officer at the time)

> *Following his capture, as much as Sporrenberg was able to divulge to Soviet intelligence and the Polish courts about the Bell was this, Witkowski said. The project had gone under two code names: "Later-nentrager" and "Chroms" and always involved "Die Glocke"—the bell-shaped object that had glowed when under test. The Bell itself was made out of a hard, heavy metal and was filled with a mercury-like substance, violet in color. This metallic liquid was stored in a tall thin thermos flask a meter high encased in lead three centimetres thick. The experiments always took place under a thick ceramic cover and* ***involved the rapid spinning of two cylinders in opposite directions.*** *The mercury-like substance was code-named "Xerum 525." Other substances used included thorium and beryllium peroxides, code-named Leichtmetall. The chamber in which the experiments took place was situated in a gallery deep below ground. It had a floor area of approximately 30 square meters and its walls were covered with ceramic tiles with an overlay of thick rubber matting. After approximately ten tests, the room was dismantled and its component parts destroyed. Only the Bell itself was preserved. The rubber mats were replaced every two to three experiments and were disposed of in a special furnace. Each test lasted for approximately one minute. During this period,*

> *while the Bell emitted its pale blue glow, personnel were kept 150 to 200 meters from it. Electrical equipment anywhere within this radius would usually short-circuit or break down. Afterward, the room was doused for up to 45 minutes with a liquid that appeared to be brine.*

Negative effects of the operation of the Bell on living things are also noted in Witkowski's descriptions - such as causing plants to decay quickly. There were also negative effects on people who were near when the Bell was powered up. It caused them to have sleep problems, spasms and was even said to be fatal to some people. From the descriptions given, Cook initially considered that the Bell was more likely some type of experiment involving radioactive material, rather than anything to do with antigravity, but he also, like me, thought the description of spinning cylinders would then seem out of place. Cook also related the "spinning cylinders" to descriptions of other experiments he had heard of. Similarly, the use of a large current or voltage in the Bell experiments is then not dissimilar to the ideas used in the experiments of Tesla or Townsend Brown.

Witkowski argued that the structure or "henge", pictured above, was some type of test rig – perhaps even for an antigravity device. He argues that because some of the green paint remained on the structure, this was evidence that it had been camouflaged. Also, Witkowski is quoted thus:

> *The ground within the structure has been excavated to a depth of a meter and lined with the same ceramic tiles that Sporrenberg describes in the chamber that contained the Bell. There are also high-strength steel hooks set into the tops of the columns. I think they were put there to support something; to attach to something. Something that must have exerted a lot of power.*

In researching for this chapter, I came across an interesting blog "UFO Hessdalen"[154] which made a connection between some of the details about the operation of the Bell (from Witkowski's information) with details included in one of the alleged MJ-12 documents. We will mention the MJ-12 documents again later, but for now we will just explain that these documents first surfaced in the 1980s and were said to reveal secrets about flying saucer retrievals in the 1940s – including, probably the craft which crashed at Roswell.

Paraphrasing what the UFO Hessdalen blog[154] says about J.P. Farrell's "Reich of the Black Sun" and "Roswell and the Reich" books, an argument is made that Witkowski's information mentions thorium and beryllium compounds and so does the MJ-12 document. There are other common elements between the two sources of information. Hence, this seems to indicate, *according to Farrell* and the author of the UFO Hessdalen blog, that whatever crashed at Roswell has some relationship to the Nazi Bell experiment, so the Roswell "crash" was probably the result of a US black programme, derived from the Bell project. (Other books have been written with this same premise.) This theme has then since been repeated in relation to the Kecksburg Crash in 1965[155], which involved an "acorn-shaped" object, also said to resemble the Nazi Bell (so, it hadn't evolved in over 20 years?)

Before we get too carried away, let us return to the Wenceslas site and some of its features. A comprehensive investigation of the site was carried out in 2005 by Gerold Schelm[156], where he took photographs, made measurements and also spoke to a Polish passer-by, who knew a little about the site. The passer by chuckled at the idea that Witkowski had implied it was a test rig for some type of antigravity device, explaining to Gerold Schelm that there was a similar "henge" structure only about 15 miles from the Wenceslas mine – in Siechnice.

Gerold Schelm writes in some detail how he did a close-up inspection of another similar "henge" structure which was part of a cooling tower at a Power Plant in Siechnice:

> *Despite the number of columns not matching (12 at Siecnice and 11 at Ludwikowice), I am sure, that even their dimensions are almost the same. The construction features are exactly the same, leading to the assumption that the cooling tower and "The Henge" once were built using the same plans, maybe even the same construction company. I had no luck in finding out when the cooling tower in Siechnice was erected, but it is in very good condition and I think it was built after WW II, maybe in the 60s or 70s.*
>
> *However, comparing the details of both "The Henge" with the Siechnice cooling tower, the purpose of the bolts mentioned by Witkowski becomes clear: The upper metal construction of the cooling tower is resting on exactly those 12 bolts, being visible just on top of every column like they can be seen at "The Henge".*

Photos by Gerold Schlem[156] - Comparison of "Henge" Structure at Wenceslas Mine (top half of each image) to a cooling tower in Siechnice

A segment on "The Henge" and the Wenceslas mine is included in Nick Cook's "UFO's - Secret Evidence" - a documentary broadcast on the UK's Channel 4 on 13 Oct 2005[157].(This documentary can perhaps be regarded as a follow up to "Billion Dollar Secret.") The documentary shows a photo of the "henge" from an allied aerial reconnaissance mission in 1944.

"Henge" seen in 1944 aerial photo

In this photo, it looks less like a cooling tower, perhaps – though it is a short distance away from something labelled "Power plant". Was it a cooling tower that was never actually brought into service or was simply unfinished when the aerial photo was taken?

So, what of the other details regarding the experiments described in Sporrenberg's testimony? Was this all disinformation and fabrication? An article by "Morice" on a French website called Agoravox[158] seems to give us some clues (although I had to use Google Translate to understand the details!)

In the discussion of "The Bell," Nick Cook's HZP book mentions two other names - Dr Walther Gerlach[159] and Dr Ronald Richter[160]. References show that these two figures both worked in Nuclear Physics. It was suspected by British Intelligence that Gerlach had worked on a Nazi atomic bomb project[161]. Richter was involved in a failed attempt to build a nuclear reactor for Argentina in the 1950s, when Peron was in power[162].

According to the French article on the Agora Website, Gerlach worked at the Goeth Institute in Frankfurt. Whilst he was there in 1933, he had experimented with the fluorescence of mercury under the influence of magnetic fields. As the article observes, mercury was said to have been inside the mysterious "bell". The article states that Gerlach had also worked on the atomic transmutation of elements – a form of alchemy (which some might say the Nazi's were very interested in).

The Agora article further asserts that "Gerlach's experimental apparatus in his research on nuclear fusion consisted of two contra-rotating cylinders containing mercury and a kind of 'jelly', all cooled by a sophisticated cryogenic system."

Tim Ventura, a researcher who runs a website called American Antigravity[68], has posted a lot of information about topics which are covered in the first part of this book (e.g. electrogravitics). He has also interviewed Nick Cook and a number of other people and posted recordings of these interviews (a useful resource!)[163]. However, care is needed, as ever, in evaluating the information on his website, as it comes from a mixture of sources. Regarding the Nazi Bell device/unit, Ventura appears to have written at least two articles[164] that reference the Bell[165]. However, rather than spending time trying to establish the facts I have given above, he chooses to write about other topics – such as Einstein's Unified Field Theory and the alleged Philadelphia Experiment. This is not to say that there isn't some useful information about the Bell unit on his website, however.

A report, posted in 2005, by Tim Ventura[164] even includes a reference to Witkowski's own characterisation of the Bell, thus:

> *Witkowski described the Nazi Bell as being very similar to a "plasma-focus", a design comparable in some ways to high-energy devices used in fusion research.*

Therefore, it seems logical to me to conclude that the Bell project was actually some type of experimental nuclear **fusion** reactor, not a fission reactor. **Neither was it an antigravity experiment**.

SARA and a Reproduction of the Bell?

This image came from Tim Ventura's online Flickr album[166], but I could not find a specific reference to what it is. From reading some of the above-mentioned report by Tim Ventura himself[164], I assume this is a photo of an alleged reproduction of the Bell experiment, done by John Dering at a company called SARA[167]. Let us take a detour, then to discuss SARA. Scientific Applications & Research Associates is seemingly a defence contractor, whose name is an eerie hybrid of two defence contractors who are known to me because of the 9/11 research I have been involved with.

Science Applications International Corporation[168] (we will be discussing SAIC later on) and Applied Research Associates[169] were both defendants in Dr Judy Wood's 9/11 Science Fraud case[170].

On their website, SARA[167] who allegedly built this "Bell" reproduction list some areas of expertise as follows:

Remote Sensing & Tactical Awareness

- Tactical Directional Seekers
- Acoustic Sensors for Unmanned Air Vehicles
- Hostile Fire Detection Sensors

Pulse Power & Directed Energy Solutions

- Integrated Pulsed Power Solutions

- Pulsed Power Products
- High Power Microwave Antennas

On page 1 of Tim Ventura's "Bell" report, there is a description of an alleged reproduction of the Bell:

> *I talked to John almost every day for about 3 months, collecting data the entire time to write a story on the Nazi Bell device. The Bell project that he'd been involved with was related to WW-II German research, but the modern replication had been financed by Joe Firmage's ISSO startup for about 1.2 million dollars and wasn't an exact replica of the original device. SARA's version was much smaller -- using only about 100 watts -- and they'd modified the design, since they didn't actually know many of the details of the original Bell's construction.*

$1.2 million – that's a fair sum! So, what results were produced?

> *SARA tested their mini-version of the Bell and found that it effectively produced a gravitational back-EMF that they couldn't shield against. That's important, because SARA does electromagnetic shielding on the B-2 bomber, and even with their expertise they tried everything imaginable before realizing that it was a time-space distortion effect. They shut the experiment down when* ***ISSO funding dried up****, although the claim is that if the cash is available, they still have the unit ready & waiting for more tests.*

This sounds like gobbledegook to me. What exactly is a "gravitational back EMF?" This is mixing together terminology related to gravity and electricity, without a proper explanation. What effects were generated? How were they measured? The report describes none of these details, nor does it provide any references to where the reader might find such details.

But what about the ISSO mentioned by Ventura? It was apparently a group set up by a wealthy figure Joe Firmage – who seemed to have some association with the Novell networking company.[171]

> *International Space Sciences Organization (ISSO). Firmage founded ISSO in October of 1998 to sponsor research and development of breakthrough insights into physics. ISSO formally launched its science operations on July 28th, 1999, and plans to release a number of book, film, and interactive public education productions.*

I wonder if Joe Firmage felt he got a "good deal" on the Bell reproduction. If I may now take a risk and be facetious, if we re-examine the unlabelled and non-explained image above, it doesn't look to me like with would "take off!" Also, no details are supplied as to how it was meant to create any kind of antigravity effect. There are no real "usable" details about the unit. Indeed, Ventura's report admits they don't really know how it worked. The role of the numerous cables is not explained either.

Disinformation

Hence, I consider that the above analysis shows that disinformation has been deliberately circulated about the Nazi Bell project, to help cover up whatever really was going on in Nazi Black programmes at the time. One can easily observe that we are dealing with an energy phenomenon – which, to my way of thinking, is one of the main reasons that these sorts of programmes – however far back in history they may be – have been covered up.

9. Hunting for Billion Dollar Secrets

We have already mentioned Nick Cook, former Jane's Defence Weekly journalist and editor, a number of times. Cook writes some of his HZP book as a journal or diary, documenting his path of discovery about hidden antigravity technology. Before Cook's book was published, some of what he discovered was included in the 1999 documentary "Billion Dollar Secret[31]," (which I will refer to as "BDS" from now on) which was aired in the UK on Channel 4 or Channel 5.

During the opening minutes of this documentary, whilst a short clip of the F-117 Stealth Bomber, flying at dusk or at night is shown, Cook states:

> *This is the plane which the world believes is state-of-the-art technology - but it's not. Somewhere, in secret, its replacement may already be flying - leaving unexplained traces in the night sky.*

A few minutes later in the documentary, Cook makes a further key observation, when holding a particular document published by the US Government, every year.

> *... This document here... is the annually published list of defense spending put out by the Pentagon. Inside this document is every single conceivable item of military equipment that America is likely to buy for its Armed Forces. But if you add up all the line items inside this budget book and you subtract the total from the annual published total of 250 billion dollars that the Pentagon reports to Congress, you find that there's a mismatch of about thirty billion dollars. Now this is what we call the "black budget" or America's billion-dollar secret - which is reserved for deeply classified programs.*

Cook, then asks a question:

> *Now I'm wondering whether inside this classified pot of money, there may be some relationship between, say, deeply classified aircraft, which is something I know about - and what people commonly report as unidentified flying objects. So, over the next three or four weeks I'm going to be criss-crossing America, talking to people [and] visiting facilities to see whether I can find the truth...*

In writing this chapter in March 2018, I am also conscious that it is almost 20 years since Cook uttered these words.

I put it to you that Cook, by making these observations and asking the questions he did, "became a blip" on someone's radar. I will discuss my reasons for suggesting this later. Before, let us first discuss Nick Cook's HZP book a little more.

The "The Hunt For Zero Point" book

In 1990 or 1991, Nick Cook became intrigued regarding the topic of antigravity research when, on his office desk, someone left a copy of the "G Engines Are Coming" article that was covered in Chapter 3. In chapter 1 of HZP, Cook describes how this sparked his interest and he tried to find documents in his own company's library regarding the topic of antigravity. He only found one article, but also found a name – George S Trimble – associated with the research. On page 11 of HZP, Cook writes:

> *In 1957, George S. Trimble, one of the leading aerospace engineers in the U.S. at that time, a man, it could safely be said, with a background in highly advanced concepts and classified activity, had put together what looked like a special projects team; one with a curious task. This, just a year after he started talking about the Golden Age of Antigravity that would sweep through the industry starting in the 1960s.*

Cook explains that Trimble became involved in the Research Institute for Advanced Studies, RIAS, a (Lockheed) Martin spin-off group. So, Cook, using a contact at Lockheed Martin, tried to track down Trimble. According to this contact, Trimble initially agreed to be interviewed, but at the end of chapter 1 of HZP, Cook recounts a late-night or early morning phone call he received from his Lockheed Martin contact:

> *"It's Trimble," she said. "The guy just got off the phone to me. Remember how he was fine to do the interview? Well, something's happened. I don't know who this old man is or what he once was, but he told me in no uncertain terms to get off his case. He doesn't want to speak to me and he doesn't want to speak to you, not now, not ever. I don't mind telling you that he sounded scared and I don't like to hear old men scared. It makes me scared. I don't know what you were really working on when you came to me with this, Nick, but let me give you some advice. Stick to what you know about; stick to the damned present. It's better that way for all of us."*

Perhaps it was this situation that propelled Cook on his journey. He continued his research "quietly" over 10 years – the result being the HZP book itself and two documentaries.

In chapter two of his book, Cook writes about how a contact, Lawrence Cross, warned him about getting involved in this field of research, stating that Cook would be impeded by "secret keepers" and that the whole field was riven with disinformation. (Both of these statements are correct, as we have already seen and we will continue to see later in this book.)

The HZP book continues by covering similar ground to what was covered in chapters 2 and 3 of this book.

Asking about Billion Dollar Secrets

In the 1999 "BDS" documentary, Cook visits a number of venues and talks to figures in government, the defence industry and even a few UFO researchers.

Early in the documentary, Cook interviews a somewhat nervous Bob Widmer, whose design for a classified aircraft, "Kingfish," was never actually put into production, but even then, strong secrecy surrounded the development and the specifications of "Kingfish" were (in 1999) still classified, decades after Widmer had worked on the design.

One notable (but short) interview is with Senator Dana Rorabacher who, at the time, served on the US Congressional Space Subcommittee. Cook asks Rorabacher about a rumoured secret aircraft development programme called

"Aurora" (again, this quickly appears in any search related to a secret space programme).

In the interview, Rorabacher states:

> *It has not been verified to me through legitimate channels that the Aurora exists and thus if it does exist, I have not been able to find out the information about the project that I would need to know as chairman of the space subcommittee, that would enable me to see that technologies that were developed in the Aurora [programme] could be part of our calculations as to how much should go into our space budget.*

Rorabacher had written to the then President Bill Clinton requesting declassification of the black budget in relation to the Aurora Programme. Rorabacher stated:

> *The only word back that I received from the president was a letter acknowledging my letter and we haven't received a letter answering my specific request about releasing information about black programs especially about Aurora.*

Cook asks Rorabacher if he thinks Aurora exists and Rorabacher states he thinks "it probably does exist" and as an "unmanned vehicle" uses a "kind of a scramjet or a ramjet engine" that enables the plane to travel "six to eight times the speed of sound." Rorabacher notes the number of credible sightings, for example. Commenting on ongoing black world programmes (after the fall of the Soviet Union, but before 9/11), Rorabacher said:

> *Now during the Cold War, it made sense to have those black programs. Now that the Soviet Union is no longer our adversary - it's still a dangerous world - but it's not like it was back in those days. If we have a functioning technology in a black program, we should not be spending billions of dollars in a program that's visible to the public trying to duplicate those technologies we already have - we can't afford that kind of waste. During the Cold War, we could, because that was part of our defense expenditure - of keeping those things secret. Well, we don't need to waste that money now.*

I wonder how much Rorabacher changed his mind after 9/11. (Yes, 9/11 was done with black world technology – perhaps because all that absolute power has absolutely corrupted those who own that technology. This is precisely what I have stated in other articles and books I have written and presentations I have given.)

Cook also interviewed Lieutenant General George Muellner and in the BDS documentary and the HZP book, Cook quotes Muellner as saying that weapons and stealth technologies are being developed in black programmes and Muellner even went so far as to say he thought he was aware of all the black programmes that were underway, but *he couldn't be sure!* Muellner also stated in Cook's interview:

> *The technologies that are developed in the black world need to continue to be matured and when the time is right brought forward in the form of **weapon systems**.*

Another significant interview (but with little information in it) is with Jack Gordon, who at the time was, head of the Lockheed Martin Skunkworks in California. This is where the U2 Spy plane and the SR71 aircraft were developed decades ago. Gordon is less forthcoming about black programmes. Cook asks Gordon to give an example of a "white," "grey" and "black" programme. Gordon responds:

> *A white program of course is the standard program that everybody is familiar with and an example of that today might be our single-stage-to-orbit work our X-33 prototype program that we're doing with NASA - and that is a program that is being managed on the internet. We're connected with a number of NASA Centres and with government organizations and with our team partners.*
>
> *A grey programme is one that has elements of secrecy associated with it but is an acknowledged program. The F-117 would be in that category today, in that everybody is aware of the F-117 stealth fighter. But of course, the actual signature characteristics and operational characteristics of the airplane are "close hold." So that would be an example of a special access grey program.*
>
> *And then a black program of course is an unacknowledged program that we just do not acknowledge that it exists at all and that it is that sensitive and is at the edge of our national defense.*

Following this Cook asks a question about existing black programmes (which, of course, cannot be answered by Gordon – so he laughs) and then he asks Gordon about whether Lockheed Martin are "responsible for any UFOs." Gordon laughs again then he says:

> *Well the skunkworks has had a long association, I suppose, in the press with UFOs - with programs like Aurora, which I believe was really a code word for some B2 funding. I know in the popular magazines there has been some speculation that there is an underground tunnel between our Helendale facility and Edwards Air Force Base - where we transport the aliens back and forth, but I can assure you that I haven't seen any of these little green folks yet.*

This tunnel/alien "story" seems a peculiar thing to mention here, though perhaps he was just referring to something that was circulating in the press/media at the time. Also, Gordon did not deny the existence of a tunnel between Helendale and Edwards AFB… In a 2017 YouTube video, Cook stated[172] that since 1999, things have changed – and as a foreign national, he would not even be admitted to the Skunkworks, let alone granted an interview. Cook also stated that he asked Jack Gordon how many different aircraft he had worked on, in his time at the Skunkworks. Gordon replied that he had worked on 15 different aircraft, only 12 of which he could talk about.

Boyd Bushman Interview

This was another important part of the BDS documentary and one in which Cook was advised to go and see another important figure, John Hutchison. Cook realised that Bushman had worked on black programmes but wasn't able to reveal any real details. We will cover more about Bushman in chapter 10. In the interview, Bushman stated to Cook:

> *I believe that nature does not speak English. What nature tells us is what must be honoured. It has been talking to us on many domains and we have data sets that we're still trying to understand. But I can't talk to all the theoreticians because there don't exist theories where I am.*
>
> *What we have is wonderful and it comes for miracles occurring, but that that we will see will not be what we have... I therefore tell my team, "If you know anything, I really don't want to talk to you - because everything we know has already gone into production. Now if there's something you kind of wonder, or if you think etc, let's take those steps to the future..." And that's what we always have to do - it is literally a kind of a lonely walk... but a very, very rewarding walk. To listen where languages are not spoken and verbalization is not used. But we have to learn its language...*

Cook then asks him about antigravity research. Bushman responds by describing an experiment which we will be discussing in chapter 10. He then describes what sounds like some of his own research:

> *I'm refining and heading towards something that all you have to do is charge - and it is lifting up it is losing weight... When we complete the technology, it should fly...*

Note Bushman's use of the *future* tense here – and I think this is about "the best Bushman could do" under the circumstances. Since the Cook interview, Bushman has revealed a little more information, which again, we will cover in chapter 10.

Cook at Area 51

The BDS documentary shows Cook visiting the border of Area 51 in Nevada. He acknowledges the high level of security and monitoring that was evident there then, as now. He meets a contact called Mark Farmer – an advocate of openness, not secrecy. Cook says to Farmer that he "feels like he's on the edge of the world." Farmer replies:

> *In some ways you are on the edge of the world - the edge of the white world. If you go over there, a couple hundred yards - is the border to the restricted area which surrounds the Groom Lake covert testing facility. And once you go just beyond the road, there, past those signs that we were just at, you're liable to arrest, search and seizure. People that are taken and arrested - they're forced to sign a statement that makes sure that they [had no] inadvertent exposure to classified material. They "do business" differently just a couple hundred yards from here... Out here, we enjoy all the rights and privileges as the citizens of the United States under the United States Constitution, but you pass that line, and all that ends.*

Let us not forget that this security and secrecy was (and is) present even in the absence of a "cold war" type scenario, which ended in the early 1990s.

Animal Mutilations and Remote Viewing

These are two other areas that Cook ventures into in the BDS documentary. He interviews Lyn Buchannan regarding his work in the "Remote Viewing"[173] programme (about which other books have been written, of course). Cook also interviews a rancher named John Harr about mutilation of livestock they owned. Harr's wife was too frightened to take part in the interview. Harr suggests that she is afraid because they are dealing with something "bigger than they can handle." Harr realises that there is a power/force operating which has no regard for the livelihood of their stock and wonders if "children would be next." Again, whole books have been written about this phenomenon and interested readers can also review Richard D Hall's documentary called Silent Killers[174].

Protecting Billion Dollar Secrets…?

It does not seem crazy to suggest that when a defence journalist starts "sniffing around" black programmes, going to see generals, senators and heads of black programme contractors, whilst also making documentaries about these topics – and also mentioning animal mutilations and UFOs, someone is going to "take notice." Perhaps this was what prompted an episode described in Nick Cook's HZP book, starting on page 140 (chapter 13). This seems to have happened when he was going to see Jack Gordon at Palmdale – the Skunkworks. The last part of his description seems to suggest that Cook had just had a "bad dream." Maybe that's all it was - I have not contacted him to ask whether this was the case. But what was Cook telling us by including this passage? Even if the episode did not physically happen, why did Cook include it?

> *It was sometime in the dead hours that I heard the sound of scratching coming from the lock … As I braced for the bang from the hammer that would drive the platinum tool into the barrel of the lock and core it out in a split instant, it happened, but so fast, all I could do was lie there, paralyzed, as the door flew in and the bolt cutters took down the chain. Then, I saw them: three shapes silhouetted momentarily in the doorway—black jumpsuits, body Armor, Kevlar helmets and semiautomatic weapons… More pressure, killing pressure, as the snub-barreled M16 was pushed down harder and harder, the force driving my head deep into the mattress… It was then that I came up for air, sucking it down in great gulps, my hands outstretched, pushing against bodies I could still see* ***but weren't there and never had been.***

Did he experience a psychotronic attack?[175]

Nick Cook's Later Research

Following the publication of the HZP book, Nick Cook received a fair amount of publicity and appeared on quite a few talk shows, one or two of which we have already mentioned.

In an interview with Tim Ventura from October 2004[176], Cook mentioned several times that he was working on a second book although looking on Amazon, for example, he seems not to have published a second book on antigravity-related topics – instead, he switched to writing fiction.[177]

In the Ventura interview, Cook talks about the fear of Al Qaida activities and what happened on 9/11 as being a boost for the black world. Sometime after this interview (perhaps around 2007) I wrote to Cook to break the bad news to him – that people in the black world either were responsible for the destruction of the WTC on 9/11, or they knew who was responsible and were covering things up in relation to this. (I never received a response to this, although he had moved to a different address.)

Nick Cook and a Free Energy Device (in 2004)

Cook understands, as I do, that the phenomena seen in experiments and techniques related to producing anti-gravitational effects are often accompanied by anomalous energy effects. Indeed, the title of his book includes the words "Zero Point," rather than "Antigravity." However, the phrase "free energy" only appears about seven times in the text of the book.

Hence, I was very interested when I heard Cook being interviewed by Art Bell on Coast to Coast. The interview took place on 14 Mar 2004[178] (just over 14 years before this chapter was written…) In this interview, Cook describes how he had become aware of an operational free energy device.

> *In my experience, there are people who are now producing small but usable amounts of energy from the vacuum of space…*
>
> *There is an individual … in America who has developed a device which, to my satisfaction, is producing small but usable amounts of energy…*
>
> *I haven't seen the device yet, but we have talked about it at length. I have sent emissaries … people I know who are scientists - to go and look at the device and they came back to me and said, "This is real. It is producing usable amounts of energy," and so at the earliest opportunity, I'm going to go out and see this guy…*
>
> *This particular device requires a small energy input to get started, but once it is up and running… it continues to churn out small levels of energy of the order of several hundred Watts…*
>
> *But it is deemed to be scalable and in time the inventor predicts that increasing levels of output will be there. So, what this tells me is that this is a reflection of many other claims out there, that there are devices which tap into the zero-point energy field…*
>
> *It's a solid-state device - it literally just sits there and interacts with zero-point field to produce its energy output. Now the guy has several other inventions which utilize other methods, including rotating magnetic fields for energy extraction, but it's this particular one that he's concentrating on and I think which seems to offer the most promise.*

Art Bell then comments:

> *That's incredibly exciting you know something like that really exists - well gee, all kinds of things are possible Nick, like this man's… I would think his life might be in danger…*

I think Art Bell was right.

"UFOs Secret Evidence" – UK Documentary in 2005

"UFOs Secret Evidence"[179] can perhaps be considered as a follow up to the BDS documentary but was more openly an investigation into certain UFO cases

– such as the Roswell Crash and Travis Walton's case (Cook interviewed Walton in this documentary).

This documentary also linked to Cook's earlier research – for example, several minutes of film was devoted to the "Bell" project that we covered in chapter 8. In this segment, Cook and Igor Witkowski visited the Wenceslas Mine – but the thrust was that the project was Saucer/antigravity related. The piece did not include the information from the site, which leads to the conclusion that it was used in a fusion energy research project of some kind. Hence, it seems Cook was used to put out disinformation.

During the production of this documentary, I sent Nick Cook electronic copies of the Douglas Aircraft Documents (discussed in chapter 5). A researcher contacted me to discuss their authenticity and I had a brief message from Nick Cook agreeing that they were indeed authentic. Hence, these documents were actually featured in [157]a short segment. However, I felt the significance of what the documents proved was played down, to some extent. Cook included coverage of the "lifter" and there was a brief segment with Tim Ventura showing the operation of the one he built.

The section on the Roswell crash was disappointing and missed out much key evidence. The documentary used a mixture of CGI and real footage – sometimes it was not clear which was which. (Footage was not captioned "simulation" or "reconstruction.")

Though there are number of interesting clips and segments, such as one of Harry Truman admitting that he had been involved in regular meetings where one of the topics of discussion was "flying saucers," the documentary, overall, seemed to want to suggest that UFOs were used as a cover for secret aerospace projects. For example, near the end of the 2005, documentary, Cook states:

> *Whether aliens exist or not, many of us want to believe in them.*

It is essentially statements like this which provide me with the motivation to write books and articles like this – to enable the interested reader to gain a deeper/fuller understanding of the true picture. Most or all of the mainstream TV documentaries are "limited hangout." Due to the way in which they are produced and the platform from which they are broadcast, it is not possible for them to be anything other than a "limited hangout." In chapter 31, I will describe specific examples of this.

I wrote to Nick Cook after the documentary aired and briefly expressed my disappointment – particularly with the segment about the Roswell crash. Nick Cook responded, implying that, to some extent, his hand had been forced and he, too, was a little disappointed.

Nick Cook Retires...?

After the 2005 documentary, Nick Cook seemed to "disappear" off the scene. For example, he made no further appearances on "Coast to Coast." Although he had created a Website www.highfrontiers.com (archived on web.archive.org[180]), which he promoted in a 2005 interview with Art Bell,[181] this website now diverts to his current website http://www.nickcook.works/ - which has been "under construction," it seems, since 2016.

In the 2005 Art Bell interview, Cook also stated that he had a follow up book to HZP in the works, which he thought would be released in 2006. However, as mentioned above, no such book appeared.

BBC Radio 4 - Brief Interview in 2010

The next time I heard Cook speaking on the subject of antigravity was on a BBC Radio 4 programme called "Punt PI" (Private Investigator).[182] In a brief clip, Cook agrees the "Henge" near the Wenceslas Mine is probably the base of a cooling tower, not a "test rig," as Igor Witkowski had suggested, but Cook also states that it seems unusual things were going on in the area.[183] There seem to be some subtle edits in this programme, which are aimed to introduce doubt or even debunk what has been suggested by Witkowski.

Since about 2005, it seems to have been a strategy, particularly for the BBC, to get a comedian or comedy writer or minor celebrity with no science background to present or participate in documentaries about the most serious and world-changing topics that exist[184].

Nick Cook on YouTube in 2017

Whilst finalising this book, I came across some short videos that Nick Cook recorded in August 2017.[185] These videos are mainly a re-telling of the circumstances of the development of his HZP book. Cook made the videos at the time of the fifteenth anniversary of the publication of one of the editions.

10. Boyd Bushman and Antigravity

Boyd Bushman was a senior research scientist for Lockheed Martin. He had worked for a number of years in U.S. aerospace and defence industries, at companies such as Texas Instruments, General Dynamics and Hughes Aircraft. He passed on, at age 78, in August 2014. We will discuss his last interview, later in this chapter.

I first heard of Bushman while watching Nick Cook's BDS documentary around the time it first aired in 1999.

Bushman was something of an enigmatic figure, but from everything that I have heard and read, I do think he was being as truthful as he could be. I think he was well aware of the reality of antigravity technology, as I alluded to in chapter 9. In the 1999 BDS documentary, he hinted that he had been working on some type of electro-gravitic technology which "should fly" (i.e. he used the future tense). It does seem rather ridiculous that Bushman, working for a military industrial contractor, had only just "cottoned-on" to the work of people like Thomas Townsend Brown. In a David Sereda interview (from 2005 or 2006), further discussed below, Bushman states:

> *I do notice in recent publications that one of the latest group of my colleagues have said that they have evidence, because of the separation of the galaxies, that there is an antigravity [force].*

Later in the interview, the term "dark energy" is used, which I think is a misnomer, as it is almost certainly vacuum energy or an aether that people should be referring to…

I would contend that he knew a lot more about antigravity research (older and newer) than he let on. This contention is, I think, supported by some of what he stated in his "last interview."

It was Bushman who suggested Nick Cook should go and see John Hutchison (covered in chapter 12) in Canada, due to the levitation effects that Hutchison had created (Cook stated in the BDS documentary that Bushman had shown him a video of Hutchison's experiments.) Of course, I have already written about John Hutchison in both of my 9/11 books, details of which are at the start of this book.

In the same BDS interview, Bushman talked about doing a simple experiment where two "balls" of the same size are made, one which contains just solid material and the other which, is the same weight, but contains two magnets which are forced or glued together with opposing poles touching. According to what he showed in the later (2005 ish) Sereda/Bushman interview[186], this experiment was done on 12 Dec 1995. (In the video of the interview, this information was shown on a certificate that Sereda held up and read out.)

> *I went back to some of the first work on gravity done by Galileo, when he climbed his leaning tower in Pisa and dropped two stones. They arrived at the ground at the same time. I climbed my leaning tower of building 500, here at the company and I dropped my two weights - that were exactly the same weight - and they did* ***not*** *arrive at the ground at the same time. I had purposely placed opposing magnets that "didn't like each other" - like two Norths - together and I had already done an iron filing test that showed that it was showering out an entire magnetic plane. I wanted to see how that magnetic plane interfaces with gravity.*

In the Sereda interview, Bushman states the magnetic field from the "ball" which had the magnets in it, extended out "three feet on each side" of the object. In the "gravity test," the "balls" were dropped from a height of 59 feet inside a building.

The experiment that Bushman describes has been reproduced by several people, as it is so simple. One such reproduction was done by William Alek, for example[187] – and a web search will likely reveal several more reproductions that have been done.

As mentioned above, the Bushman/Sereda interview was included in Sereda's film "From Here to Andromeda[188]" (although the interview was broken up). The date that this interview was recorded is not clear, but it was sometime between 2003 and 2006 I understand. Sereda's film, however, promotes the Apollo Hoax in some places (covered in chapters 14-18), as so many other films and videos do. In the interview, we "don't get the best" out of Bushman, as Sereda keeps interrupting him and changing the subject.

Bushman gives more details about the "opposing magnets/gravity" experiment, mentioned above. He states that "gravity goes through anything that is solid and anything like iron… but it has to have a magnetic component, which may be 'cancelling out within itself' when in materials other than metals. He then points out that, in the opposing magnets/falling object experiment, the behaviour of gravity is changing, due to the presence of the constant magnetic field from the magnets.

Bushman then drops a magnet through a copper pipe and compares this to dropping the magnet through an aluminium pipe. The magnet falling through an aluminium pipe falls more slowly. Bushman mentions that this is to do with free-moving electrons in the aluminium. This demonstration is quite reminiscent of what Eric Laithwaite demonstrated in the 1960s and 1970s[189] – the effect of aluminium in a changing magnetic field (this was the basis for Laithwaite's maglev units). This is not "mysterious" – it is due to induced "eddy currents."[190] I remember seeing this effect demonstrated either in a science lecture, or perhaps even in my physics class at school, in the 1980s.

Bushman also mentions what he was told by a Roswell witness and relates a story about the craft being "shot down," with some kind of "Tesla Weapon." However, Sereda keeps interrupting and it becomes rather unclear what Bushman is saying. The discussion switches topics rather quickly with little verifiable detail discussed.

Bushman talks about Einstein and suggests what motivated the statement "you cannot communicate at beyond the velocity of light using photons." He facetiously suggests that if Einstein had been blind, he might have said "you cannot communicate at beyond the velocity of sound." That is, Bushman was saying you have to consider the way your own senses work. To follow this up, Bushman alludes to the idea that perhaps we should be considering that "you can't travel faster than the velocity of thought." I will just note here, again, what Wilbert Smith said in his Top Secret Memo from 1950, which we referenced in chapter 4:

> *"I was further informed that the United States authorities are investigating along quite a number of lines which might possibly be related to the saucers such as mental phenomena ..."*

We will return to this theme, when we talk about Ben Rich in chapter 21.

Boyd Bushman on Bob Lazar

I wanted to include a discussion of Boyd Bushman's thoughts on the Bob Lazar story, so I will now try and summarise that story.

Bob Lazar in a 2014 Interview

Much has been written about the disclosures of Bob Lazar, which were revealed in 1989. This was another account which some deemed to be an extremely significant disclosure and others judged that it was "too good to be true." There are several copies of an interview which appeared in a Sci-Fi channel documentary in the 1990s, so I recommend readers listen to that to hear Lazar's own words.[191]

While there are some serious questions about Lazar's account, there are, to my mind, very interesting statements he makes which I will try to summarise here.

Lazar's Story

In one of the interviews in the 1990s (referenced above), Lazar stated:

- He met Edward Teller at a lecture Teller gave in 1982 (Teller had seen Bob Lazar's work on a dragster engine).
- Was hired by Los Alamos National Labs (his name appeared in the phone book).
- Drove to and parked at a Special projects EG&G building at Las Vegas McCarran Airport and flew from there into Area 51.
- Travelled by bus to S4 Lab, allegedly near Papoose Lake.
- Worked there, sporadically, from early 1989 for about 6 months.
- Went into hangars, set against a mountain, with sand-coloured sloped doors (he did not notice any underground levels) and he went through security checks there.

- Was hired to replace someone who was killed when working on a reactor at Nevada test site.
- He went into a briefing room and told to read about what was going on in other projects. Saw some autopsy photos of aliens.
- Craft originated from Zeta Reticuli star system (same as the craft that Betty and Barney Hill encountered, according to Stan Friedman and Marjorie Fish).[192]
- He directly worked on a craft in a hangar there, which he went inside - and realised it was extra-terrestrial and he was there to "back-engineer" it (i.e. determine how it worked and how it had been built so that it could be duplicated).
- He had "majestic clearance" 38 levels above Q clearance and he worked with about 20 other people on "Project Galileo"
- Craft produced a gravity wave similar to what the earth produces.
- Craft had a "reactor" and 3 gravity amplifiers which generated its own gravity field(s). Amplifiers used different configurations (delta and omicron) to lift, then propel the craft in a given direction (by creating a distortion which the craft "falls into")
- Reactor which he was looking at was powered by element 115 – a stable element. He was told they had been given 500lbs of element 115 (a few atoms have since been synthesised). It was put into the top of the reactor and there is apparently a small particle accelerator in the base – like a cyclotron. Particle is accelerated, aimed up a small tube and aimed at element 115. That produces radiation and antimatter. This then reacts with a gas to create energy. This is <u>heat</u> energy which is converted to electrical power in the reactor itself using a thermoelectric converter which is "100% efficient." This is used to power other subsystems on the craft (but there is no wiring for this). The reactor sets up a gravitational wave – from the element 115 being bombarded.
- There is no thermal radiation.
- Gravity wave is guided (similar to the way microwaves are guided) through "tuned tubes" into amplifying cavities and through the projectors at the bottom of the craft.
- He witnessed one "low performance," piloted test flight, inside the hangar, at sundown.
- Craft lifted off ground silently with a corona discharge at the bottom and moved slightly from side to side. There is no exhaust.
- Craft presumed to be metal.
- When asked about whether he stole element 115, it was the only question on which he would not comment.
- Was told about test flight schedules and they were on a Weds. Night – when, statistically, there was less air traffic in the area.

- He took John Lear and Gene Huff to witness a test flight on 22 Mar 1989 and they saw a craft.[193]

Lazar comes across very well, and his statements about being a physicist and engineer are true. For years, he has run a company called United Nuclear[194] and sold equipment and materials which scientists and educators use. He has also designed a safe hydrogen storage system which can be used to run an ordinary car.[195] He states, frequently, that he would rather people disbelieved his Area 51 story. He has to be persuaded to tell this story and be interviewed about it. He has not written any books about it, nor has he been on the "UFO speaking circuit."

Several people, such as Stanton Friedman[196], have looked into Lazar's background and cannot verify, for example, where he was educated. Lazar gave a presentation in 1993 at the Little Ale Inn in Nevada and when questioned, could not remember the name of any of his college tutors/lecturers.[197]

Paul La Violette on Lazar

PLV has a useful section on the Lazar case in his 2008 book "Secrets of Antigravity." On page 284-5, PLV writes:

> *...at one UFO seminar in 1993, Lazar disclosed his belief that gravity is electromagnetic in nature but that it is an electromagnetic wave of a particular microwave frequency, which he did not wish to disclose at the time. Yet, in my opinion, it is a major error to assume that gravity is electromagnetic in nature or to suggest that the electric or magnetic field itself produces gravitational effects. It would instead be more reasonable to postulate that electric and gravity potential fields are coupled and that electromagnetic waves and electric shock discharges are accompanied by a distinct gravity wave component. To refer to the craft's microwave emissions as gravity waves per se and to claim that such gravitational effects manifest only at a specific frequency, in my view is rather outlandish.*

He also suggests that the technology that Lazar is describing does not work exactly as Lazar describes – and that it could actually be part of another black programme that PLV was told about – Project Skyvault. (This was alleged to be a project to launch a craft using a high powered microwave beam. It is covered in chapter 7 of PLV's book.)

Boyd Bushman's Comments

In an October 2004 interview on the American Antigravity website with Tim Ventura[198], Bushman discussed what he thought of Lazar's story.

> *We did collectively investigate everything that Bob Lazar [said], but not by calling Bob, or going through the process. I'm accustomed on the inside of the military - with high levels of security that I've had – of realizing that at least every time we were doing an investigation, that we knew would be top-secret or above, we always have what we called the story - which is what we were really doing - and then we had the cover story. In the event that anyone talked to us, [the cover story] was "what could we tell them we were doing." Bob Lazar's stuff fits beautifully with it as a "cover story." As a story, it stank - but as a cover story, it's fantastic. The items that we identified by independent*

investigation were that his description regarding his work assignment was absolutely proper. That, they allowed [it] to be part of the cover story, because they knew that we would be able to verify that. His description regarding element 115 being a potentially stable element was ***allowed*** *to be part of the cover story.*

There turns out to be this zone around 115 that we've known for years - 50 years at least - that there is a stable plateau that points down to the large elements - about 114 to 118 is good. It was interesting to me that as soon as Bob Lazar 'did his thing...' In order to have a new element you have to spend about ... six million to a billion dollars (?) ... When Bob Lazar leaked his information, or should I say, gave the cover story for his information, we only had on the chemical chart element 109... Because people had positive enough reaction to Bob's stuff, the political people permitted... a great deal of money to be expended in the United States and they also gave a great deal - and I'm talking about larger than billions - of dollars to the CERN collider as well as on into Russia. Between Russia, CERN and us we now have on your periodic tables, up to element 118. Now, it's interesting to me that they said that they got something around 118 to 116 - to be semi stable. But not in the chemical chart, but in the physics chart of elements, you'll find that there are whole batches of isotopes. I suspect that the individuals found there to be several isotopes which are stable and usable and I would suspect they're probably busy with stable isotopes today, conducting further experiments, which are obviously not going to be released because that's part of the story, not the cover story. The cover story is released not the story...

Bob has acted perfectly properly for a person that is working on a cover story. I've had to use cover stories many times, in my life as well. He had no contact with anyone of the standard UFO people... Bob Lazar contacted none of them. I think that they were fishing to get some of us to add some more understanding to their information - breaking through to the story. I, nor, any of my friends decided to give them a call – you understand who the "great them" are?

We will consider the significance of Bushman's analysis at the end of this section.

The Cover Story?

Similarly, Michael Schratt[199], an aerospace expert and SolidWorks draftsman based in Tucson (Arizona), has compiled a detailed report on the Lazar case[200], which seems to indicate that there are a few too many holes in Lazar's story. Schratt has recorded several interviews with Kerry Cassidy which can be found via the Project Camelot website.[201] (Like many other similar sites, though, the "Project Camelot" website is also used to circulate disinformation – with little discernment employed by the people that run the site.) Most interestingly, however, Schratt highlights a small detail in one of Lazar's sketches, which were shown to some people. This sketch may indicate that Lazar had seen something classified when he was working in Area 51. Schratt proposes this might have been something codenamed "Blackstar" – a two stage to orbit spaceplane also nicknamed "Speedy," "Black Magic" and "Xov."

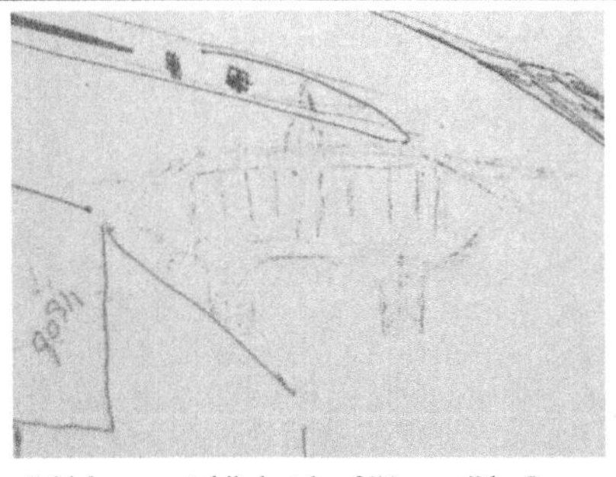

"chicken scratch" sketch of "Aurora" by Lazar

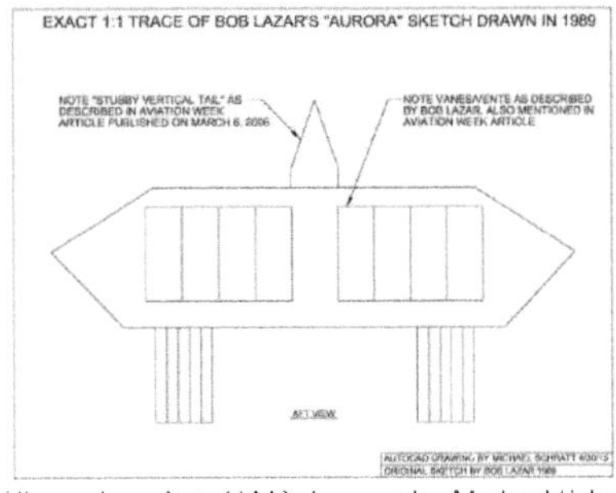

Cleaned up AutoCAD drawing by Michael Schratt

In his report, Schratt includes a suggested diagram/schematic of the "Blackstar" craft and also notes that a story about it was featured in the March 6, 2006 issue of Aviation Week[202].

What was Lazar Doing in Area 51?

We can, for a moment, consider Bushman's analysis in relation to the fact that Lazar had built a particle accelerator at his home[203] and he had used this to actually develop a new material (a hydride) for his hydrogen storage system. This means that Lazar must have a very good understanding of nuclear physics and he is also good at engineering. Also, let us note that he was allegedly employed by Los Alamos National Laboratories – which was where they had designed the atomic bomb in the 1940s. Similarly, he said he was employed after a meeting with Edward Teller – regarded as "the father of the Hydrogen Bomb"[204]. Hence, Teller was also a nuclear physicist.

Was Lazar working in Area 51, near Papoose lake, not at the S4 Lab, but at another lab? (The S4 Lab is at the Tonopah Test Range.) Neither was he working on reverse engineering a flying saucer's propulsion system. Rather, he was working on some black programme which was dealing in nuclear physics of some kind? Did he get a glimpse of other secret programmes inside Area 51 and was it these glimpses that resulted (a) in the sketches Michael Schratt analysed in his report and (b) Lazar knowing the timing of the test flights?

In other words, as Boyd Bushman implied, Lazar never disclosed what he was *really* working on at the time – even if he did relate some correct/true information about what he saw while inside Area 51. By Lazar circulating a false story about reverse engineering a craft (which apparently uses a nuclear fuel according to him), this helps to keep the details of actual programmes covered up – and attention is deflected away from some of the other aspects of these programmes. Perhaps he is basically a truthful person and was forced to put out some disinformation, which is why he said "it would be better if people don't believe" his story. That is, he didn't want people to believe something which wasn't true, but he was forced into a position where he had to put out some disinformation.

In the David Sereda/Boyd Bushman interview, Bushman showed Sereda a photo of a nuclear-powered plane.

Bushman's Nuclear Powered Plane (date unknown)

Bushman also showed Sereda a schematic of an alleged nuclear-powered saucer (from the 1960s or 1970s according to Bushman). Was Lazar actually working on a successor to one of these aircraft, which used some type of nuclear fuel? Perhaps, initially, he only knew about the fuel – perhaps he was "drafted in" to work on some nuclear physics or chemistry project.

It could, perhaps have been some form of nuclear thermal or nuclear-electric propulsion? (We will mention these technologies again later.)

I always felt that the element 115 story, in relation to it being some kind of fuel, for the ET craft was disinformation. It is also the case that neither Bob Lazar, John Lear nor Gene Huff can give any real details about the alleged theft of this material.

So, perhaps Lazar's story is that he was indeed working on something related to propulsion – but it wasn't reverse engineering a flying disk. His use of the word "reactor" in his story, though, may be a clue to what he was really working on. It is interesting to note that Stanton Friedman, a strong critic of Lazar,[205] also worked on secret Nuclear Powered aircraft projects[206], before he became renowned for his UFO research efforts.

Weighed against this, however, Lazar does sound sincere. If he wasn't working on reverse-engineering a flying disk, why did he come out with his story in the first place? It's possible there had been another security leak around the same time. Lazar's story could then be used to discredit the real one if it surfaced somewhere. So, Lazar was instructed to put out a scripted story. This is entirely speculation on my part, but we don't have any corroboration of the details of what Lazar said he was actually working on, even though his description (apart from the nuclear reactor bit) sounds plausible to me.

This is not to say that ET craft have not been reverse-engineered. In chapter 24, we will hear of an account which is somewhat similar to Lazar's (though it predates Lazar's story). Perhaps Lazar heard about such stories while he was working in Area 51. Could it even be the case that the "Blackstar" craft that was sketched by Lazar and identified by Michael Schratt used some type of nuclear fuel?

Boyd Bushman's "Last Words"

In 2014, shortly before he died, Bushman recorded an interview[207], presumably at his home, which was easy to "make fun of." Were it not for what Bushman had said in earlier interviews, perhaps even I would not have taken what he said seriously. In my opinion, Bushman does not come across particularly well in this interview – perhaps due to failing health. However, we must also consider why someone of his age and situation would even bother to make such a video. What did he have to gain by misleading others, so close to the time of his death? Many of the photographs he shows are difficult to see and therefore, for the most part, rather unconvincing. However, I think he was being truthful.

Near the start of the interview, Bushman says:

> *I do have a top secret clearance. I choose however for their purposes [his contacts], not to use it, because the intelligent ones and me and actually believe that a great deal of information should be lifted up from those dark recesses of Area 51 and moved over, so people can see it. So, that's how it began.*

One of the "most ridiculed" areas of this interview is where Bushman shows photos of an alleged alien being. It is not difficult to see why some claim that the photos he shows are of a plastic dummy or doll. If one draws that conclusion, then it is only logical to write Bushman off as someone who has finally "lost his marbles" in old age. However, I had seen this photo of the alleged alien in about 2008 or 2009 when I received it from someone that Bushman communicated with – i.e. John Hutchison[208] (whom we will discuss in chapter 12).

Much of the information that Bushman discusses in the interview (including most of the photographs) appears to be second-hand and so, due to the sensitive nature of this information, it is wise to consider he may have been fed disinformation deliberately by the secret keepers or by well-meaning but misinformed contacts.

However, towards the end of the 30-minute interview, round about the 25:30 mark, Bushman appears to be talking about first-hand information, thus:

> *They would give me pieces of UFOs and I would come to my laboratories - I have the laboratories at Lockheed and every place else... So, I would do that, but would get pieces of UFOs for example. I have three little pieces of UFOs.*

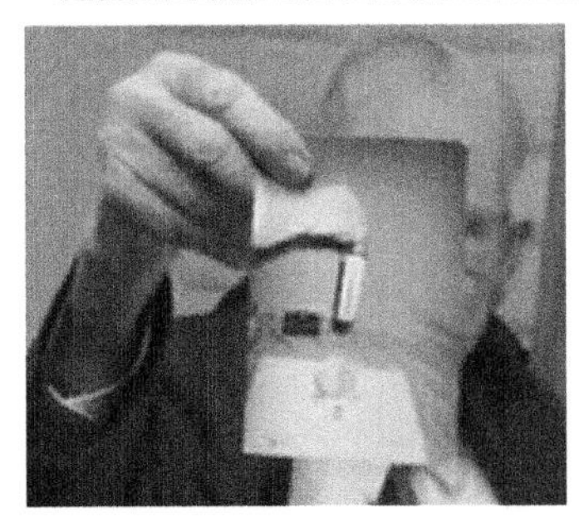

> *Down underneath - the thing that looks like a quartz crystal. notice that is my hand and it is giving weight. It says that it weighs 100 grams. Now without those three items, when I take the UFO material away, notice that the three UFO items are gone. But also notice that the weight of the crystal is now 650 grams. the implication here is that if I were to take the pieces of UFO and put them underneath an average man's scale, he would weigh 45 pounds. Basically, they're working in antigravity and therefore, our efforts are in antigravity. Not only that, but we take these same pieces of items and I go to a standard voltmeter and notice that in my voltmeter I am getting a total voltage of 249 volts out of one rock, one piece of UFO. Not only that, but the team took a piece of UFO and connected it up like this - and was able to have a complete laboratory operate for six months - with heavy equipment and it goes all the way up to 12 amperes...*

So, was Bushman making this up? I somehow doubt it.

It appears that the FBI investigated Bushman due to the fact that someone was worried he might "talk too much."[209] An FOIA to the NSA regarding Bushman was essentially refused on the grounds of… you've guessed it – the results of years of 9/11 propaganda (i.e. anti-terrorist considerations).[210] What a shame the operatives in the NSA won't be tried for acts of terrorism (after all, who keeps files on them…?) I hope I live to see the day that those who have helped cover up crimes are either accordingly prosecuted or they simply have the clarity of mind to see that they have been brainwashed or deceived and they then adopt an approach like Bushman's – i.e. an approach of study and research and then disclosure.

11. GRASPing for a "Greenglow" Propulsion Breakthrough

This chapter covers some "similar territory" to that covered in Chapter 9 of Nick Cook's HZP book. Cook spoke to some of the people mentioned below.

Millennial Gravity Research

In 2003, when I first started the "intellectual odyssey" which has lead me to compile this book, I began to come across some relatively recent articles and research relating to gravity control, as I have already alluded to elsewhere in this work. I actually came across a PDF file that someone posted on a website which contained over 1000 pages of information[211]! I only read a few pages of this document at the time (it was far easier to watch videos and listen to MP3 files by then!)

Perhaps it was the case, that as broadband internet services became more and more widely available, the rate at which information was being shared also increased exponentially, so there was a new opportunity for people to explore suppressed topics – such as free energy and antigravity research – much more easily.

A catalyst seemed to be the work of a Russian chemist, Dr Eugene Podkletnov. The events unfolded fairly predictably – there was an initial interest and some fair descriptions of the research, but then it was claimed the results of experiments could not be reproduced and the original scientists were either mistaken or cranks etc.

Dr Eugene Podkletnov – Gravity Shielding and Beam

Dr Evgeny (Eugene) Podkletnov in a 2016 BBC documentary.

According to an article on BBC news posted in July 2002[212], Podkletnov was born into a highly educated family in the Soviet Union in the mid-1950s. His father was a scientist and a professor in St Petersburg, while his mother was a researcher in medicine.

Dr Podkletnov earned a degree in Chemistry from the Mendeleyev Institute in Moscow, and then went to work at an institute within the Russian Academy of Sciences.

In the late 1980s, Dr Podkletnov went to pursue his research at the Tampere University of Technology in Finland. It was from here that he received his doctorate in materials science.

It was in 1992 that "the trouble" started, when he conducted a successful experiment into what he called "gravity shielding". This resulted in the publication of a paper in a journal called "Physica C: Superconductivity and its Applications." The paper he wrote was called: "A possibility of gravitational force shielding by bulk $YBa_2Cu_3O_7{-}x$ superconductor."[213] The abstract of this paper reads:

> *Shielding properties of single-phase dense bulk superconducting ceramics of $YBa_2Cu_3O_7{-}x$ against the gravitational force were studied at temperatures below 77 K. A small non-conducting and non-magnetic sample weighing 5.48 g was placed over a levitating superconducting disk and the loss of weight was measured with high precision using an electro-optical balance system. The sample was found to lose from 0.05 to 0.3% of its weight, depending on the rotation speed of the superconducting disk. Partial loss of weight might be the result of a certain state of energy which exists inside the crystal structure of the superconductor at low temperatures. The unusual state of energy might have changed a regular interaction between electromagnetic, nuclear and gravitational forces inside a solid body and is responsible for the gravity shielding effect.*

As indicated above, the ceramic disc was apparently made of a YBCO compound[214] containing — Yttrium[215] (a group 3 rare-earth metal) Barium and Copper Oxide.

Later, Dr Podkletnov submitted a paper about his research to the Institute for Physics in London and this was to be published in October 1996[216]. However, in a scenario somewhat reminiscent of what happened to Pons and Fleischmann in 1989 regarding their research in "cold fusion," in September 1996, the UK Sunday Telegraph published an article discussing the possibilities for "anti-gravitation" technology[217]. The BBC article said that the Sunday Telegraph story was published after a leak. This generated a lot of negative reaction against Podkletnov and so the paper was withdrawn from the Journal of Applied Physics before it was published.

Various claims were made in the press that Podkletnov's results could not be reproduced. There was talk of teams in Toronto (Canada) and Sheffield (UK) attempting to reproduce his work[218], but none have ever published results. The Sheffield team were not able to construct a large enough disc, nor replicate the exact method of rotation (due to limited resources). Podkletnov suggested that the other reason these teams hadn't come forward was because of the fear of ridicule. The lengthy PDF file that I downloaded in 2004[211], however, does actually contain a paper/article by the Sheffield researchers, starting on page 126.

The same document, from page 326, includes a 1997 paper by Podkletnov entitled "Weak gravitation shielding properties of composite bulk Y Ba2Cu3O7-x superconductor below 70 K under e.m." This paper contains instructions which describe how Podkletnov's 275mm superconducting disk was made and it includes 9 schematic diagrams of his experimental configuration. Reading this gives a better understanding of the scientific and engineering challenges presented in constructing a similar disk.

In 2003, he published a paper[219] with a colleague regarding their investigations into an electrical discharge phenomenon which occurred through large, ceramic super-conducting electrodes placed in a gas at low pressure.

From 2004 onwards, Podkletnov was heard to discuss further research[220] he had been doing on a "force beam," which was able to knock small objects over – with several kilogrammes of force. This was also mentioned/discussed in a later 2012 paper that Podkletnov published with G. Modanese[221], where they described an experiment, using a two-million volt discharge, to scatter laser light. They claimed some highly interesting results – such as the propagation speed of a gravity impulse/wave as being 64c (64 times the speed of light), although Podkletnov pointed out this does not conflict with modern interpretations of Relativity Theory if viewed correctly.

In 2012 and 2013, Podkletnov was interviewed again by Tim Ventura and the interviews were posted on Ventura's American Antigravity website[222]. In these interviews, Podkletnov described further details of his superconductor/gravity-related experiments. He also talked about using rotating magnetic fields without any superconducting materials – to generate an antigravity effect. This, of course, sounds quite similar to what Wilbert Smith talked about, as we discussed in chapter 4.

In 2016, Podkletnov was interviewed in a BBC documentary – and this included him showing a very brief clip of one of his experiments on a laptop screen[223] – where some kind of panel can be seen flapping up and down when it is placed over a rotating disk. (We will discuss this documentary further in chapter 31.)

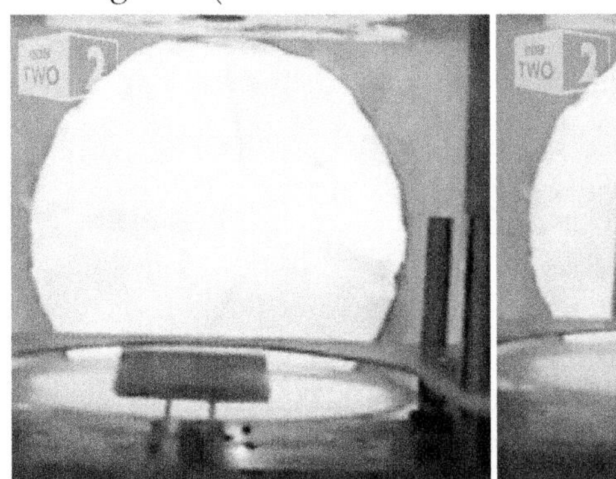

Later version of Podkletnov's disk rotation experiment.

Some further technical details about some of Podkletnov's experiments can be found in a paper published as part of the 2011 Space, Propulsion & Energy Sciences International Forum.[224]

As we shall see in the following sections, in the 1990s, Podkletnov aroused considerable interest – within NASA and Boeing.

Boeing GRASPs for Antigravity

On 29 Jul 2002, only a few years after the announcements regarding Podkletnov's research, Jane's Defence Weekly published an article written by

Nick Cook (who, himself, had published the HZP book in 2000). The article was titled "Antigravity propulsion comes 'out of the closet'"[225]. Cook wrote:

> *Boeing, the world's largest aircraft manufacturer, has admitted it is working on experimental antigravity projects that could overturn a century of conventional aerospace propulsion technology if the science underpinning them can be engineered into hardware.*
>
> *As part of the effort, which is being run out of Boeing's Phantom Works advanced research and development facility in Seattle, the company is trying to solicit the services of a Russian scientist who claims he has developed antigravity devices in Russia and Finland. The approach, however, has been thwarted by Russian officialdom.*
>
> *The Boeing drive to develop a collaborative relationship with the scientist in question, Dr Evgeny Podkletnov, has its own internal project name: 'GRASP' — Gravity Research for Advanced Space Propulsion.*
>
> *A GRASP briefing document obtained by JDW sets out what Boeing believes to be at stake. "If gravity modification is real," it says, "it will alter the entire aerospace business."*
>
> *GRASP's objective is to explore propellent-less propulsion (the aerospace world's more formal term for antigravity), determine the validity of Podkletnov's work and "examine possible uses for such a technology". Applications, the company says, could include space launch systems, artificial gravity on spacecraft, aircraft propulsion and 'fuelless' electricity generation — so-called 'free energy'.*

An article echoing similar themes and statements also appeared on 29 Jul 2002 on the BBC News Website.[226]

Cook wrote that the "GRASP" paper that he had described Podkletnov's "force beam," (discussed in the previous section), but talked about its potential use as an anti-satellite weapon. Cook stated in his article that NASA had tried to reproduce Podkletnov's experiments (see later) and that "Boeing recently approached Podkletnov directly, but promptly fell foul of Russian technology transfer controls." Cook also stated that the GRASP briefing document said that BAE Systems and Lockheed Martin had also contacted Podkletnov and "had some activity in this area" (whatever that meant). Interestingly, Cook suggested that Boeing had alluded to the existence of "classified activities in gravity modification ". Cook further stated that the GRASP document noted that "Podkletnov is strongly anti-military and will only provide assistance if the research is carried out in the white world."

It seems that Cook's article "rattled someone's cage," because on 31 July 2002 - only two days after the Jane's "GRASP" article was published - a "not so fast" type article was posted on the "Space.com" website.[227]

The article by Jim Banke was entitled "Gravity Shielding Still Science Fiction, Boeing Says." Banke's article reported that Nick Cook's analysis was:

> *Almost true, Boeing told SPACE.com Wednesday in a prepared statement.*

> *"We are aware of Podkletnov's work on 'antigravity' devices and would be interested in seeing further development work being done. However, Boeing is not funding any activities in this area at this time," the statement said.*
>
> *"The recent report that we are is based on a misinterpretation of information. For instance, GRASP is not a codename for a current project but rather an acronym for a presentation entitled "Gravity Research for Advanced Space Propulsion," in which a Boeing engineer explains Podkletnov's theory and proposes that we should continue to monitor this work and perhaps even conduct some low-cost experiments to further assess its plausibility. No steps have been taken beyond this point by Boeing."*

In a 2017 YouTube video[228], Nick Cook explains how the publication of his article about the Boeing GRASP programme triggered a series of events which resulted in the programme's cancellation.

However, almost one year later, in May 2003, it appears that Boeing were still interested, as a PowerPoint presentation by G.V. Stephenson seemed to strongly indicate.[229] This presentation seemed to relate to communication technologies – not propulsion or weapons, as its title was "The Application of High Frequency Gravitational Waves to Communication." Page (slide) 27 of this presentation (a "reserve slide") shows a simple schematic of Podkletnov's experiment – and it is reproduced below.

Of additional interest, perhaps, is the fact that this presentation was given to the "International HFGW Working Group" - HFGW meaning "High Frequency Gravitational Waves." From what I could find, this HFGW Working group seemed to have been short-lived.[230]

BOEING **GW Valve Concept – Spinning HTSC in an Alternating Magnetic field (Podkletnov)**

Hypothesis: Rotating a type II superconductor through an alternating magnetic field causes shielding via gravitational wave annihilation due to gravito-electric to gravito-magnetic frame transformation.

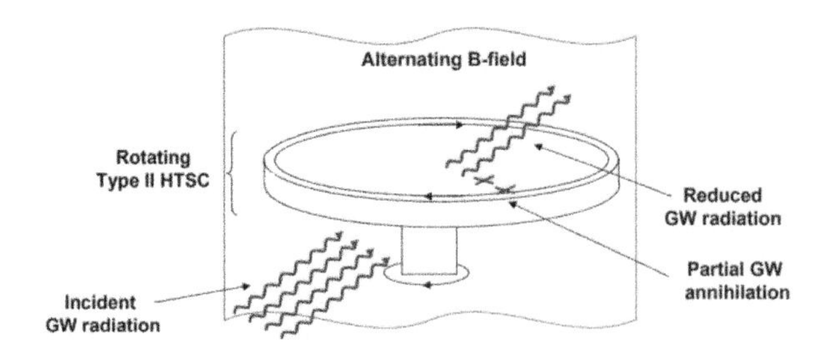

Project Greenglow (BAe Systems)

This initiative, again, seemed to start in the mid-late 1990s and seemed to be "kicked off" by Dr Ron Evans. Evans has since written a book, published in 2015, called "Greenglow: & The Search for Gravity Control,"[231] which has positive reviews[232]. Dr Evans, according to the "author information" on his

Amazon page worked for BAe Systems on advanced military aircraft projects. It was his work on aerodynamics and radar stealth technology, and his knowledge of repeated patterns in nature that led to his interest in the possibility of gravity control.

A website was set up in 1998[233], but in 2018 is no longer active (though the domain name still works). Four years later, a BBC article[212] that we referenced earlier, stated this about the project:

> *In 2000, the British defence contractor BAe Systems confirmed that it would be funding research into a device said to modify the effects of gravity.*

Dr Ron Evans - in a 2016 BBC documentary.

Project Greenglow – Concept Drawing (or was it some pulp-science fiction novel book-cover!)

On the website, there was not much information about areas of research that were conducted.[234] It essentially just seemed like a the project was a discussion group of some kind. It was described thus:

> *What is Project Greenglow?*
>
> *British Aerospace is sponsoring a speculative research programme in the realm of gravitational physics with the aim of initiating some new areas of research and priming technology development, with the hope that subsequent implementation, could lead to significant advances in the aerospace industry. For historical reasons we have called this programme project Greenglow.*
>
> *We see Project Greenglow as the beginning of an adventure which other enthusiastic scientists from academia, government and industry might like to join, particularly those who believe that the gravitational field is not restricted to passivity and who have new theories, to the contrary, which they would like to propose. The network provided by Project Greenglow could provide the arena for publicising, debating and pursuing these ideas. We feel that emphasis should be given to those theories which can be tested in a laboratory at moderate cost and which, if successful, could be developed into technical applications for the aerospace industry.*
>
> *Our approach to the research programme will be much along the lines of the newly established NASA Breakthrough Propulsion Physics Programme which has, as its central theme, the goal of developing propellant-less propulsion.*

We will discuss the NASA Breakthrough Propulsion Physics Programme in the next section.

Browsing through the Internet Archive[235] to locate later versions of the site, it seems that only two links were ever added to the site. One link was to a site called "Quantum Cavorite"[236] and another was to a site called Electrogravity.[237] The Electrogravity site is still active, but neither of these sites belong to Boeing, NASA or BAe systems – they appear to be "hobbyist" type sites only.

The only other item I could find on the Greenglow website was a page about a pair of lectures at the University of Lancaster Physics Building in September 1998.[238] One was given by Dr. Costas Kyritsis, of the National Technical University of Athens – which was about understanding electromagnetism and gravity. It was titled "A unified derivation of nonlinear electromagnetism and gravitation. Implications in electromagnetic propulsion."

The second lecture was by Mr. Stavros Dimitriou, of the Technological Education Institute in Athens and was about the propulsive effect observed in massive plane capacitors. It was entitled "Thrust from time-derivatives of the electric charge." The lectures were sponsored by BAe. (Does this remind you of anything in chapter 3?)

The website was never updated after 1999 and "Future plans for the Project Greenglow web site"[233] don't seem to focus on any particular research area (such as electrogravitics or superconductor-related research).

NASA Breakthrough Physics Propulsion Programme

Marc Millis in 2016 BBC Documentary

It was again in the mid to late 1990s that this programme was initiated and it was co-ordinated by an aerospace engineer named Marc Millis of the NASA Lewis Research Centre (LeRC). The project brief stated[239]

> *NASA is embarking on a new, small program called Breakthrough Propulsion Physics to seek the ultimate breakthroughs in space transportation.*

The brief that he came up with seemed to be more focused than that of "Greenglow," and included 3 main areas of study:

- Propelling a vehicle without propellant mass.
- Attaining the maximum transit speeds physically possible.
- Creating new energy production methods for propulsion systems.

The program brief said that it represented "the combined efforts of individuals from various NASA centres, other government labs, universities and industry." According to Nick Cook's HZP book, Millis was given a budget of $500,000 and initially awarded this in 5 contracts.

Also noteworthy was that the program was "supported by the Space Transportation Research of the Advanced Space Transportation Program managed by Marshall Space Flight Centre(MSFC)." In this respect, it seemed to be more significant than the Greenglow project, having apparently had the support of several offices and groups, not just a "small discussion group."

Marc Millis prepared a White Paper, which was dated 02 Oct 1998.[240] In this paper, he notes the obvious limitations of rocket technology writing:

> *Propellant mass rises exponentially with increases in payload, destinations, or speed. This limit cannot be overcome with engineering refinements: it is based on the very physics of rocketry. To dramatically reduce the expense of near-Earth journeys or to journey beyond our Solar system in a reasonable time, new propulsion physics is required.*

It is interesting to see the document mention vacuum fluctuation energy, the Casimir effect, wormholes, warp drives, quantum tunnelling and "anomalous weight reductions over spinning superconductors." It states that "gravity, electromagnetism and spacetime are coupled" but the "coupling is still not fully understood."

It then states that if there were breakthroughs in any of the three goals already listed above, it would "revolutionize space flight" and then it states:

> *Achievement of any one of these would usher in a new era, where people could explore deeper into space, reaching more destinations, in less time and with less infrastructure. Achievement of all three would enable human voyages to other star systems.*

In discussing the approach to research and acknowledging that much of it will "fail", it then says:

> *The emphasis on credibility is because such long-range ambitions are often tainted by non-credible work or even "pathological science" and since genuine progress can only be made with credible work.*

It then says that because of this, and related factors "scepticism is actively sought." It also notes that up to the time that this Whitepaper was written (about 2 years after the inception of the project), about 200 researchers expressed interest in assisting. It then refers to a number of experiments which had already been undertaken that might inform progress to reaching one of the above goals.

In 2000, Space.com[241] reported that the BPP project had tried to replicate Podkletnov's experiments, according to Ron Koczor, then assistant director for science and technology at the Space Science Laboratory in NASA's Marshall Space Flight Centre:

> *Koczor assembled a team that worked together with scientists at the nearby University of Alabama at Huntsville, to build a device partially simulating the one Podkletnov had used. But the researchers were unable to replicate Podkletnov's results, and the partnership fell apart last year with bad blood between the two sides.*

It seems that the BPP was supported by NASA until 2002, but may have "run on" at least until May 2004 when Marc Millis posted a very interesting summary of the research completed[242]. It does mention things such as Lifters and the work of Townsend Brown but writes off the effects as purely the result of an ion wind:

> *3.2.7. Biefeld-Brown and Variants. In 1928 a device was patented for creating thrust using high-voltage capacitors [50]. Since then, a wide variety of variants of this "Biefeld-Brown" effect, such as "Lifters" and "Asymmetrical Capacitors" have claimed that such devices operate on an "electrostatic antigravity" or "electrogravitic" effect. One of the most recent variants was patented by NASA-MSFC [51]. To date, all rigorous experimental tests indicate that the observed thrust is attributable to ion wind.*

Sadly, it does not note that Brown himself had ruled out this ion wind as being the sole cause of the effects he generated – and his research had been done perhaps six decades before Millis' report.

The summary we referenced above gives more details about the attempted reproduction of Podkletnov's experiments:

> *3.2.4. Podkletnov Force-Beam Claims. Through undisclosed sponsorship, Podkletnov produced a new claim - that of creating a force-beam using high-voltage discharges near superconductors. His results, posted on an Internet physics archive [45], claim to impart between $4x10^{-4}$ to $23x10^{-4}$ Joules of mechanical energy to a distant 18.5-gram pendulum. Like his prior "gravity shielding" claims,* ***these experiments would be difficult and costly to duplicate****, and remain unsubstantiated by reliable independent sources.*

So, it appears they couldn't reproduce them due to **cost** – not because Podkletnov was wrong! (This is confirmed in chapter 9 of Nick Cook's HZP book.) Such is the problem with the quality of journalism – often no better than "Chinese Whispers."

Interestingly, the summary notes:

> *3.1.8. Explore Vacuum Energy. Quantum vacuum energy, also called zero point energy (ZPE), is a relatively new and not fully understood phenomenon. ... It has been shown analytically, and later experimentally, that this vacuum energy can squeeze parallel plates together. This "Casimir effect" is only appreciable at very small dimensions (microns). Nonetheless, it is evidence that space contains something that might be useful. The possibility of extracting this energy has also been studied. In principle, and without violating thermodynamic laws, it is possible to convert minor amounts of quantum vacuum energy.*

We will learn later in this book (assuming you don't already know) why this project never really "got off the ground." All similar projects will fail, as their starting point is based on hopelessly wrong assumptions.

Ning Lee and AC Gravity

Dr Ning Lee (Centre)
Photo by Philip Gentry

Dr Ning Lee is perhaps the least well-known of the "antigravity research" names I have mentioned so far in this book. It seems that information about her research is quite limited. Dr Lee, who was educated in China, was highly qualified. She had a B.S. in Semiconductor Physics, an M.S. in Space Plasma Physics and an M.E. in Electrical & Computer Engineering and a Ph.D. in Plasma Physics.[243] Her biography also notes that:

> *Dr Li's original development of **the gyro magnetically produced gravitomagnetic field** was published in Phys. Rev. D in 1991, in Phys. Rev. B in 1992, and in Found. Phys. in 1993.*

So, it appears she had been doing research in gravity fields for quite a few years. In the abstract of a 2003 paper, that we will reference again later, she wrote:

> *In the Lorentz gauge, the gravitational-generalized Lienard-Wiechart retarded potential shows that there are two types of gravitational fields. One is DC gravity, which is local and static such as the Earth's gravity. Another is AC gravity, which is radiation and can transport far away without energy decay such as gravitational waves (GW).*

Her name became known in 1999 when she was doing research in Huntsville Alabama. An article in Popular Mechanics called "Taming Gravity"[244] by Jim Wilson, described a rotating "HTSD" – High Temperature Superconducting Disk, which sounds very similar to what Dr Podkletnov was using. In the article, Wilson wrote:

> *...the most important component of their proof-of-concept demonstrator... is a 12-in.-dia. high-temperature superconducting disc (HTSD).*

In chapter 9 of the HZP book, Nick Cook writes:

> *In 1993, the Advanced Concepts office at the Marshall Space Flight Centre was handed a copy of a paper written by two physicists, Douglas Torr and Ning Li, at the University of Alabama at Huntsville. It was called "Gravitoelectric-electric coupling via superconductivity"[245] and predicted how superconductors—materials that lose their electrical resistance at low temperatures—had the potential to alter gravity.*

The paper referenced is theoretical. Six years Later, however, Lee was apparently aiming to build a device that might have been called a "force-field machine," which when complete, could cause a bowling ball placed anywhere above to stay exactly where it was left. The article continued:

> *Prospects for the Alabama HTSD are attracting serious attention because this particular disc was fabricated by Ning Li, one of the world's leading scientists. In the 1980s, Li predicted that if a time-varying magnetic field were applied to superconductor ions trapped in a lattice structure, the ions would absorb enormous amounts of energy. Confined in the lattice, the ions would begin to rapidly spin, causing each to create a minuscule gravitational field...*

> *Using about one kilowatt of electricity, Li says, her device could potentially produce a force field that would effectively neutralize gravity above a 1-ft.-dia. region extending from the surface of the planet to outer space.*

Again, this sounds remarkably similar to Podkletnov's experiment. Some further explanation is given:

> *"The first thing to understand about Li's device is that it is neither an antigravity machine nor a gravitational shield," says Jonathan Campbell, a scientist at the NASA Marshall Space Flight Centre who has worked with Li. "It does not modify gravity, rather it produces a gravity-like field that may be either attractive or repulsive." Li describes her device as a method of generating a never-before-seen force field that acts on matter in a way that is similar to gravity. Since it may be either repulsive or attractive she calls it "AC gravity." "It adds to, or counteracts, or re-directs gravity," explains Larry Smalley, the former chairman of the University of Alabama at Huntsville (UAH) physics department.*

The article also states that the effects which seem to be happening are predicted by the theory of relativity, but the effects are normally too weak to measure on earth. Indeed, a 2004 probe launched by NASA sought to measure Gravitomagnetism[118]. According to my understanding, they found that the results of their measurements were as predicted[246] – i.e. that there indeed *is* an interaction between magnetism and gravity (even though it is extremely weak in "everyday" situations). In the Wilson article, Dr Ning Lee explains how this effect is amplified in superconducting materials:

> *Li explains that as the ions spin they also create a gravito-electric field perpendicular to their spin axis. In nature, this field is unobserved because the ions are randomly arranged, thus causing their tiny gravito-electric fields to cancel out one another. In a Bose-Einstein condensate, where all ions behave as one, something very different occurs.*
>
> *Li says that if the ions in an HTSD are aligned by a magnetic field, the gravito-electric fields they create should also align. Build a large enough disc and the cumulative field should be measurable. Build a larger disc and the force field above it should be controllable. "It's a gravity-like force you can point in any direction," says Campbell. "It could be used in space to protect the international space station against impacts by small meteoroids and orbital debris."*

The Popular Mechanics article then notes that Ning Lee had left the University of Alabama where this research had been conducted and collaboration with NASA ended. Ron Koczor, then assistant director for science and technology at the Space Science Laboratory in NASA's Marshall Space Flight Centre suggested Dr Ning Lee was more interested in "proving the Science" than building a device:

> *Koczor said the project fell apart not because of incompetence, but because Li was primarily interested in proving her theories of why the "gravity shield" would work. That differed from NASA's goal of simply building a working device, he said.*

> *"She wanted the research to focus on her particular theory. Our intent was simply to show there was a gravity effect, without saying 'theory A is right' or 'theory B is right,'" he explained.*

However, it seems Lee was concerned about who would end up controlling the research. The Space.com article from 2000[241], that we referenced above states:

> *Li said she dropped the NASA collaboration and decided to work independently after the agency "wasted" the project's money and resources.*

Similarly, the Popular Mechanics article reported:

> *This summer, Li left UAH. She and several colleagues are striking out on their own to commercialize devices based on her theory and a proprietary HTSD fabrication technique.*
>
> *Li's next step is to raise the several million dollars needed to build the induction motor that individually spins the ions in the HTSD. "It will take at least two years to simulate the machine on a computer," says Smalley, who plans to join Li's as-yet-unnamed company after he retires from UAH. "We want to avoid the situation that occurred in fusion where extremely expensive reactors were built, turned on, and didn't work as intended because of unforeseen plasma instabilities." Li says she has turned down several offers for financial backing. It is less about money than control. "Investors want control over the technology," she says. "This is too important. It should belong to all the American people."*

Gravitational Waves Conference in May 2003

Following this, it seems Lee all but "disappeared." Thanks to a blog[247], I was able to track down this document about a "Gravitational Wave" conference in May 2003[248]. This document shows about 20 other people were involved with the conference including George Hathaway (who we will likely mention again in chapter 12), Gary V. Stephenson (who was working at Boeing) and Marc G Millis – of the NASA BPP programme mentioned earlier. Another name that is mentioned is Dr Hal Puthoff. The blog I referenced above makes an excellent observation:

> *"I would point out, however, that once again we see Dr. Hal Puthoff involved in this conference. Everywhere there is a mystery involving science, this man is involved somehow."*

The conference document contains abstracts of papers that were discussed, including Dr Ning Lee's – which was about "The possibility of generating and measuring an AC gravitational field."

It seems that since around the year 2000, we have heard more about Gravity waves, but mainstream science has focused on measuring waves created by cosmic events – for example the alleged detection of gravity waves from the collision of two black holes in the LIGO experiment in 2016.[249] There is far less talk of generating gravitational waves.

Conclusion

The effects that are mentioned in this chapter are tiny, difficult to create and perhaps almost as difficult to measure. I think the research that was done around the turn of the millennium reveals that white-world science is not capable of creating any antigravity breakthroughs. As we shall see when we study parts of the "regular" (rocket-based) space programme, the whole basis of white-world antigravity research is flawed – perhaps in large part because it is hamstrung by standard "white world" theories and models of how matter and electric and magnetic fields should behave. The scientists concerned have "snatched a glimpse" into what I call "black world physics" but if they get any deeper insights, black world interests are never very far away – and will "clamp down" on their research and projects.

In the next chapter, we will learn more about the afore-mentioned Canadian researcher John Hutchison. As Boyd Bushman stated in 1999, Hutchison achieved much more significant antigravitational effects working outside mainstream science, although he needed a large amount of equipment to do this.

12. The Hutchison Effect

The text in this chapter is mainly derived from 2 sources (a) a press release I wrote in 2008, in relation to Dr Judy Wood's 9/11 research[250] and (b) an affidavit I helped John to compile in 2008 in relation to Dr Judy Wood's 9/11-related Science Fraud case against NIST's contractors.[251] The people named in this chapter were named by John in his affidavit, and so we can assume he was quite certain of what he said. Also, important named individuals such as Col John Alexander have openly spoken of their work with John Hutchison.

Nick Cook covers John Hutchison's story (up to the end of the 1990s, that is) in his HZP book. Cook also went to meet John, although Cook apparently did not witness any of the antigravity/levitation effects happening at that time.

John Hutchison - early 2000s

John Hutchison is a Canadian inventor and experimental scientist who has been working with "field effects" since the 1970s. John Hutchison's life changed drastically in 1979 when, upon starting up an array of high-voltage equipment (which he was using to investigate phenomena produced many years previously by Nikola Tesla), he felt something hit his shoulder. He threw the piece of metal back to where it seemed to have originated, and it flew up and hit him again. This was how he originally discovered some sort of method for shielding gravity. When his Tesla coils, electrostatic generator, and other equipment created a complex electromagnetic field, heavy pieces of metal levitated and shot toward the ceiling, and some pieces shredded.

Hutchison uses radio frequency and electrostatic sources – such as radar equipment, microwave generators, a Tesla coil and/or a Van de Graaff Generator. He has also used a radioactive source to help trigger certain effects.

The Hutchison Effect occurs in a volume of space where the beams intersect and interfere. The results are levitation of heavy objects, fusion of dissimilar materials such as metal and wood, anomalous melting (without heating) of metals without burning adjacent material, spontaneous fracturing of metals (which separate by sliding in a sideways fashion), and both temporary and permanent changes in the crystalline structure and physical properties of metal samples. Upon analysis and thorough investigation, the Canadian government dubbed this phenomenon the Hutchison-Effect.

The reason for including a discussion of John's research in this book is due to the levitation effects he has repeatedly created in a fair number of his experimental runs. A collection of videos, audio recordings and documents can all be found on a website I created for John in 2009.[252]

John has been visited by many TV and News crews over the years and participated in several documentaries made by various production companies. He has hundreds of videos posted on his YouTube channel.[253]

A Note to and about "Debunkers"

From the evidence you will find discussed here and online, you should conclude that what John Hutchison has been doing is real but is not well understood in the realm of conventional science. It is untrue to say that mainstream scientists dismiss John's results - you will see here documents which show that people like Ken Shoulders, George Hathaway and Hal Puthoff have all taken a keen interest in and endorsed John's experiments. However, as with other real, but unusual phenomena, whatever amount of evidence is presented to a sceptic or debunker, it is dismissed, as the cognitive dissonance and highly restrictive world view take precedence over the open-ended study and analysis of the phenomena.

Often, the first refuge of the Hutchison Debunker is the "Toy Flying Saucer" video.[254] This video did indeed use *a wire*, attached to the toy UFO, to help it "levitate." The truth behind this experiment is that it was not a levitation experiment in the same way some of his other experiments were. This was for a high voltage experiment – with the voltage being delivered through a wire. It was actually similar to experiments performed by Thomas Townsend Brown, discussed in chapter 3. In another video, John Levitated the same (or a very similar) toy using his field interference method.[255]

A second debunker's refuge is to write John Hutchison as a "crank" or a "nutcase." For example, John has openly posted videos of himself dressed up in a blonde wig, pretending to be "Karla Kniption" a female journalist.[256] Whatever John dressed up in, or whatever voice impersonations he does, the photos shown here and the videos of his experiments all show real effects and events.

Hutchison Effects

Perhaps the most obvious of the effects repeatedly created and observed that present the biggest problem for "debunkers" are anomalous metal samples John has created over approximately a 20-year period – about 500lbs total weight.

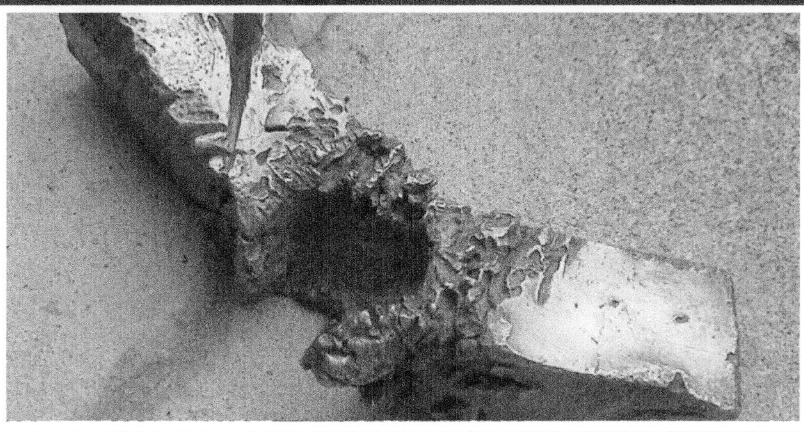

John has witnessed the tops of steel bars turn to dust and white powders as well as chrome plating being "blown off" other samples. At various times since 1980, John has witnessed anomalous effects of foaming water in some experiments. John has other samples of dissimilar materials, such as wood and metal, that have fused together during experiments. Similarly, a recent video of some of the metal samples[257] shows anomalous regions which are magnetised – some type of ferrous material is embedded in copper. (The green film allows one to view lines of flux from magnetised samples).

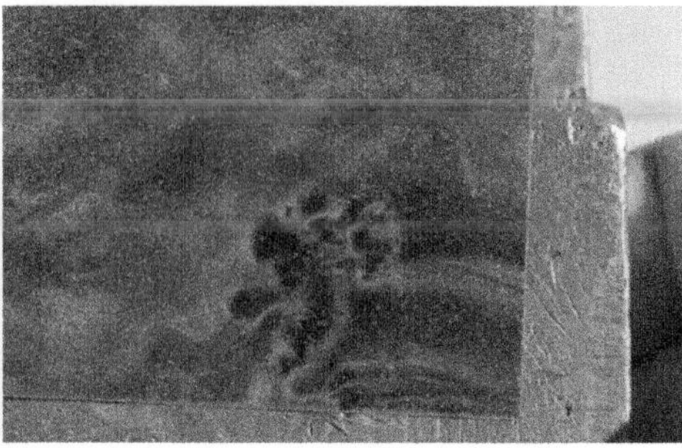

Other effects observed in his experiments include weird fires and transmutation. Material scientist George Hathaway observes that the Hutchison Effect causes either lift or disruption of the material itself. Not surprisingly, these effects came to the attention of a number of people in the US Military – over a period of 5 – 10 years. Col. John Alexander and others from the U.S. military visited[258] John Hutchison in 1983 and filmed his experiments with a team from Los Alamos National Laboratories (LANL)[259] (see letter below).

I wrote the original press release in 2008 in relation to the clear and obvious similarities between effects seen in the aftermath of the destruction of the World Trade Centre Complex on 9/11 to the effects seen in John Hutchison's experiments. Some of these are summarised in the table below. For full details, please see Dr Judy Wood's book "Where Did the Towers Go?"[76]

Phenomenon	"The Hutchison Effect"	Anomalies at the WTC
Weird Fires The fires seen near the toasted cars don't seem to ignite the paper. Some photos show firemen walking very near or even through them. Are they "cold" fires?	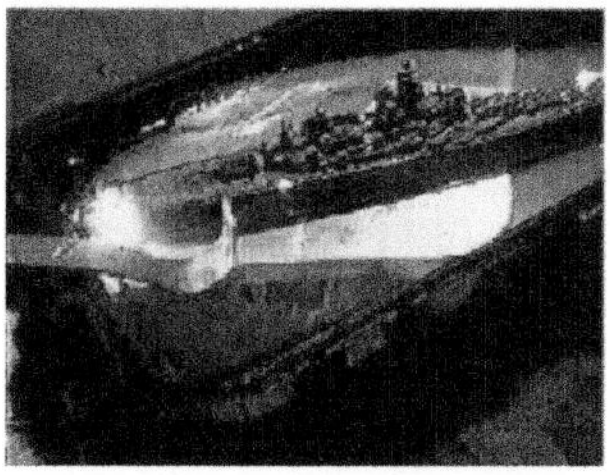	
Bent Beams Samples that John Hutchison has produced show very unusual effects on the metal – sometimes severe bending occurs		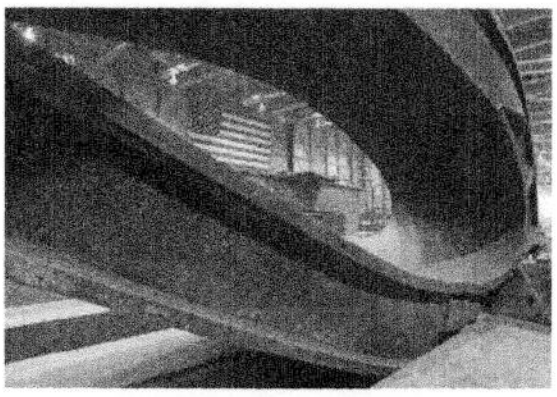
Jellification Sometimes the metal "jellifies" - other effects are also seen.	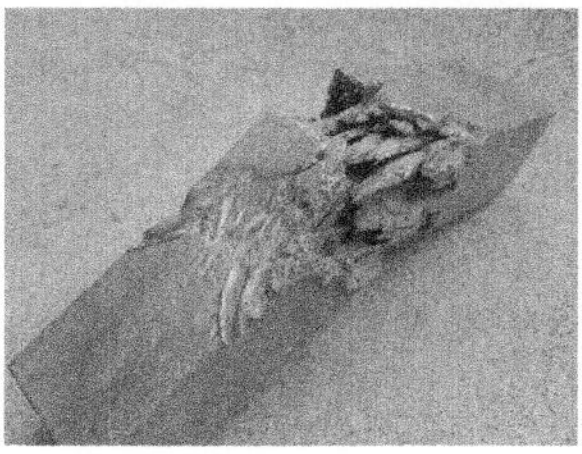	
Cars/Lift and Disintegration Some WTC pictures show cars that are upside down. (How?) One of the key effects John Hutchison has reproduced many times is a "levitation" or "antigravity" effect.		
Holes Samples seem to end up with "voids" in them, following the experiments. Could this effect have created holes in WTC6 and other buildings?	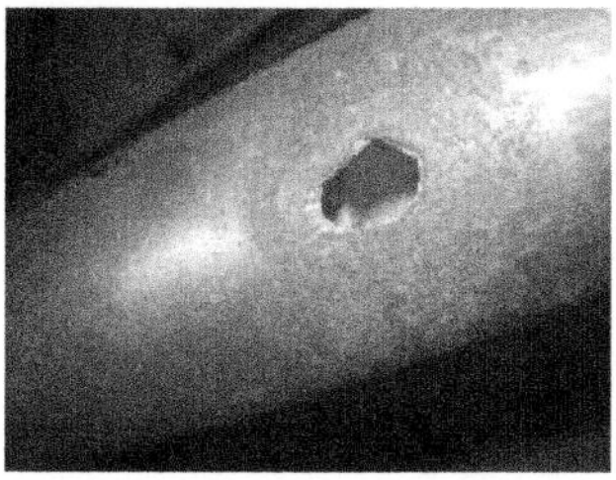	

DEPARTMENT OF THE ARMY
UNITED STATES ARMY INTELLIGENCE AND SECURITY COMMAND
FREEDOM OF INFORMATION/PRIVACY OFFICE
FORT GEORGE G. MEADE, MARYLAND 20755-5995

REPLY TO
ATTENTION OF:

(Files) FOI March 13, 1991

Axiom General Systems
911 Dublin Street
New Westminster
Vancouver, B.C.
Canada

Dear Mr. Hutchison:

Re: your letter December dated the 18 1990.

We understand that you received some slides of your work taken by Los Alamos and U.S.A.I 1983 June.

As you know the report is classified.

We suggest you go threw the proper foia act, The F.O.I.A. Title 5 Code 552.

However your project (The Hutchison Effect) holds merits in future developments

The best of luck in your research.

Sincerely,

Mr. S Brammer Files (DCII)

ENCLOSURES

9.9.99. Dorian - you may like to try to obtain the Los Alamos + mentioned above, made by the Los Alamos after their visit to Hutchison's lab. John H. himself has been able to ... promptly classified!

Letter regarding LANL's classified report about John Hutchison's research

LANL Team and Col John Alexander

In 1983, a group from Los Alamos National Laboratories (LANL) contacted him on behalf of the US Government. At that time, the following individuals were involved: Flynn Marr (Lawyer) Joanne Maclusky (Lawyer for Pharos Technologies), George Hathaway (Scientist), Alex Pezzaro, Col. John Alexander, Bob Friedberge and John Rink. Also, at that time, he gave private demonstrations to Edna Drake and Bill Ross (who were civilians).

During the work with the Los Alamos Team, Starlite Lighting, who resided in the industrial premises adjacent to the lab witnessed effects in their own warehouse. Experiments performed with the Los Alamos team were videotaped, but John has been told, following requests, that these tapes are no longer available.

During experiments performed with the Los Alamos team, large effects took place outside the target areas such as mirrors breaking, fires starting and there were unusual effects with the lights and other things. Col. Tom Bearden said the project with the Los Alamos team was sabotaged by either Bob Freeman or Bob Friedberg.

Around 1984, information was presented to Ed Murand of the Iranian embassy and Henry Kissinger of the USA government in an effort to get financial support for Mel Winfield at a level of $500,000, to fund research into the effects already observed in other experiments. But Col. John Alexander stepped in to prevent funding being given.

In 1985, Hutchison completed 2 days of controlled experiments for Col. John Alexander, who was representing interests in Washington D.C. Again, video recordings were made but this time, the report he produced was not classified, and John has copies of the report and the video recordings. Also, in 1985, John gave demonstrations for George (Jack) Houck of McDonnell Douglas and around this time, John witnessed some metal samples giving off a grey mist during some experiments.

Levitation

In 1979, Mark Murphy and another individual, Mr. Murphy of Canadian Pacific Air, lived in the same building as his lab at 1458 East 29 Street, Lynn Valley, North Vancouver and they witnessed levitations of and effects on samples. During 1980 Bill Ross (a civilian) and Mel Winfield (of Vancouver) also visited his lab at 1458 East 29 Street and they observed and took photos of levitations. These photos were later published in Dr. Winfield's book "The Science of Actuality".

Levitation/lift effects have been seen to occur with objects of only a few grams (ounces) in weight to approximately 750 kg (1500 lbs). This has been in cases where the total energy input to his set of test equipment has been between 75 watts and 4000 watts (4kw), at 110 volts, depending on the equipment and the test being done. The results have been reported on CTV (Canadian Television) news and by Government personnel.

Many of his experiments have caused effects at a distance — such as levitation of objects, vibration of objects and other effects. On some occasions, effects have been seen at around 300 feet away from the main area of experimentation, one such case being in Dec 2007 when plumbing on the ground floor of the building, in which John's experiments were being performed, failed and caused minor flooding. That part of the building then had to be evacuated.

George Hathaway and Norman Hathaway witnessed or filmed further demonstrations or experiments at around this time. Mr. Dennis Edmonson of NASA discussed the possibility of obtaining $5 million in research funding being made available from NASA or Boeing. Also, in 1985, CKVU T.V. filmed experiments for the news hour and a mid-air "aurora" type effect was also witnessed.

Independent Lab Tests of Samples

In 1986, the Max Planck institute in Germany undertook tests on his metal samples. Their tests showed anomalous properties of the crystal structure of the samples. The crystalline structure was "changing very rapidly over time".

In Germany, Siemens Laboratories, BAM labs and Berlin University have also done tests on his work, but that communication between them was very limited. By this point, many articles had been written discussing aspects of his work, such as those in Esquire Magazine and in other publications in Germany, Japan and the UK.

Germany Trip and Lab Equipment Confiscation

In 1989, John was contacted by Prince Hans Adam Liechtenstein expressing an interest in his experiments and research. John has been in communication with him since that time. Also, in 1989, John had taken a trip to Germany, to do some work at the Max Planck Institute. John planned to have equipment shipped over there, but during his time away, government officials took the lab equipment while it was in transit and while it was in the previous lab installation. Lawyers Hogson and Kowasky worked on his behalf to get the equipment shipped safely by Rolf Kiperling.

Judge Paris set up a B.C. Supreme Court order to protect the lab equipment. The Vancouver Sun ran the story of the lab equipment seizure on the cover of the Feb 22, 1990 edition. At this time, Henry Champ of NBC got involved in covering the story while John was in Germany.

Ongoing Interest From the 1990s Onwards

Around this time, George Hathaway, Col. John Alexander and some people at Sandia Labs (USA) and Washington D.C. were still involved in analysing his experiments, but they did not intervene in the seizure of the lab equipment. George Lisacase formed a company called Pinnacle Oil International and he stole some of the energy cell technology John had developed and used it for oil prospecting. George Lisacase also has/had videos of levitation demonstrations as well as photos of free energy tests and experiments.

Also, in 1993, Nobuo Yokoyama, as part of the Tokyo Free Energy Project, published a Japanese book about the Hutchison Effect.

Harkening back to chapter 11, it seems that Marc G Millis of NASA's Breakthrough Propulsion Initiative had become aware of John's research, and he asked some fair questions…

[illegible]

National Aeronautics and
Space Administration
Lewis Research Center
Cleveland, OH 44135-3191

Reply to Attn of 5340

December 7, 1995

Mr. John Hutchison

CANADA

Dear Mr. Hutchison:

I received your Fax of Dec 2, 1995, but had a little trouble reading the handwriting. Can you please mail me a more legible copy?

I am already aware of some of your work, in particular from the writings by Hathaway from a few years ago. I have also seen some of your videos of objects set into vibrations, objects split apart, and objects being abruptly lifted up.

From what I have already read and seen, I am under the impression that it is difficult to reproduce the effects at will. Have you made any progress toward making any of your observed affects completely repeatable? Have you been able to isolate the key factors which make the effects happen or which keep the effects from happening? Are you able to carefully record the conditions of your experiments, including for example the magnitudes and frequencies of electromagnetic fields you use, the geometry of your set up, or any other parameters that my shed light on how these effects are occurring?

Do you still work with Hathaway or anyone else who can help you write about your work in a clear, logical fashion? Do you have anyone who can help you with your experiments to give the experiments the systematic rigor needed to trace down how the effects work?

Sincerely,

Marc G. Millis
Aerospace Engineer

In 1997, more levitation experiments were demonstrated to members of the company "Harry Delighter Productions" and they put these in a video called "Free Energy — The Hunt for Zero Point". In year, 2000 his premises were raided by the local New Westminster Police Department and other people with them photographed things in his apartment during the raid.

John has documentation that "Car Check" of Canada was established as a front-company for the Canadian based research, and all results were classified and remain so.

In the period 2004-2005, Bruce Burgess of Blue Book Films worked on another film project with Hutchison. Also, in the period 2004-2005, John made attempts to retrieve copies of official reports about his experiments through FOIA requests in Canada and the USA, but nothing of value was released to him.

In April 2005, John spoke on Art Bell's US "Coast to Coast" radio show and discussed the likelihood of the replication of his technology by other groups within the military[260].

> *Art Bell: John is it possible that your effect has been not only discovered, but perhaps perfected by the military and that we do have craft that can virtually change the makeup of their mass and virtually disappear - to the human eye - or go - wherever they go - when the effect that you stumbled on is applied?*
>
> *JKH: I tend to believe, Art that they have replicated it. I get signals and messages from, well from the early 90s, about that - that basically we left them too close. This stuff was presented in the Pentagon in 1981, 1980 and they came and investigated. Then the Canadian government got involved - also to investigate it... But I'm trying to retrieve the video stocks. One especially from the Canadian government. Those people don't exist although I have all their letters and all their names - you cannot contact them they deny they were ever here and I'm trying to go through the Freedom of Information Act to the Canadian government...*
>
> *It's interesting that you're talking about this, because in the last 10 months, there's been a really heavy traffic and phone calls to me - emails and letters to me from NASA,* ***from the Pentagon from - SAIC ...*** *I have all this stuff printed out. It comes to almost two inches of material. It seems to go in waves and ... they say, "Well, can you send us videos..." So, I send videos - to the Airforce division of the Pentagon. I don't want to mention names...*

John mentioning SAIC is significant because this company helped to co-ordinate the production of scientific/technical reports (compiled by NIST) into the destruction of the WTC complex on 9/11. Those reports were released in August 2005 – not long after this Hutchison/Bell interview. We will discuss SAIC a little more in chapter 28.

Also, around this time Harold Berndt of Surrey, BC filmed some demonstrations of effects and posted a video on the American Antigravity Website.

Reproducibility

This is another refuge of sceptics – to make them feel comfortable that what John Hutchison has achieved is either not real or it is irrelevant. Reproduction in the "white world" is tricky – for example, it requires two tons of equipment, carefully configured and arranged. As of writing this in March 2018, an organisation in Germany is planning to attempt reproduction[261], though they need John's help to do this and he has not been able to travel to Germany to assist the group. This situation is quite similar to the reproducibility of Podkletnov's experiments – that is to say, the experimental setup and configuration is difficult or expensive to copy, emulate or build from scratch.

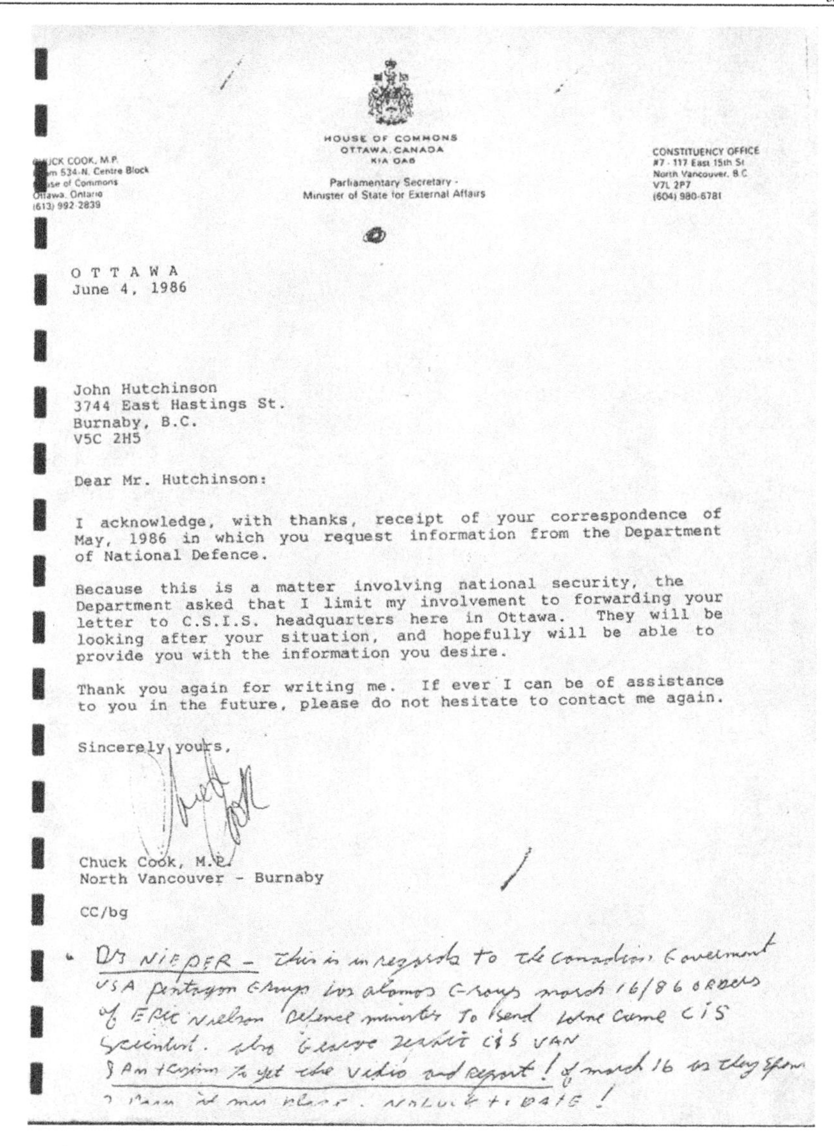

HOUSE OF COMMONS
OTTAWA, CANADA
K1A 0A6

Parliamentary Secretary -
Minister of State for External Affairs

...UCK COOK, M.P.
...m 534-N, Centre Block
...se of Commons
Ottawa, Ontario
(613) 992-2839

CONSTITUENCY OFFICE
#7 - 117 East 15th St
North Vancouver, B.C.
V7L 2P7
(604) 980-6781

O T T A W A
June 4, 1986

John Hutchinson
3744 East Hastings St.
Burnaby, B.C.
V5C 2H5

Dear Mr. Hutchinson:

I acknowledge, with thanks, receipt of your correspondence of May, 1986 in which you request information from the Department of National Defence.

Because this is a matter involving national security, the Department asked that I limit my involvement to forwarding your letter to C.S.I.S. headquarters here in Ottawa. They will be looking after your situation, and hopefully will be able to provide you with the information you desire.

Thank you again for writing me. If ever I can be of assistance to you in the future, please do not hesitate to contact me again.

Sincerely yours,

Chuck Cook, M.P.
North Vancouver - Burnaby

CC/bg

DR NIEPER – this is in regards to the Canadian Government
USA pentagon Group los alamos Group march 16/86 ORDERS
of ERIC Nielson Defence minister To send some came CIS
scientist. also [illegible] CIS VAN
I Am trying to get the vidio and report! of march 16 as they [illegible]
[illegible]

Letter from Canadian MP in response to a query about the Canadian Government's investigation of his research.

Conclusion

What John Hutchison achieved is, by some measures, miraculous – in the way that Boyd Bushman described in his 1999 interview with Nick Cook. We have clear evidence that the agents of the Military Industrial Complex know all about the Hutchison Effect – they essentially spent years studying it. I can only conclude, based on what was seen in the destruction of the WTC on 9/11 that they have perfected its use and this is a very closely guarded secret – because of its importance. I have written about this several times before, for example, in my two 9/11 books which I have referenced already. The key aspects of the Hutchison Effect seem to be that:

- It allows access to large amounts of energy – with a small energy input.
- It causes or facilitates levitation – antigravity, and therefore is a "gateway" to new propulsion technologies.
- It causes certain temporal effects (for example, background radiation counts have been shown to drop).
- It seems to create transparency or even invisibility – so it has obvious "stealth" applications.

Theoretical work by Ken Shoulders regarding "charge clusters" is probably the most appropriate in explaining how the Hutchison Effect works.[262] The fact that Hutchison Samples show strong evidence of elemental transmutation in many (or all) cases shows that the process that occurs has some relationship to that other suppressed area of research – LENR – more commonly known as Cold Fusion.

13. John Searl and the SEG

This chapter is based on an article I wrote and posted in April/May 2010.

In April or May of 2010, a film/DVD was released about the life and work of a British inventor named John Searl[263]. The opening caption in the film says: "The claims made by this inventor are open to debate." and "Here is the evidence as it was presented to us".

Alleged Searl Flying Disk – 1970s

Searl in approximately 2007

SEG Reproduction by Fenando Morris (left) (circa 2010)

Often referred to as "Professor John Searl," his biography does not list any period when he studied at a University and there aren't any details about how he might have been awarded an Honorary Degree or what subject his "professorship" was related to.[264]

Searl claimed to have invented a device that produced a large amount of excess energy and was even able to levitate! PLV discusses what became known as the "Searl Effect" in chapter 10 of his "Secrets of Antigravity" book:

> *It was during an indoor test that British inventor John Searl watched his permanent magnet generator levitate off its bench and ultimately bump into the ceiling. As he had envisioned in his dreams, his generator was not only able to propel itself with an over-unity power efficiency, but it was also able to defy gravity.*

The SEG (Searl Effect Generator) used an arrangement of concentric rings which held magnets of different kinds.

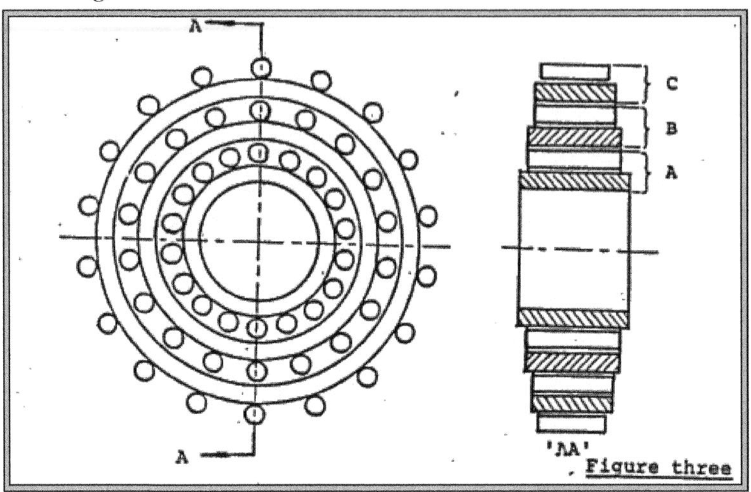

Schematic of SEG[265]

Searl states that the design of the generator was based on an idea he had regarding the "Law of Squares" – which he discovered as a child, perhaps as a result of playing the game of Hopscotch.

In his "Secrets of Antigravity" book, PLV describes how, in 1952, Searl and a friend built one of these 3-foot diameter generators and they tested it outdoors.

> *They set the rotor in motion using a small engine that applied torque through a clutch mechanism. Even though it was rotating at a relatively slow speed, the device produced an unexpectedly high potential at the rotor's periphery, on the order of 100 kilovolts or more. The large potential was indicated by the field's characteristic crackling sound and ozone smell and the effect it had on surrounding objects.' After the rotor had passed a particular threshold speed, its rate of rotation began to accelerate. Operating on its own power, it began to lift off and break the union to its starter engine. It is said to have risen to a height of some 50 feet, where it hovered for a while, still speeding up. It surrounded itself in a bluish pink halo similar to the glow discharge phenomenon seen when air is ionized in a moderate vacuum. Its pulsing field caused nearby radios to turn on. Finally, the whole generator is said to have accelerated upward at a fantastic rate and disappeared, presumably flying into space.*

It was also claimed that the device got extremely cold and that, if a load was attached to the generator, its output would increase even further. I am not aware of any photos or videos of a SEG powering other devices, however.

Though we have some black and white photos of Searl's alleged disk flights, these seem to be at least 40 years old and no one has reproduced a version of the generator which behaves like the one described above.

As shown in one of the photos above, by 2010, Fernando Morris had constructed a reproduction of the SEG and this unit was inspected by Richard D Hall[266]. He observed, from oscilloscope readings, that the magnetic field produced by the generator was unusual, but the device was not, at that point, an "over unity one". It didn't levitate or take off either. It appears that Morris is associated with a company called Searl Magnetics[267]. Their website stated[268] (in 2016):

> *Although we have made great strides in recent years, there is still a lot of work to do in the areas of materials, magnetization, and optimization before we shift gears and move into a full production phase. Nevertheless, we are closer than ever to fully realize Professor Searl's vision of a world filled with clean, portable, abundant energy for all.*

Here again, they refer to Searl as a Professor, when he has neither a "regular" nor an honorary degree. At least one other website appears[269] to be devoted to developing Searl's technology – and this is associated with Searl Magnetics.

Scepticism Effect Generator

When the "Searl Story" film came out, I was already sceptical of claims about the device. This was partly because I'd already investigated these claims and had not found much substance. My scepticism regarding Searl's research/developments was also increased following conversations I had with respected veteran UFO researcher Omar Fowler, of Derby, UK. Omar ran a group at which I was a regular speaker until his death in 2017. He also distributed a booklet called "OVNI" to readers all over the world.[270]

Omar Fowler in 2012 in Derby (Allenton) Victory Club[271]

Omar described to me how he attended a press conference held by Searl probably in the 1970s. He said the conference was packed and there was some anticipation that Searl was going to show something quite revolutionary. Omar said that Searl told the assembled crowd that he had built a working flying saucer, but then proceeded to lift the cover off what was... a cardboard model. At that point, Omar stated, approximately half of the assembled crowd got up and left.

My scepticism increased slightly when I noticed people like Joanne Summerscales wanting to promote the new film even though I explained to her (and others) why it was almost certainly not worthy of promotion. The same feeling returned when I attended the BEM conference in Hilversum two years later in November 2012, where the Searl Research was heavily promoted[272] in more than one talk (and there was no new evidence then, either).

Richard Vere Compton and John Searl

In the "Searl Story" video, there is a segment of an interview with engineer Richard Vere Compton, who worked with Searl in the 1970s. This interview seems to be heavily edited. In the video, Vere Compton reported that initially he helped Searl, by converting some electric drills to run on 12 volts rather than 240 volts, so that they could be used outside, to help construct a disk. Vere Compton stated:

> *Ken Pirelli turned up at my doorstep and told me about Searle. I went to see him and saw what he was doing and was rather amazed that he was building a disk in somebody's garden... this could be fun – let's see what happens. And of course, very soon, one began to feel that nothing was going to happen.*

Vere Compton understood what Searl was claiming:

> *Basically, the Searl effect was meant to turn the ether, which is all around us which contains energy, and tap into this energy which is euphemistically now be known as zero-point energy and various other people who followed Searl's footsteps have also pottered around and they've tapped into it by mistake...*

Also covered in the 2010 "John Searl Story" film is the fact that, years earlier, Searl was convicted of stealing electricity while he was living in (Cleobury) Mortimer, West Midlands, UK. This appears to be related to a story Vere Compton tells in the film, where Searl claimed to demonstrate running devices off the generator.

> *He said what he was going to do was to switch the Searl generator on by cutting off the electric power and it would then be running on the Searl effect generator. Well, as I said before ... the jury's still out as to what John actually did. Were there puffs of smoke and mirrors? Did he switch a straightforward inverter-to-inverter on? Was there any dip in the voltage? As far as we were concerned, one minute the lights were on and the next minute the lights are still on and there was no dipping, there was no fading there.*

Vere Compton also stated when asked the question about him witnessing the SEG directly:

> *No, I have not seen the Searl effect work.*

John Searle refuses to give up his "Secret"

As part of the run of promotion of the Searl effect in 2010, referenced above, this film/DVD was discussed on the US talk radio show "Coast to Coast AM".[273] I listened to this broadcast, with Bradley Lockerman (who, I think, financed and produced the film) and John Searl himself. One of the key points, for me, was raised by a caller – when Searl was asked about others who were working to reproduce his technology – and why he had not encouraged this.

Searl basically said that he cannot reveal how his technology works (fully) because "terrorists might get hold of it". John Searl was not aware that terrorists have already deployed a weaponised version of fee energy technology. As I have mentioned already, several times, in this volume, and discussed at length elsewhere, in 2008, it was disclosed by Dr Judy Wood that the evidence found in the destruction of the WTC tied up closely with a set of evidence produced by the experiments of another free energy researcher – John Hutchison[274] (see chapter 12). Clearly, John Searl and other free energy activists and researchers – and, in fact, everyone - need to be made aware of what happened on 9/11.... i.e. terrorists *already have* the technology – and they have *already used it*!

Conclusions

In "The John Searl Story" video, which covers about 50 years of experiments and research, only one working model of any device is shown. This particular model also appears to stop - it doesn't keep "going and going" as he previously claimed. This model was developed by a wealthy philanthropist who is clearly a good mechanic - and yet it still needs energy input to get it going - and then it stops. To me, the whole "Searl flying disk" issue is rather incredible - we have about 7 or 8 stills, but no films of disks flying - despite all the press attention he speaks of. With all that effort expended to make and build a disk, no one thought to take an 8mm home camera to film it – flying off into space? Why don't we have more stills of it? (Apparently several disks were built and flown, according to Searl's own words.)

It is certainly possible that one or more of the websites promoting the SEG has (or have) been created to lead people away from more important lines of research, to a dead-end. I certainly know the underlying ideas that Searl is trying to exploit are valid - but what is shown in the "Searl Story" film (and elsewhere) leads me to believe we will not be getting anything of value out of John Searl. Even taking him at his own words, he cannot pass anything onto anyone else "because it will be stolen by greedy people" – which brings us to the "catch 22" situation – "the technology cannot be shared because it's too world-changing - and it will be stolen". Which, again, brings me back to the destruction of the World Trade Centre…

The SEG didn't really warrant a chapter in PLV's "Secrets of Antigravity" book. It appears that though there is some anomalous effect being seen in Fernando Morris' version, it cannot (currently) be exploited to produce "free energy" nor can it be used in an antigravity type propulsion system. If Searl ever did create a levitating or flying disc, there is no evidence left and it cannot be reproduced, over 40 years later.

Of course, Searl's goals seem laudable, but as with all free energy device research, the goals are unreachable, without a full and true knowledge of the current "state of affairs."

Part 2

Big Fireworks

14. Apollo and "Truth's Protective Layers"

Introduction

Many videos have been made about the Apollo Hoax and quite a few books have been written. Whatever people say about the evidence, there are a number of important facts that are beyond dispute:

- The last time a human being was alleged to have travelled beyond low earth orbit was in December 1972. As of writing this book, that was over 45 years ago.
- The termination of the US Space Shuttle programme in July 2011 brought the US manned space programme to an end, since when, all publicly disclosed manned space missions have been conducted outside the US.

These facts on their own are extremely peculiar, considering the claims made by NASA and the advances in technology since manned missions first began in 1961.

In this chapter and the following four, we will review evidence which might help to explain why, in the case of manned lunar and planetary landing missions, technology appears to have "gone backwards" – and we are now having to look for what Dave McGowan called "The Lost Technology of the 1960s."

Before going into more detail, I might suggest that for a fuller discussion of evidence and history, consider consulting the sources mentioned below. Remember, this book isn't about the Apollo Hoax specifically, though it forms a significant part of the evidence which strongly suggests or even proves that a secret space programme exists.

David Percy's website www.aulis.com and his tremendous two-part video/film "What Happened on the Moon."[275] . Similarly, his book, co-authored with Mary Bennett, called "Dark Moon[276]," is worthy of detailed study.

Bart Sibrel's hugely important films "A Funny Thing Happened on the Way to the Moon"[277] and "Astronauts Gone Wild"[278] reveal voluminous important evidence. (It was Sibrel's films which were the first to illustrate to me that the Apollo missions were, indeed, faked.) A few other sources worthy of study (in no particular order) are:

- Ralph Rene's book "NASA Mooned America."[279]
- Bill Kaysing's book "We Never Went to the Moon."[280]
- Jarrah White's "Moonfaker" videos[281].
- "Moon Hoax Now" video by Jet Wintzer[282].
- Jim Collier's video "Was it Only a Paper Moon"[283].

In the UK, Marcus Allen has, for many years, given talks about the Apollo Hoax and discussed problems with the photos and other aspects – you will find quite a few interviews and presentations by him online too.

In compiling these five chapters, I have noticed that many of the more obvious aspects of fakery or studio filming seem to have occurred on the last mission – Apollo 17. Is it the case that the hoaxers had become over-confident by this time, and made too many obvious mistakes?

Apollo – A History

In this section, we will briefly examine some of the strange aspects of Apollo History. However, I will leave you to do your own more detailed study of Apollo symbolism and nomenclature, should this be of interest to you. In his Book "Dark Mission,"[284] Richard Hoagland spends some time noting how many of the dates, names and the symbolism used in the Apollo programme have occult significance (though Hoagland thinks Apollo craft did successfully land there and he claims he proves this in his book).

Hoagland claims that the astronauts are secretive about what they saw on the moon (i.e. artefacts). He goes into some detail with regard to which constellations were rising or setting at significant points during the missions.

Others have noted the recurrence of Masonic themes and numerology – for example Apollo 11 was the first "successful" mission. Whilst earlier missions had modules named after cartoon characters, later names were based on more symbolic and mythological names, such as Columbia, Aquarius, Eagle and so on. Videos by Texe Marrs and several others make some interesting observations, which readers can form their own opinions on.

From V2 to Saturn V

Much has been written elsewhere about Project Paperclip, which involved bringing Nazi German scientists to the USA at the end of the second world war. This project was revealed in the 1980s and was given the "paperclip" name because files kept on various WWII scientists by the US Intelligence Services were apparently "marked" with a paperclip to indicate they were of special interest.

The German scientists were then brought to the USA to work in a number of programmes – including the rocketry programme and various CIA programmes including what would become MKULTRA. The latter involved research into the effects of different methods of torture and pain induced trauma on the human mind. Various books have been written about this subject, including those by Walter Bowart, Jim Keith and Neil Sanders.

Here, we will concern ourselves (briefly) with Werner Von Braun, who was the main figure behind the development of rockets which culminated in the enormous Saturn V booster. Von Braun was also responsible for Germany's V2 programme, which resulted in the launching of rockets from Germany such that they caused devastation in London and killed many people.

In 1999, Aron Ranen was paid by the State of Ohio to investigate the Moon Landing Claims. He produced a 1-hour documentary he called "Did We Go?"[285] In this documentary, he discussed some of the history of Von Braun and he showed documents proving that Von Braun authorised and oversaw the use of slave labour at the Mittelwerk site, where the V2 rockets were constructed using this labour. Rana also spoke to Eli Rosenbaum, then Director of the US Justice Department's office of Special Investigations, an office responsible for locating and, where possible, prosecuting former Nazis for war crimes. Rosenbaum said:

> *The key individuals responsible for the great accomplishment of Apollo 11 - the landing of human beings on the moon - were deeply involved in Nazi crimes during WWII. This isn't just guilt by association. It's not just that some of them were in the SS or the Nazi Party. They were personally complicit in Nazi crimes.*

In the documentary, Aron Ranen stated:

> *Arthur Rudolph was Werner von Braun's head of rocket manufacturing under both Adolf Hitler and John F Kennedy. The Justice Department brought war crime charges against him.*

Eli Rosenbaum then states:

> *Rudolph was very much a de facto deportation he was removed unceremoniously from the United States and the Justice Department exposed him for what he was - which was a Nazi criminal.*
>
> *Unfortunately, von Braun died before the US government, belatedly in 1979, created this office and seriously undertook to investigate Nazis living in the United States. Had he still been alive if there is no doubt that he would have become a major subject of federal investigation.*

Dr. Werner von Braun (left), the NASA Director of the Marshall Space Flight Centre, and President John F. Kennedy at Cape Canaveral, Florida on November 16, 1963. (Photo: NASA)

Some might wish to excuse people like Von Braun who, it can be argued, would have been imprisoned or killed had he not co-operated with the Nazi regime and worked to further their evil goals. Whatever one's view of this situation, if knowledge about Von Braun's history had been made public at the time, perhaps it would not have been possible for the rocketry programme to progress as quickly as it did whilst maintaining public approval.

It can also be surmised that, if Von Braun went along with the Nazi's murderous plans in WWII, he would have no problem going along with a programme which was, in fact, an enormous deception (but didn't involve killing large numbers of people).

Moving away from ethical issues to technical ones, Von Braun wrote a book, published in 1953 called "Conquest of the Moon."[286] At the time, the mission plan for getting to the moon was different to the one which NASA finally settled upon. We will discuss this, after considering what Von Braun wrote in his book:

> *It is commonly believed that man will fly directly from the earth to the moon, but to do this, we would require a vehicle of such gigantic proportions that it would prove an economic impossibility. It would have to develop sufficient speed to penetrate the atmosphere and overcome the earth's gravity and, having traveled all the way to the moon, it must still have enough fuel to land safely and make the return trip to earth. Furthermore, in order to give the expedition a margin of safety, we would not use one ship alone, but a minimum of three ... each rocket ship would be taller than New York's Empire State Building [almost ¼ mile high] and weigh about ten times the tonnage of the Queen Mary, or some 800,000 tons." (footage from "Apollo Zero")*

It was pointed out to me that Von Braun's initial idea was that there would only be a single craft which landed on the moon and returned, rather than having two craft travel there, separate, re-connect and then return. (i.e. the idea of using the Command Module and Lunar Excursion Module had not been thought of in 1953). Whilst it is true that landing a smaller craft on the moon would use less fuel, the figures Von Braun quotes for the size of the craft needed are orders of magnitude larger than that of the final Saturn V rocket. Hence, the total amount of fuel needed for a round trip to the Moon certainly seems to be a fundamental technical hurdle that was difficult to overcome.

Douglas Gibson also advised me of Von Braun's appearance in an educational film made by Walt Disney and aired in December 1955.[287] (This was part of a longer educational film when it was first shown.) We will discuss elements of this film.

Near the start of the segment, Von Braun, standing near a large model rocket, describes something similar to what he described in his book and states:

> *A voyage around the moon must be made in two phases. A rocket ship taking off from the Earth's surface will use almost all the fuel it can carry just to attain a speed great enough to balance the pull of gravity. Unpowered, it will then keep circling the earth in an orbit outside of the atmosphere. This is a first phase. However, if we can refuel the ship in this orbit with fuel brought up by cargo rocket ships, it can set out on the second phase - the trip around the Moon - and back. To facilitate this refuelling operation, we will establish an advanced space [station] in the orbit a thousand miles above the earth. This advanced space or space station will be headquarters for the final ascent to the moon...*

After a mission description, there follows, a dramatization of the "trip to the moon." About three quarters of the way through the dramatization segment, the proposed ship starts orbiting the moon. (It can be noted that the model of the moon that is used in this sequence is quite detailed. Considering the Disney film was made 14 years before the first alleged landing, the "model moon" looks quite realistic.) During the orbit sequence, the following dialogue is heard, and around the same time the image, below, is shown:

> *Officer: Captain! I'm getting a high Geiger count - **33 degrees**!*
>
> *Captain: My sonolation counter indicates a high degree of radioactivity on the same bearing. Contour map shows a very unusual formation at about 15 degrees southern latitude and Meridian 210*
>
> *Captain: Get some flares in that area quick!*

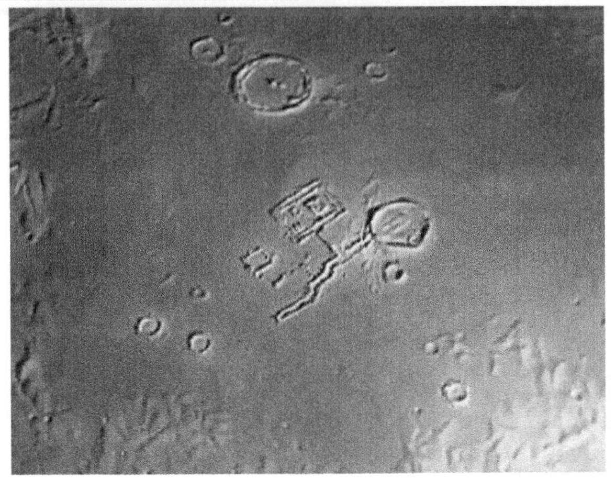

Still from Disney's "Man and the Moon" 1955 film

Of course, this proves nothing about the validity of the actual mission plan and doing a complete set of calculations to determine how much fuel would be needed for a return trip would be highly complex, so we must leave this as just another unanswered question. Later, we will show evidence that more obviously demonstrates the fakery involved in the Apollo missions, however.

Gus Grissom and Apollo 1

Left to Right - Ed White, Virgil (Gus) Grissom and Roger Chafee

Any self-respecting astronaut, whilst on the one hand being happy to be involved in "making history" would surely have concerns about the safety and the technical progression of the programme. Gus Grissom was the commander of Apollo 1 and was also slated to be the first man to walk on the moon.[288] However, he was also a man of integrity and was apparently not afraid of speaking his mind. It was no secret that he was not happy with the progress that was being made in the Apollo programme. According to a NASA website posting[289], which refers to Page 182 of the Betty Grissom and Henry Still book[290] "Starfall":

> *On January 22, 1967, Grissom made a brief stop at home before returning to the Cape. A citrus tree grew in their backyard with lemons on it as big as grapefruits. Gus yanked the largest lemon he could find off of the tree. Betty had no idea what he was up to and asked what he planned to do with the lemon. "I'm going to hang it on that spacecraft," Gus said grimly and kissed her goodbye. Betty knew that Gus would be unable to return home before the crew conducted the plugs out test on January 27, 1967. What she did not know was that January 22 would be "the last time he was here at the house."*

During a test which was being run on the same day, Grissom was heard to say, in a recorded conversation[291]:

> *"I can't hear a word you're saying. Jesus Christ... How are we going to go to the moon if we can't even communicate between two or three buildings?!?"*

Soon after this, a fire started in the Capsule, killing Grissom, Roger Chafee and Ed White – the Apollo 1 crew. The fire was almost certainly the result of a 100% oxygen atmosphere being in use inside the capsule. The pure Oxygen was at a much higher pressure than had been used in previous similar situations with the Gemini tests.

In Nov 2017, a Newsmax report by Christopher Ruddy stated[292]:

> *Virgil I. "Gus" Grissom, the astronaut slated to be the first man to walk on the moon, was murdered, his son has charged in the Feb. 16 edition of Star magazine.*

> *In another stunning development, a lead NASA investigator has charged that the agency engaged in a cover-up of the true cause of the catastrophe that killed Grissom and two other astronauts.*
>
> *The tabloid exclusive by Steve Herz reports that Scott Grissom, 48, has gone public with the family's long-held belief that their father was purposefully killed during Apollo I.*

(This story, perhaps unsurprisingly, has been deleted from the Newsmax website, so only a few re-postings remain.) The article repeats statements made years earlier by Grissom's son, Scott:

> *"My father's death was no accident. He was murdered," Grissom, a commercial pilot, told Star. Grissom said he recently was granted access to the charred capsule and discovered a "fabricated" metal plate located behind a control panel switch. The switch controlled the capsule's electrical power source from an outside source to the ship's batteries. Grissom argues that the placement of the metal plate was an act of sabotage. When one of the astronauts toggled the switch to transfer power to the ship's batteries, a spark was created that ignited a fireball.*

(Though the story above was posted in November 2017, I first heard Scott Grissom's comments in the Apollo Zero documentary[293] which came out in 2009.)

Thomas Ronald Baron and Apollo

Thomas Ronald Baron in the mid 1960s

In November 1966, some months before the Apollo 1 fire which killed the three astronauts, a quality control inspector called Thomas Ronald Baron had produced a report detailing safety concerns[294]. Baron, who worked for North American Aviation (NAA), the company responsible for building the Command Module, wrote the paragraphs below, and these are posted on one of NASA's own websites:

> *North American Aviation, has not, in many ways, met their contractual obligations to the United States Government or the taxpayer. I do not have all the information I need to prove all that is in this report. I just hope someone with the proper authority will use this information as a basis to conduct a proper investigation. Someone had to make known to the public and the government what infractions are taking place. I am attempting to do that, someone else will have to try to correct the infractions…*
>
> *When I was hired by NAA, I was assigned to the Quality Control Department. I was told of the vast importance of my task, and of the great responsibility associated with it. I was told how the slightest infraction could be detrimental to the objectives of the program. I, along with others, was told how important our job was when it came to manned launches. We were told to report every infraction, no matter how minor we felt it was. Unfortunately, this is not practiced by the Company.*

According to the same NASA page referenced above[294], when the tragedy occurred, Baron was apparently in the process of expanding a 55-page paper he had written into a 500-page report. He had finished this report when he testified before congress during the Investigation into the Apollo 204 Accident Hearings, held before the House Subcommittee on NASA Oversight held at Cape Kennedy, Florida, on 21 April 1967. There is a transcript of Baron's meeting with Congress on the Clavius website[295]. Baron stated he did not think the Apollo programme would succeed in getting a man to the moon in the projected timescale. There were just too many problems and safety concerns. From reading some of this transcript, it appears some members of the committee wanted to try and characterise Baron as having "one too many personal or health problems."

The NASA page about the Baron report also notes what happened to Baron only three months after the Apollo 1 fire:

> *Ironically, Baron and all his family died in a car-train crash only a week after this exposure to congressional questioning.*

Is it just a coincidence, then, that two people who had spoken to the press and been sceptical of the progress of the Apollo programme were both dead within about three months of one another?

"Baby We're Just Landin' in the Dark..."

Some years ago, I heard John Lear on a radio interview make a strange observation about Apollo 13. He pointed out that if we assume this mission really was going to the moon, then according to the information published by NASA, when Apollo 13 landed on the moon, it would have landed in total darkness. I was able to test this claim and it is true. A press release about the proposed Apollo 13 landing[296] (before the "accident" or "problem" report to Houston Mission Control) contained the following statement:

> *Lunar surface touchdown is scheduled to take place at 9:55 p.m. EST April 15... The Apollo 13 landing site is in the hilly uplands to the north of the crater Fra Mauro. Lunar coordinates for the landing site are 3.6 degrees south latitude by 17.5 degrees west longitude...*

This then gave us the time and proposed landing site for Apollo 13. Here is information from Britannica.com about the location of Fra Mauro,[297] which gives a slightly different location than the press release, but as we will see, this doesn't really matter:

> *Fra Mauro, crater on the Moon that appears to be heavily eroded; it was named for a 15th-century Italian monk and mapmaker. About 80 km (50 miles) in diameter, Fra Mauro lies at about 6° S, 17° W, in the Nubium Basin (Mare Nubium) impact structure.*

We can now take the stated time and location and generate a simulated view of the Moon in Celestia, a freeware 3D-Desktop Planetarium package.[298] I have included 2 screenshots below and I encourage people to check this for themselves.

I set the view to show the moon at 14:55 UT (9:55pm adjusted to EST) on 02 April 1970. Even if we take the time as being ±5 hours to account for possible errors in conversion to different time zones, this doesn't make much difference to the sun's position during the lunar day – which is about 28 terrestrial days in length.

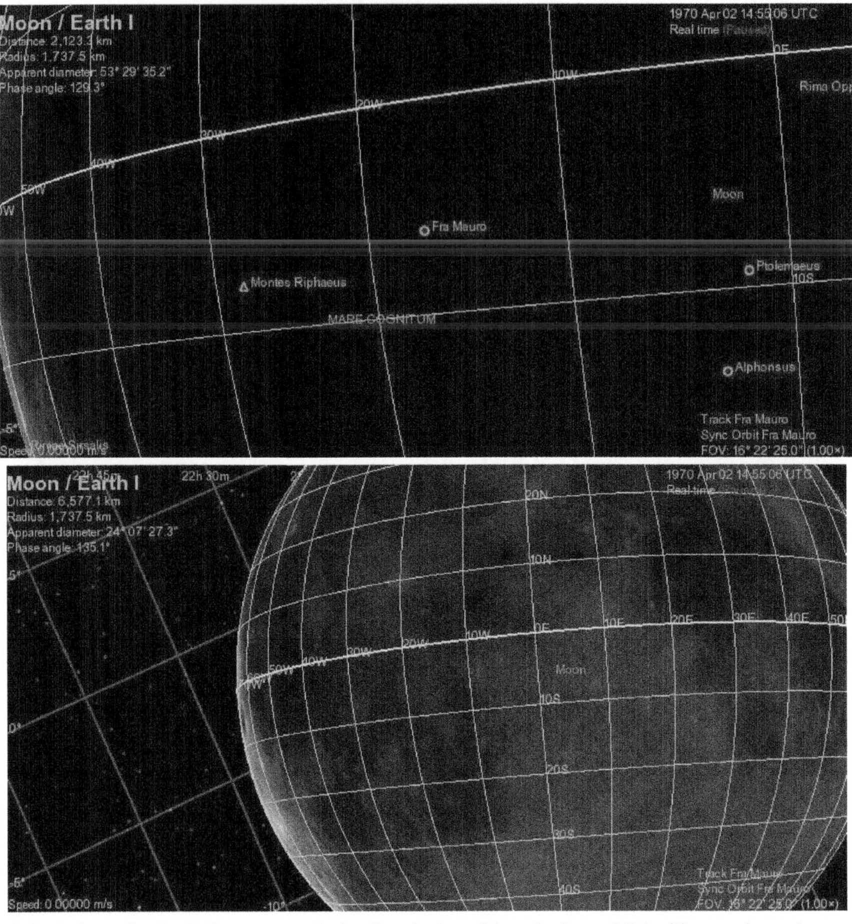

In both these simulated views, we see the landing site is well into the Lunar Night. The terminator is way over on the left hand side. Years ago, I simulated the same scene in the "Redshift" desktop planetarium package, with the same results.

Here we can clearly see that if Apollo 13 had landed at the time and place given in the press release, they would need to have taken a substantial lighting rig and attached it to the Lunar Excursion Module (LEM), to allow them to avoid obstacles when attempting to land. They would have also needed it to illuminate their activities while on the surface. Does this sound a likely or even possible part of the mission plan?

Is it the case, then, that someone "messed up" in stating what the landing site was and they had to concoct a Space Drama story to cover this up? A problem was turned into an opportunity! In the 1970s, only a tiny number of people would have been able to determine where the landing site was, and how it was illuminated. Even today, with free software like that identified above, few would bother to check these details. (This is yet another example of the cover up being "as good as it needed to be.")

Ah Yes, Those Moon Rocks…

It is my understanding that many of the "moon rocks" were ground into a fine powder and distributed around the world. Apart from this, a few are on display in museums with captions saying, "here is a rock from the moon". How many people would be able to show (by laboratory tests) that the caption was accurate? In at least one case, we know, now, that the rock presented was fake…

In August 2009, the BBC and other outlets posted a story entitled "Fake Dutch 'moon rock' revealed"[299] Here is the story in its entirety:

The lump of 'moon rock' had been on display for decades.

A treasured piece at the Dutch national museum - a supposed moon rock from the first manned lunar landing - is nothing more than petrified wood, curators say. It was given to former Prime Minister Willem Drees during a goodwill tour by the three Apollo-11 astronauts shortly after their moon mission in 1969. When Mr Drees died, the rock went on display at the Amsterdam museum. At one point it was insured for around $500,000 (£308,000), but tests have proved it was not the genuine article. The Rijksmuseum, which is perhaps better known for paintings by artists such as Rembrandt, says it will keep the piece as a curiosity. "It's a good story, with some questions that are still unanswered," Xandra van Gelder, who oversaw the investigation that proved the piece was a fake, was quoted as saying by the Associated Press news agency. "We can laugh about it." The "rock" had originally been vetted through a phone call to NASA, she added. The US agency gave moon rocks to more than 100 countries following lunar missions in the 1970s. US officials said they had no explanation for the Dutch discovery.

My friend Lloyd Pye, who was somewhat uncomfortable with "letting go" of the official story of Apollo, pointed out that this story on its own does not prove that the Dutch museum was always in possession of a fake Moon Rock. It's possible they were originally given a piece of one of the meteorites or other *real* moon rocks that had somehow been obtained, by other methods (not the Apollo landings though). Then, this original rock could have been stolen by someone and replaced with a look-a-like or even a substitute. Consider that in the photo above, the rock is a brownish colour – not a colour we normally associate with the grey/black appearance of the lunar surface.

In 1999, Aran Ranen visited California Institute of Technology (Caltech) and interviewed Professor Gerald Wasserburg[300] regarding his study of rocks allegedly brought back from the Moon. Wasserburg said the rocks he studied, could not have come from earth. Wasserburg stated:

> *I don't care if it's a Hollywood set. The rocks are not a Hollywood set. They're not from the Earth. They're not from any place near the earth or made on the Earth. They're not synthetically made. They cannot be generated within the earth and they came back in a box marked "From the Moon."*

In the "Did We Go" documentary, Ranen also asks Wasserburg about meteorites originating from the Moon being found in Antarctica. Wasserburg responds:

> *Well I got a phone call once [from] a fellow with the Smithsonian. He said, "Jerry we have now a meteorite which I just brought back from Antarctica, which is from the Moon." I said, "Bryan that's nonsense." Of course, I said something stronger than that... A whole assembly of scientists attacked this object and at the end of a year and a half, the conclusion was it was "Uncle Sam." It was a Moon rock. There was no question that it was a moon rock. It was unambiguously a Moon rock. Just as later, one could identify pieces of Mars...*

Exploring Antarctica (1967) - Intrigued by exploration in space and on Earth, Dr. Von Braun participated in an expedition to Antarctica. This photo was made on or about January 7, 1967.[301]

Retro Reflections

"But, wait! Wait! Those reflectors left on the Moon! That proves it!" We can fire a laser and get a reflection back!

The problem with this claim is that (a) the Russians sent the Lunokhod rover up to the Moon and it had a retro-reflector, so NASA's reflectors *could* have been remotely landed. It must also be borne in mind that when the laser being fired at the moon reaches the surface, the laser diverges so that the area the beam covers on the surface of the moon is *several km wide.*[302]

Even more important in settling this matter is an article in National Geographic Magazine from December 1966.[303] "The Laser's Bright Magic by Thomas Maloy, Howard Sochurek," Page 874-6. It states:

> *Four years ago (1962), a ruby laser considerably smaller than those now available shot a series of pulses at the moon, 240,000 miles away. The beams illuminated a spot less than two miles in diameter and were reflected back to earth with enough strength to be measured by ultrasensitive electronic equipment. The beam of a high-quality searchlight, if it reached that far, would spread out to several times the moon's 2,160-mile diameter."*

So, it seems we cannot be 100% certain that a laser beam fired at the moon is being reflected by anything that was allegedly placed there by astronauts.

15. Astronauts, Astronots or Actornauts

A study of what the Apollo astronauts have actually said is quite revealing. Indeed, a whole book could be written on this area alone. In this chapter, we will cover a few of the very peculiar things they have said. As I write this in 2018, I can say that soon, any secrets they are keeping will die with those who still remain. Those who want to study this area more can read the various books written by the astronauts themselves, listen to interviews they have given, or watch and study more "challenging" material such as Bart Sibrel's "Astronauts Gone Wild" and Bill Kaysing's "We Never Went to the Moon."

One especially noteworthy source of Apollo astronaut testimony is the Post Apollo 11 Mission Press Conference video.[304] We will discuss part of this, below.

Left to Right - Edwin R (Buzz) Aldrin, Neil Armstrong and Michael Collins at the 12 August 1969 Press Conference after the Apollo 11 Mission.

Neil Armstrong – First Man on the Set?

On July 20, 1994, the 25 Anniversary of the supposed Moon Landings, Armstrong addressed the crowd and said[305].

> *Wilbur Wright once noted that the only bird that could talk was the parrot, and he didn't fly very well. So I'll be brief.*

Knowing what we know about Gus Grissom and the Apollo 1 crew, and what happened to them, could we ever interpret this strange remark as a reference to the horrible events which happened in 1967? I leave the reader to ponder this idea and decide if this is likely. Shortly after this cryptic introduction, Armstrong stated[306]:

> *"Today we have with us a group of students, among America's best. To you we say we have only completed a beginning. We leave you much that is undone. There are great ideas undiscovered, breakthroughs available to those who can remove one of the truth's protective layers. There are places to go beyond belief..."*

Also in 1994, Armstrong's first marriage – of 38 years – ended in divorce.[307] Was the emotional strain of this situation a factor that influenced him to make (what I consider to be) this disclosure?

In 2005, Armstrong was interviewed by Ed Bradley of CBS. The interview contains some interesting remarks.[308]

> *Ed Bradley: You said once to a reporter "How long must it take before I cease to be known as a space man." Why did you make that comment?*
>
> *Armstrong: I guess we'd all like to be recognized not **for one err.. piece of fireworks** but for the ledger of our daily work.*

Bradley then describes Armstrong as "easy to talk to, but hard to know." Bradley asks about his experience of walking on the moon. Armstrong's response seems peculiar.

> *"It's a brilliant surface in that sunlight. The horizon seems quite close to you because the curvature is so much more pronounced than here on earth. It's an interesting place to be. I recommend it."*

Armstrong rarely gave interviews – only about five or six are available online, which were done decades apart. In 2012, Armstrong was interviewed by Alex Malley of the CPA (Certified Practicing Accountants) of Australia.[309] Malley asked Armstrong what he thought about the idea of people not believing that Apollo Landed on the Moon. Armstrong replied:[310]

> *People love conspiracy theories. I mean, they are very attractive. But it was never a concern to me because I know one day, somebody is going to go fly back up there and pick up that camera I left.*

Well, Mr Armstrong, Commander Armstrong, or whoever – I don't "love conspiracy theories." What I *like* to have is evidence – which is kind of the "opposite" of theory. Also, "conspiracy" is not what I focus on either – I want to know whether people are telling me the truth or not.

Following this interview, Armstrong showed a video where he compared a "Google Simulation" of the final 3-minutes of the Apollo 11 landing, run side by side with the "actual film" of his landing. Does this prove he "landed the Eagle" in July 1969?

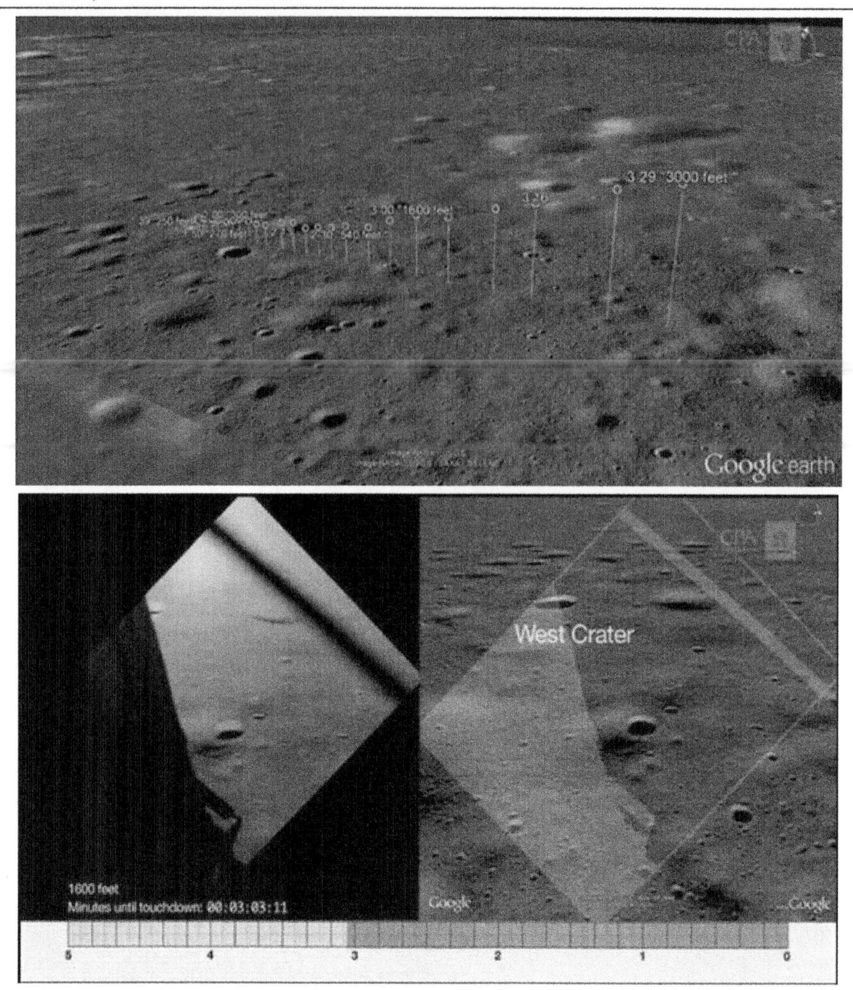

Stills from a video shown by Neil Armstrong to the CPA Audience in 2012. The left image is the alleged film from Apollo 11. The right image is from a google simulation.

I would say "no it does not." What it does show is that the lighting conditions used in the Google image simulation are remarkably close to those shown in the Armstrong version – which means that whatever (unnamed) lunar probe took the images used in Google's simulation had identical or near identical lighting to the alleged 1969 mission and the 3D projection/image wrapping used in the simulation is almost perfect! Why did Armstrong even want to show this video? Did he need to prove to the audience he walked on the Moon?

Statement Analysis

In 2017, Richard D Hall decided to ask[311] professional statement analyst Peter Hyatt[312] to analyse a 1970 interview between Patrick Moore and Neil Armstrong[313]. Peter Hyatt's method involves breaking down the responses of interviewees and considering their use of phrases, pronouns and other linguistic elements. Hyatt's technique is used by investigators to hone their questioning in order to detect when someone is attempting to deceive the interviewer.

A portion of the interview and an analysis is reproduced below:

> *Moore: When you were actually walking about did you have any difficulty in distance judging, because I, I think I heard you say once that ... far things looked quite near?*
>
> *Armstrong: Yes, we had er, some difficulties in perception of, distance... for example our television camera, ... we judged to be from the cockpit of the lunar module only about er fifty to er sixty feet away, yet we knew that we had pulled it out to the full extension of a one hundred foot cable. Er, similarly we had difficulty er guessing how far the hills out on the horizon might be. Err, [a] peculiar phenomenon is the closeness of the horizon, due to the greater curvature of the Moon than we have here on Earth - of course four times greater, and the fact that errr, it is an irregular surface with err crater rims overlying other crater rims. Er you, you can't see the real horizon you're seeing hills that are somewhat closer to @. Er, there was, a large crater er which we overflew during our final approach which was .. had hills of the order of a hundred feet in height, and er we were only eleven, twelve hundred feet west of that hill and we couldn't see a hundred foot high hill...*

Hyatt commented on this segment, thus:

> *The topic is sight. The one being interviewed is singular. We look for him to answer with "I saw" in some form. Sensory is very individual. When it is discussed one may then say the thoughts or perceptions of another, but here, he is interviewed alone and is being asked for first person, eye witness account. He does not give a first person singular account.*
>
> *Perception: He does not speak for himself including in the realm of "perception." This is to address the brain's interpretation of what was seen or experienced. He consistently, in the interview, tells us what "we" saw, "we" thought, "we" perceived, as well as what "you" saw, and so on. Although many of these topics were likely discussed, he should still be speaking for himself. This use of "we" is most unexpected. We must now consider that if he is not deceptive, why does he have a reason to "join a crowd" - even in personal experience and perception?*

In Hall's video, he shows Hyatt's conclusion regarding the detailed analysis of the 8-minute interview between Moore and Armstrong, Hyatt concludes:

> *Neil Armstrong does not linguistically connect himself to the lunar landing. This is evident in his consistent 'distancing language' including intuitive pronouns and passivity. What may have caused this?*
>
> 1. *Security Mandate*
> 2. *Baseline [i.e. Armstrong talks like this all the time.]*
> 3. *Deception - The lack of commitment mimics deceptive language. The lack of personal commitment, in context, is stark.*
> 4. *Unknown*

Edwin "Buzz" Aldrin

Aldrin has often appeared in public and, as illustrated in the "Apollo Zero" documentary, seems to have made a very good living in relation to his participation in the real or fake space programme. In my book "Secrets in the Solar System: Gatekeepers on Earth," in chapter 15, I discussed his comments about Mars' moon Phobos. Since roughly year 2000, he seems to have appeared in quite a number of news programme items, interviews and commentaries. He was also present when Barack Obama (Barry Sotero) gave a presidential speech at NASA in 2010, which is covered in a later chapter.

There are some interesting notes about and quotes from Aldrin in Bill Kaysing's book, "We Never Went to the Moon." Kaysing writes:

> *In June of 1971, Aldrin rejoined the Air Force and was assigned to the Edwards Air Force Base in California, a windy, high desert locale. He had been on medication for his nervous disorder: one Ritalin pill per day.*

Kaysing then quotes from pages 278 (where Aldrin mentions Ritalin) to 280 of Aldrin's first book (co-authored with Wayne Warga) "Return to Earth."[314]

> *I looked great. There was only one problem. I believed my confidence to be rooted in reality. Early in June an event was scheduled that we regarded as a new beginning for us and as such, we looked forward to it a great deal. I wanted to be at my very best... The occasion was a big meeting of the Lancaster Chamber of Commerce. They had invited members of the Society of Test Pilots; I was one of the guest speakers. What intrigued me most was that 1 would not be giving a speech. Instead, Roy Neal, the NBC news- caster, would interview me in a most informal way. The afternoon of the banquet, I stopped by Roy's motel and asked if he wanted to run over any questions with me. He assured me they would be very easy to answer and that no preparation at all was necessary.*
>
> *As banquets go, this was a large one. The base commander, General and Mrs. White attended. I began to be more and more apprehensive as the time for my interview grew near. The first question that Roy Neal asked me was, 'Now that almost two years have gone by,* ***why not tell us how it really felt to be on the moon****?'*
>
> ***If any one question was anathema to me, that was it. Roy, I suppose, felt he had no choice. Yet it has always been almost impossible for me to answer that question with any sort of decent response. My throat went dry and I became dizzy. Carefully I picked my way through a reply, thinking that all the test pilots in the audience would burst out in laughter.***

Regarding the question of "what it felt like to walk on the moon," Aldrin, like Dr Edgar Mitchell (discussed in chapter 27) experiences difficulties in answering. A possible reason for this reaction is discussed later.

Kaysing also observes the following about the end of Aldrin's 1975 book (page 338 of that hardback edition):

> *Aldrin: "When I began this book I had two intentions. I wanted it to be as honest as possible."*

Kaysing: (Why not just plain "honest"?)

Once a Freemason… Always a Freemason

Some years after I had realised the truth about the Apollo Hoax, and I had already heard about the connections of NASA to Freemasonry, I came across a posting by Terry Melanson about Aldrin's "Masonic Moon Machinations."[315] The posting proves that Aldrin lied about issues related to Masonic practices in the Apollo 11 mission. This is shown by comparing statements Aldrin made to fellow liar Alex Jones on one of Jones' "Infowars" broadcasts on 17 August 2009[316] (which was around the time Aldrin released another book called "Magnificent Desolation"):

> *AJ: We know there's Masonic influence in the founding of the country ...what is the Masonic influence on NASA?*
>
> *Aldrin: As far as I can tell, zero. There were some Masonic brothers of mine in Texas that wanted me to take some kind of a Masonic emblem to the moon, and some gesture of – I don't know what it would be a gesture of – but I told them that it was not within my ...my authority to do such a thing.*

Quoting now from Terry Melanson's 2013 article, we read:

> *That's funny, because there was an article in the New Age Magazine, December 1969 (the official organ of the Scottish Rite Southern Jurisdiction), replete with pictures of Aldrin presenting to the Scottish Rite headquarters in Washington, just such an "emblem" which he had carried with him to the moon and back.*

Melanson then includes a copy of the letter, as shown below, the text of which is included here (emphasis added):

> *Dear Grand Commander*
>
> *It was a great moment in my life to be so cordially welcomed to the House of the Temple on September 16, 1969, by you and Grand Secretary General Kleinknecht, 33, and also the members of your staffs. My greatest pleasure, however, was to be able to present to you on this occasion the Scottish Rite Flag which* ***I carried on the Apollo 11 Flight to the Moon—emblazoned in color with the Scottish Rite Double-headed Eagle, the Blue Lodge Emblem and the Sovereign Grand Commander's Insignia.***
>
> *I take this opportunity to again thank you for the autographed copy of your recent book, entitled "Action by the Scottish Rite, Southern Jurisdiction, U.S.A.," which is filled with a wealth of information about your Americanism Program sponsored by the Supreme Council, participating activities and related activities of the Rite. Cordially and fraternally.*
>
> *Edwin E. Aldrin, Jr. NASA Astronaut*

So, it appears that Aldrin lied about a trip to the moon and then, later, lied about Masonic connections/influence.

NATIONAL AERONAUTICS AND SPACE ADMINISTRATION
MANNED SPACECRAFT CENTER
HOUSTON, TEXAS 77058

IN REPLY REFER TO:

September 19, 1969

Illustrious Luther A. Smith, 33°
Sovereign Grand Commander
Supreme Council, 33°
Southern Jurisdiction, U.S.A.
1733 16th Street, N.W.
Washington, D.C. 20009

Dear Grand Commander:

It was a great moment in my life to be so cordially welcomed to the House of the Temple on September 16, 1969, by you and Grand Secretary General Kleinknecht, 33°, and also the members of your staffs. My greatest pleasure, however, was to be able to present to you on this occasion the Scottish Rite Flag which I carried on the Apollo 11 Flight to the Moon--emblazoned in color with the Scottish Rite Double-headed Eagle, the Blue Lodge Emblem and the Sovereign Grand Commander's Insignia.

I take this opportunity to again thank you for the autographed copy of your recent book, entitled "Action by the Scottish Rite, Southern Jurisdiction, U.S.A.," which is filled with a wealth of information about your Americanism Program sponsored by the Supreme Council, participating activities and related activities of the Rite.

Cordially and fraternally,

Edwin E. Aldrin, Jr.
NASA Astronaut

Aldrin's 1969 Letter to Luther A Smith.

Aldrin and the Phobos "Monolith"

As I wrote in chapter 15 of Secrets in the Solar System[104], on 22 Jul 2009, C-Span, a US cable news channel, broadcast an interview/report with the tag line "Buzz Aldrin Reveals Existence of Monolith on Mars Moon."[317] In a peculiar interview, where a renewed programme of manned space exploration was being discussed, Aldrin then side-stepped unexpectedly, stating the following:

> *... if there's something very important to be developed from the moon... I'm not sure what it is right now and I think we should identify what it is for America to make such gross expenditures again for human habitation on the moon. We can help - we can join with, together... We can explore the moon and develop the moon. We should go boldly where man has not gone before... fly by the Comets, visit asteroids, visit the moon of Mars. There's a Monolith there - a very unusual structure, on this little potato shaped object that goes around Mars once in seven hours. When people find out about that they're gonna say, "Who put that there? Who put that there?" Well, the universe put it there. If you choose, "God put it there."*

No image of this monolith was actually shown in the C-Span report, though images of such a monolith had been discussed on some websites previously. It is not really clear to me what prompted this statement, other than possibly it was some type of media spin from NASA to obtain more funding for Mars missions.

Michael Collins

If you watch Bart Sibrel's "Astronauts Gone Wild," you will be able to see Collins' peculiar reaction to Sibrel's questions and requests. This makes for uncomfortable viewing, but we can also consider Collins' contradictory statements regarding his viewing of stars during the hours he allegedly spent orbiting the moon and on other space flights he made.

In the post Apollo 11 press conference Collins "chimed in" with a statement following a question to Neil Armstrong by UK Astronomer Patrick Moore. At approximately 1 hour and 4 minutes into the Conference, Moore was asking about the visibility of the Solar Corona and the visibility of stars.[318] As Bart Sibrel observes in the commentary of "A Funny Thing Happened on the Way to the Moon," it is the only question that Armstrong responds to with a "lack of memory."

> *PATRICK MOORE: When you looked up at the sky, could you actually see the stars in the solar corona in spite of the glare?*
>
> *ARMSTRONG: We were never able to see stars from the lunar surface or on the daylight side of the Moon by eye without looking through the optics. I don't recall during the period of time that we were photographing the solar corona what stars we could see.*
>
> *COLLINS: I don't remember seeing any.*

Interestingly, NASA's transcript of this event contains a correction regarding this very statement by Collins – which was attributed (in the transcript) to Aldrin.[319] Only when the video of the event was posted was a correction made, it appears.

Some have tried to say that Armstrong and Collins were specifically referring to not seeing stars *in the solar corona*, which is, after all, what Moore asked about.[320] However, Armstrong's answer is ambiguous and Collins' answer is too. Why didn't they describe their views of the stars more clearly?

In Collins' book "Carrying the Fire," (Farrar, Straus and Giroux, New York copyright by Michael Collins 1974) Collins says as he is getting ready to exit Gemini 10 for a spacewalk (page 221):

> *"My God, the stars are everywhere: above me on all sides, even below me somewhat, down there next to that obscure horizon. The stars are bright and they are steady. Of course, I know that a star's twinkle is created by the atmosphere, and I have seen twinkle-less stars before in a planetarium, but this is different, **this is no simulation**, this is the best view of the universe that a human ever had."*

Are we therefore to believe that, from the Command Module, he saw no stars? (Remember, Collins was not on the surface of the moon and would, if the astronauts' stories are true, have spent several hours in darkness when he was not travelling over the sunlit portion of the moon's disk.) Armstrong's answer also implies that they *could* see stars when they were on the unlit side of the moon, though this question is not discussed in the press conference. The reaction to the question about stars, overall, is not answered with enthusiasm, detail or comfort.

No Stars in the Photographs Either?

While we are discussing this area, I will note that arguments rage on the internet about the lack of stars in any of the photographs. Most of these arguments are started by people who don't understand how cameras expose images. That is to say, all the published Apollo photographs are taken with exposure for sunlit scenes and so stars, relatively speaking, are too dim to be seen with this exposure setting. However, there was nothing to stop them taking some of the best astro-photographs ever if they had simply set a camera on a tripod –but, as far as I know, they did not do this.

Dr Edgar Mitchell

Another frequent public commentator on the Apollo missions was the late Dr Edgar Mitchell, who was allegedly the "sixth man" to walk on the moon. When I finally realised the moon landing had been hoaxed, I was disappointed to find out that Mitchell was just another gatekeeper. In 2004, I wrongly assumed that Mitchell was disclosing real information about the ET/UFO cover up. At that time, he had spoken out and was reported in the press as saying "The Aliens Have Landed."[321] These issues are discussed further in chapter 27.

Don Petit - 2016

He made what I consider to be one of the most bizarre statements that is currently in the public record. In an article "NASA astronaut Don Pettit: Next logical step is to go back to the Moon - then Mars and beyond," dated 21 Sept 2016, on the IBTimes Website, he is quoted thus[322]:

> *"I'd go to the moon in a nanosecond. The problem is we don't have the technology to do that anymore. We used to, but we destroyed that technology and it's a painful process to build it back again."*

The quote is accurate – and can be heard - in a short video posted by IB Times[323] - for anyone that wants to listen.

Astronauts, Hypnosis and Richard C Hoagland

In Richard C Hoagland's "Dark Mission" book, he claims he was told by an unnamed MD who worked at NASA that astronauts were hypnotised. On page 176 he writes:

> *Early in his low-key lunar ruins investigation, a NASA insider (an MD directly involved in the medical aspects of the Program ...) confirmed to Hoagland and Johnston anecdotally that during their "debriefings" by NASA, all of the Apollo astronauts had been hypnotized - ostensibly to help them remember more clearly their time on the Moon. In actual fact, it seems more likely that these sessions were used to make them forget what they had seen, as evidenced by the astronauts own post-Apollo behavior.*

Hoagland then goes on to claim that Mitchell and other "Lunarnauts" must have seen the remains of enormous but shattered glass structures, which Hoagland claims as evidence of the work of a lost civilisation. (He doesn't describe how these structures may have been avoided during landing and provides no clear images or other evidence of these structures in any of his books or website postings.) It is therefore easy to dismiss Hoagland's claims as fantasy – including the story about the astronauts being "hypnotised". Also, of note here is what I also included in my "Secrets in the Solar System" book. On page 172 of Hoagland's Dark Mission book he writes (emphasis added):

> *As we discussed in the Introduction, this idea (most recently advanced by a few well-known **self-promoters** such as David Percy, Bill Kaysing and the late James Collier) had its origins as far back as the Apollo 11 mission itself. This myth is based on the simple—albeit **naive and absurd**—notion that the Apollo missions and subsequent Moon landings were "faked."*

So here we see Hoagland "playing the man, not the ball." That is, he attacks (or criticises) the messenger, not the evidence.

He continues:

> *That said, one thing they [NASA] did not do, **unquestionably**, was fake the Moon landings. In fact, most of the charges made by these "Moon Hoax" advocates are so absurd, so easily discredited and so lacking in any kind of scientific analysis (and just plain common sense) that they give **legitimate conspiracy theories** (like ours) a bad name (which is more than likely now the real objective—see below). The comedy of errors and willful ignorance represented by the Moon Hoax advocates is too extensive to detail here. We instead refer our readers to the "Who Mourns for Apollo?" series on Mike Bara's Lunar Anomalies web site for a detailed, claim-by-claim dismantling of the whole Moon Hoax mythos.*

This is no different from what he accuses other people of doing. If you have read this far, you should be able to understand that Hoagland is wrong – and he offers no evidence in his book to counter what has been posted by the researchers he projects his opinion onto. Also, Hoagland claims he has a "legitimate conspiracy theory." I think it would be far more sensible of him to refer to "evidence" than "theories." Those wishing to review a study by Hoaglands' co-author, Mike Bara, called "Who Mourns for Apollo?" series can now find it on a blog he posted.[324] Why hasn't Hoagland himself authored such a study?

Returning to the "astronaut hypnosis" issue, we can now consider a peculiar story which was posted on the Smithsonian Magazine's website in November 2014. The title was "How NASA Censored Dirty-Mouthed Astronauts. NASA really didn't want astronauts swearing on air." The article reports[325]:

> *During the early days of the space race the public relations handlers at NASA had an image to uphold. America's astronauts were the new face of a nation: they were the bold, brave explorers of the beyond. But this image didn't always align with the more rough-and-tumble types who got the job.*

It then quotes from an earlier article on a website called "Vintage Space" by a space history writer, Amy Shira Teitel[326]:

> *One [astronaut] in particular had the unfortunate habit of filling space when his mind wandered with profanities. This posed a problem for NASA - with the world watching astronauts walking around the lunar surface, how could the organization be sure his transmissions from the Moon would be family-friendly?*
>
> *In preparing for his mission, NASA had the astronaut hypnotized. Rather than curse, a psychiatrist put the idea in his head that he would rather hum when his mind wandered. The hypnotized astronaut is rarely named, but only one man can be heard humming as he skips across the lunar surface. Transmissions from Commander Pete Conrad are punctuated with "dum de dum dum dum" and "dum do do do, do do" making him the likeliest candidate.*

Whilst this may seem like a harmless and amusing anecdote, we must remember that around the same time as Pete Conrad was allegedly "hopping around" on the moon, public enquiries into the CIA's mind control programme MKULTRA were being held. We will mention this again in chapter 27, when we will discuss Edgar Mitchell's case – he also admitted to being hypnotised.

Having observed the astronauts' behaviour in interviews, it appears that most of them believe they did, indeed, walk on the moon. However, in my view, the reactions of some of the astronauts seem to indicate some type of "internal emotional conflict" – like they know something is wrong, but they can't express this.

16. A Brief Review of some of Apollo's Photographic Evidence

One of the most studied areas of the Apollo Hoax is the collection of thousands of photographs that were taken on these missions. Most of that analysis has been done on those photographs from the surface of the moon, rather than the ones taken in orbit etc. You may find better or more thorough or more complete photo analyses elsewhere. I will, however, point out what I consider to be the most glaring problems with the photos and why they prove (to me at least) that said photographs were not taken in the harsh, hot and airless moon environment. Instead, they were taken in a studio. Doubtless, you will be able to find alternative "explanations" for most or all of the images shown here. On internet forums, most of the "explanations" will be doled out by anonymous apologists (similar to how I described them in the Preface of this book). One famous debunker is James Oberg. NASA originally tasked him with writing a book rebutting moon hoax evidence, but they then withdrew their request.[327] I hope he reads this book and learns at least some small thing he didn't know.

That "First Boot Print" … Hmmm…

It is one of a number of iconic images from the alleged Lunar Landings.

Can we really be sure that image AS11-40-5878 is a genuine photo of a foot print on the moon[328]?

Knowing now that Neil Armstrong could not possibly have landed on the moon, I would say the material has a consistency more like flour or cement. Lunar dust should not bind together like this – as there is no surface moisture. However, in a studio, moisture would be ever-present – in the air. Was this really taken with a chest-mounted camera? Maybe it was, as these probably would have been used in the studio to take most of the images.

Shadow/Shade Analysis and Counter Argument

Much has been made of the anomalous shadows and lighting in the photos, such as the photo below, of Buzz Aldrin allegedly descending onto the surface from the Lunar Excursion Module (LEM).

I have not discussed other issues with the shadows, such as their direction "changing" based on an objects location in an image. Some examples of the shadows "being wrong" are more obvious (or contentious) than others.

Apollo Image Atlas: AS11-40-5868[329] – Aldrin Descends from LEM

The argument with photos like this (and there are many) is that Aldrin is well illuminated, even though he is completely in the shadow of the LEM. Additionally, a close-up examination of the parts of the image reveal highlights which imply a second source of light or a reflector was used (but they carried no additional lighting with them).

Though I think these arguments are correct, they are equivocal, as demonstrated by a simple experiment by someone called Thomas Bohn. On his (now defunct) Website, Thomas Bohn wrote[330]:

> *"I tried my lighting experiments on a surface of black asphalt. For this experiment I went to my local auto supply store and bought a new toy: a handheld floodlight. 1,000,000 candlepower … I took it to a tennis court late at night and hung it from the fence so it was pointing down at some common household objects. The idea was to simulate the sun on the moon. I would shine the floodlight on the backs of these objects and see whether I could see their fronts."*

Here are some of his photos.

Objects put on Tennis Court – Camera is near top, centre of picture.

Scattering of light gives enough illumination to see objects which should be in total shadow.

A few years ago, I wrote to Bohn and asked him about some of the other photos which I present here. He never replied. Perhaps he got fed up of avoiding difficult questions or correcting other points which were sent to him by people who were mistaken about aspects of the Apollo Hoax.[331] In any case, his interesting photographic analysis of the shadow argument has been taken offline.

Apollo LEMmings and Parallax

On his website, David Percy has posted an interesting article called "Serious problems in the Valley of Taurus Littrow"[332]. The article shows a comparison of two sets of composite photos from the Apollo 17 mission (reproduced below). AS17134-20437 to AS17134-20443[333] form the top composite and AS17147-22494 to AS17147-22521[333] form the bottom composite. In the upper image, we can see the LEM is some distance away from the camera, over to the right-hand side (it's quite small). In the lower composite, the LEM is much closer. However, the field of view is almost identical! Indeed, it is possible to take the two images and do a "fade" between them – and only the foreground changes (I have shown this "fade-between" demonstration in multiple talks and videos I have given over the years). This is another serious problem with the imagery and I will show other examples of similar problems shortly (it's to do with parallax). You can easily see that the distant landscape features (labelled B, D, E) are in exactly the same relative place!

When I posted this image on a forum (or maybe it was on my website), some anonymous person wrote to me saying there was "nothing wrong" with this image – it was simply that the camera was further away from the LEM and this could be proved by checking that the individual rocks shown in the images match what is on the highly detailed surface maps. I don't remember the exact references this fellow gave, but I wasn't convinced by his explanation. In the article I linked above, from which these images are derived, a further set of examples and information is/are given showing similar problems.

Two Composite images from the Apollo 17 Mission – see description above.[332]

Also, on the "Aulis" website, we can find an article by Oleg Oleynik, Ph.D.c - Previously of the Department of Physics and Technology, Kharkov State University, Ukraine. He has completed a study entitled "A Stereoscopic method of verifying Apollo lunar surface images."[334] He shows an example photo of a scene taken on earth and using some maths, works out the distance to the object (a power station) in the distance, based on the position change of objects in the foreground when two pictures that were taken along the same horizontal line, a few metres apart.

In the example below, Oleynik calculated that the Power Station is about 4km away and the woodland is about 2km away.

Note the changes in position of objects between the two images – e.g. of the thickest pole on the left and the pylon on the right. If you know how far the camera has moved, you can work out the distance to objects in the image.

In the same way, pairs of Apollo 15 photos, taken close to each other, are used and the "parallax" between the pairs is determined, to work out the distance to the objects in the background.

This pair of images is AS15-86-11601[335] and AS15-86-11602[336] from EVA-1 near the LM

Oleg Oleynik made an animated GIF file using the marked sections of each image[337], to give the same field of view. The effect he describes – a sharp line of demarcation between the foreground and background detail – can easily be seen. Similar lines of demarcation appear in quite a few Apollo lunar surface images. These have been pointed out by researchers such as Jay Weidner in his "Kubrick's Odyssey" videos. (Once you know what to look for, you will be able to identify demarcation lines quite easily in some images.) I have indicated the approximate position of the line in the image below.

An example of a line of sharp demarcation between foreground and background (look for more in other Apollo surface images)

In his article/paper, Oleg Oleynik writes:

> *Nearby objects: the LM, the rover, and astronaut Jim are shifting relative to each other. The Apennines and the crater St. George are also moving as a whole. (Moreover, the shadow is changing on the mountains and the crater.) This finding indicates that it is less than 300 metres to the background (the 'mountains') instead of 5 kilometres!*
>
> *In addition, the Apollo 15 stereoscopic photos feature a clear separation line between the 'mountains' and the foreground. Based on the distance between*

> *the camera and rover, the distance to the panorama of the 'lunar' scape cannot be more than 150 metres.*

The rest of the article shows about six or seven different examples, along with various parallax calculations. It is definitely worth your attention! Oleg Oleynik concludes:

> *The Apollo 15 photographic record contradicts the stereoscopic parallax verification method. The apparent change in the relative positions of objects by moving the camera when the camera angles are separated by several tens of cms show that:*
>
> - *the distance to distant objects such as mountains is not tens of kilometres but is no more than a few hundred metres;*
> - *the landscape is not continuous, but with clear lines of separation;*
> - *there is movement between nearby sections of the panorama, relative to other sections.*
>
> *Thus, based on the above examples, this study concludes that the Apollo 15 photographic record does NOT depict real lunarscapes with distant backgrounds located more than a kilometre away from the camera.*
>
> *These pictures were, without doubt, taken in a studio set – up to 300 metres in size. A complex panorama mimicking the lunarscape shows degrees of movement, such as horizontal and vertical changes to give an impression of imaginary distance to the objects and perspective.*

In a short afterword, the following is also noted:

> *Two years have passed since the original publication of this article in Russia. During that time, NASA decided to create a series of stereo photographs for 3D red-cyan glasses (anaglyph images), superimposing overlapping parts of Apollo surface photos. Reports slip out now and then that some of the photos on NASA's Web sites have been replaced by retouched counterparts.*
>
> *An article entitled "The method of correlative calculation of parallax and camouflage" was published (in Russian). I criticized the article stating that: "The merging of frames is carried out in the application for creating 360 degrees panoramas PTGui, which erases parallax, and eventually the distance to background objects artificially increases. Please double check the algorithm of the application". More here (In Russian).*
>
> *There was no answer from NASA. Instead, in the Russian Wikipedia, late 2009, the following paragraph was added (and removed on July 31, 2011) to The Moon Hoax article: "Also, analysis of the lunar surface images, taken during the missions shows that distance to background objects is indeed vast and cannot be achieved in a soundstage with trick photography", referring to "The method of correlative calculation of parallax and camouflage" publication.*
>
> ***Any attempts to change or correct the information in Wikipedia, and to point out the serious errors in the Wikipedia article did not succeed, the moderator continued to erase the link.***

Now you know why there are no Wikipedia references in this book.

Not enough Dust on the Feet of LEM

Another obvious problem is illustrated on several photos of the feet of the LEM. They show little or no dust on them following the landing[338]. In general, the landing site shows almost no evidence of dust being blown around during the landing[339]. There should be streaks/lines of dust stretching for many meters around the landing side. The photo here is AS16-107-17442 from Apollo 16. It is worth inspecting high resolution versions yourself.

A Genuine Photo on a Fake Moon...

An argument which seems to me to be irrefutable is one which has nothing to do with shadows or angles of shadows. If I speak about a single piece of evidence in relation to the Apollo Hoax, this is frequently the piece I speak about. It is the impossible family photo of astronaut Charlie Duke. With an average temperature of over 100°C on the Moon's surface in daylight[340] (although different figures are given from different sources[341]), the photo would curl up almost immediately and the plastic bag it is contained in would shrivel or wrinkle. Instead, we see a photo which seems to have experienced only mild curling, perhaps due to the heat in an enclosed studio set.

Apollo 16 Photo AS16-117-18841[342] of Charlie Duke's family – this was taken in a studio. The ambient temperature must be low – we observe only a very limited effect on the plastic or paper in the photo.

You can easily demonstrate what should be seen in this photo by putting a photo inside a plastic bag and putting it in your oven, after setting a temperature of 100C. You will quickly see the photo curl up.[343]

We will see this photo again, in a slightly different form in a later chapter.

Orange Juice and the Reseau Plate

In his film "Moon Hoax Now," Jet Wintzer highlights an episode in the Apollo 16 mission for which the only logical conclusion can be, as he clearly illustrates, that the sequence was filmed in a studio. Those interested in understanding this should watch the sequence of the film and check the evidence he discusses.

Wintzer points out how a sequence of Apollo 16 photos, starting with image AS16-114-18444[344], each show a smeared pattern across them[345]. This pattern was created, according to NASA and the astronauts, because the film was changed inside the LEM and in the process, orange juice was spilled on the lens of the camera. Wintzer proves that this story cannot be true, due to inconsistencies between what is written in the Apollo 16 Lunar Surface Journal[346] and what is shown in the film/video[347] of the actual events - and the images - themselves. Wintzer shows that the orange juice must have leaked out of the space suit and gone onto the camera's reseau plate[348], while the astronaut was *outside* of the LEM. This proves that the space suit was not water-tight and therefore was not air-tight either.

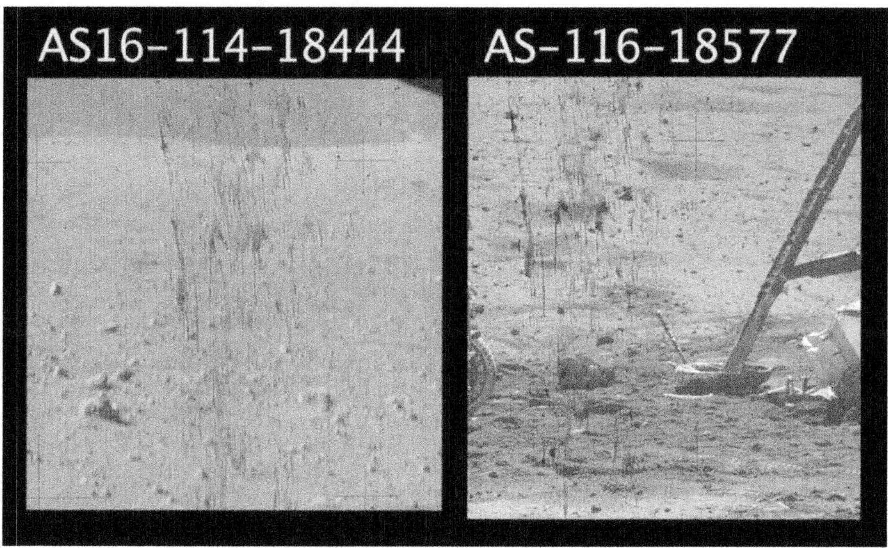

Still from Jet Wintzer's film "Moon Hoax Now"[349] When the circumstances relating to this image sequence are studied, one can only conclude these images were taken in a studio.

As with the Charlie Duke "family photo," the only conclusion I can draw from the analysis of the evidence is that the images we are being shown from the Apollo 16 mission images from a simulation.

Trackless Rover Photos?

As with most other photos I discuss, it is easy to find "explanations" for them. However, in view of the non-photographic evidence I have discussed in this book, I think these "explanations" cannot be correct.

In the following images, consider the appearance of the rover tracks. To me, it seems like the rover has been placed in a set… Of course, as we don't have a full 360° image of the rover, it's hard to be sure.

In image AS15-88-11901[350], it sure is difficult to see the rover's tyre tracks… Is it just the camera angle…?

And in image AS17-140-21354[351], the tracks have just been covered by the Actornaut's Footprints, right…?

Image AS17-135-20544HR[352] is, again, just another bad camera angle… of course…

In Search of Apollo Artefacts…

The most obvious way, surely, of proving that the Apollo missions did land men on the lunar surface would be to photograph the landing site… We will look at two possibilities regarding this. However, we must realise that we cannot get an independent view of this – because the instruments that would be able to photograph the landing sites are currently under the control of NASA…

The Hubble Space Telescope (HST)

Quite a few years ago, I was asked if the HST could see the Apollo craft on the moon. I assumed the width/diameter of the LEM, as viewed from above, to be 4.2 metres[353], but you could arguably say it was 9.5 metres if you include the feet and measure the widest dimension. I then needed to know the resolution of the HST. I found that figures given for this vary slightly – from about 0.05[354] to 0.014 arc seconds[355]. A "Hyperphysics" page on the University of Georgia Website states:

> *It has a resolution of 0.014 arc seconds, compared to an angular size of 1800 arc seconds for the moon. An arc second on the moon corresponds to about 1.92 kilometers on its surface, so Hubble has a resolution of about* ***27 meters on the moon's surface.***

So, this means the HST would not be able to see the artefacts on the moon – we will have to "look elsewhere…"

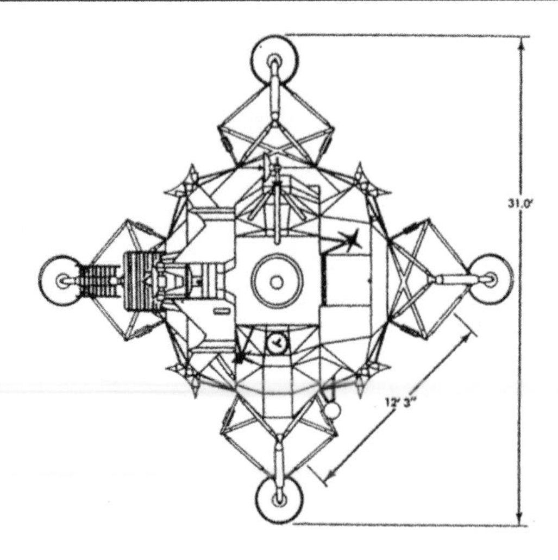

LEM is 31 feet or 9.5 metres at the widest.

Lunar Reconnaissance Orbiter (LRO)

I mentioned LRO many times in the "Secrets in the Solar System" book. It's time to use data from LRO again, to illustrate some strange Apollo anomalies…

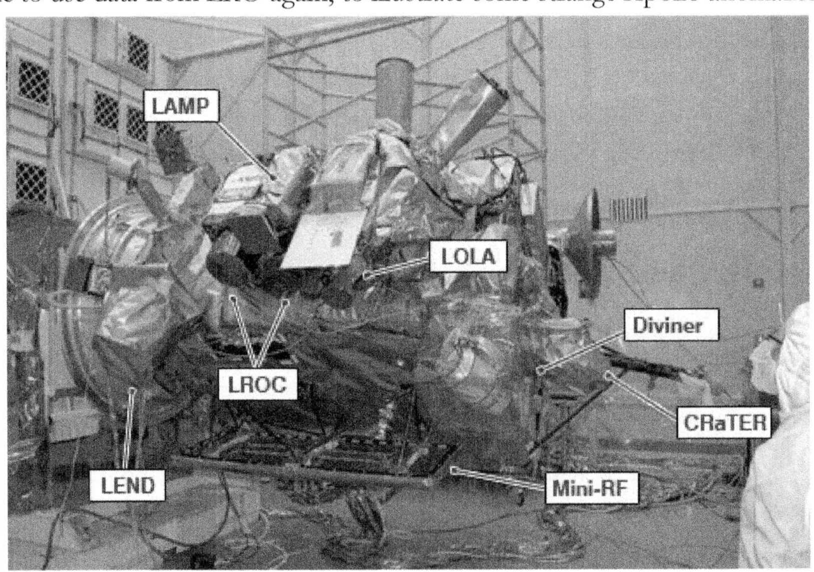

LRO before launch

LRO was launched on June 18, 2009, by an Atlas V from Cape Canaveral Air Force Station.[356] It was one of two payloads carried by the rocket, the other being the Lunar Crater Observation and Sensing Satellite (LCROSS). The objectives of the Lunar Reconnaissance Orbiter (LRO) were to find safe landing sites for future crewed missions, locate potential resources, characterize the radiation environment, and demonstrate new technology.

We will now try to establish if the LEM can be *clearly* seen on the moon by LRO. A posting on NASA's website, dated 9 November 2009 (that's 9/11/9 or 11/9/9 depending on where you live!) has a title "LRO Gets Additional View of Apollo 11 Landing Site."[357]

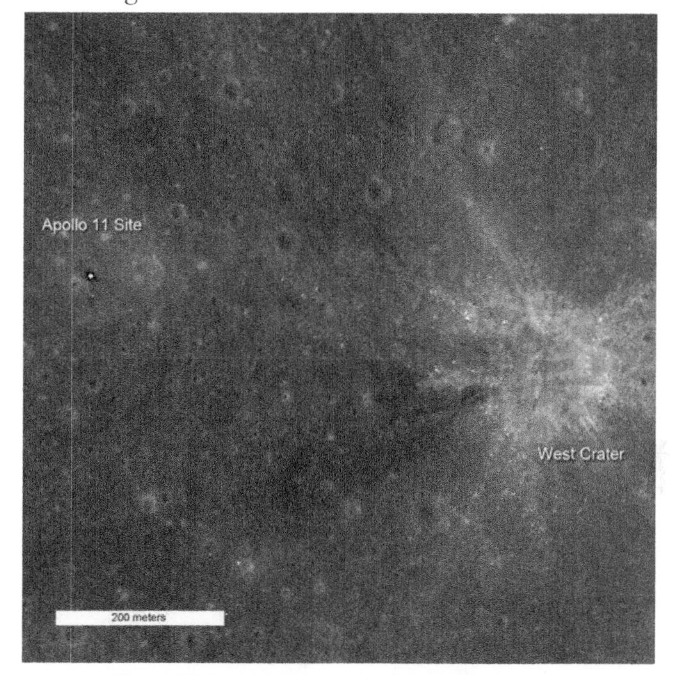

As the Apollo 11 Lunar Module (LM) neared the surface, Neil Armstrong could see the designated landing area **would have been** in a rocky area near West Crater. He had to change the flight plan and fly the LM westward to find a safe landing spot. This image is 742 meters wide (about 0.46 miles). North is towards the top of the image. Credit: NASA/GSFC/Arizona State University[357]

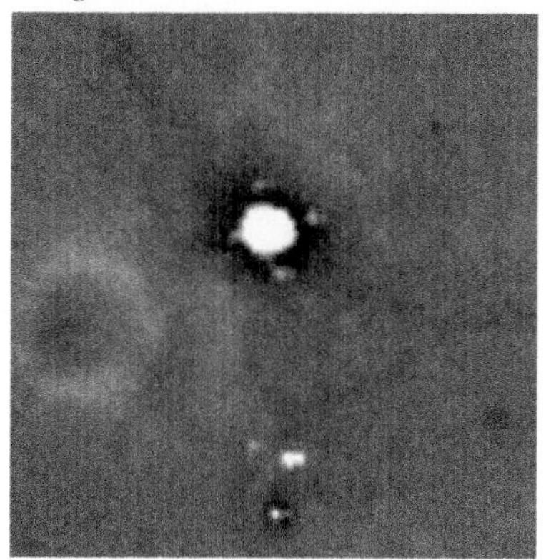

Enlarged section of image, where LEM is supposed to be.

I measured that 200 metres = 374 pixels in the large LRO image, shown above. In the Plan View/Elevation of LEM, we see a measurement of 31 feet from one landing pad to the other – a distance of about 9.5 metres. With 200 metres for 374 pixels, this would mean that each pixel covered 0.53m. Hence the LEM should cover approximately 9.5 / 0.53 pixels = 17.92 or 18 pixels. Zooming in on the LEM part of the image, and enlarging and contrast enhancing it, we get this:

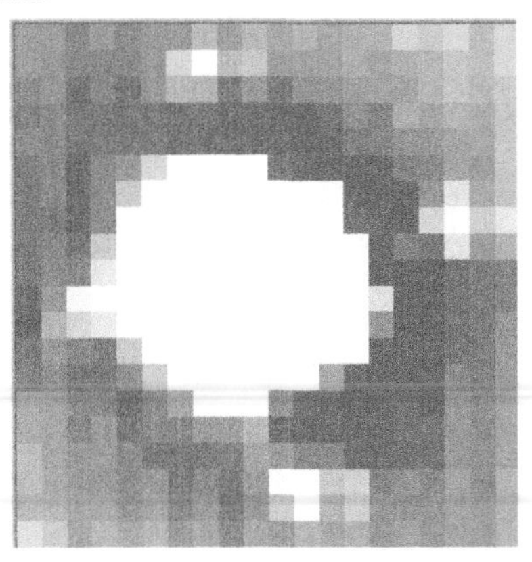

Zoomed in on LEM - from LRO image taken 09 Nov 2009

The above image is actually 20 x 20 pixels – and so is arguably a little too small – but it's basically the right size.

Now, consider that this was a high resolution camera, with *no atmosphere* to blur or add haze to the image of the LEM. Let's try to get an idea of how this image should have appeared. We will do this by using a high-resolution image of the LEM in space and then reducing it to the same size as the image shown on the left. Fortunately, there is one such image – from Apollo 9. It's not quite a "plan view", but fairly close.

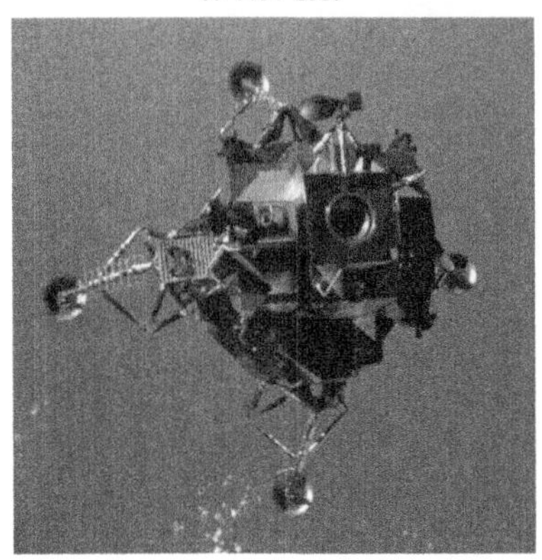

Apollo 9's LEM in Earth Orbit - – image AS9-21-3181 – 07 Mar 1969.

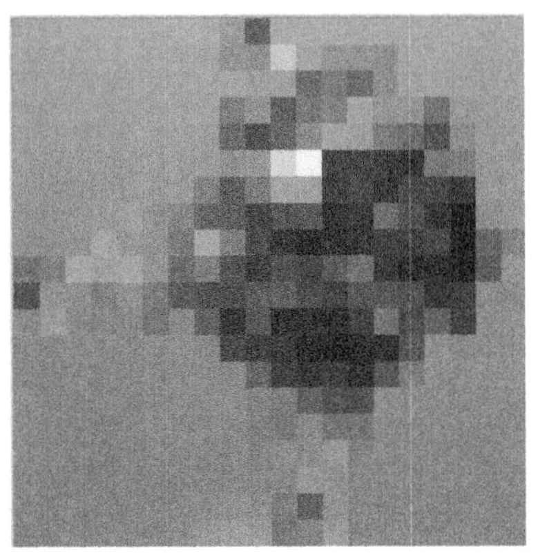

Apollo 9 LEM image re-sized to 20 x 20 pixels and converted to greyscale.

So, compare the level of detail in these 2 images – consider the amount of different shades of grey within the body of the LEM itself.

This is *not* a totally fair comparison, because LRO images were taken from orbit, but the LEM images used were taken close-up. However, there is no atmosphere on the moon and the optics on LRO should be better than those used in the 1960s to take the LEM images.

But hey, in 2009, it was only NASA's first attempt at imaging the Apollo 11 site – that's why the images were so bad! They got better, OK?[358]

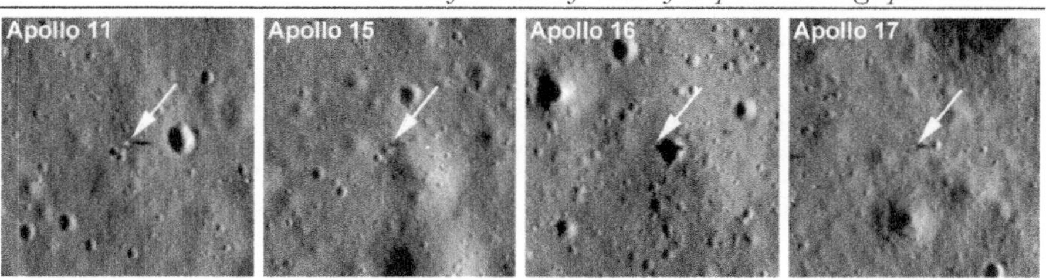

Let's have a closer look at "Apollo 16 Footsteps" – and Lunar Roving Vehicle (LRV) tracks and remains![359]

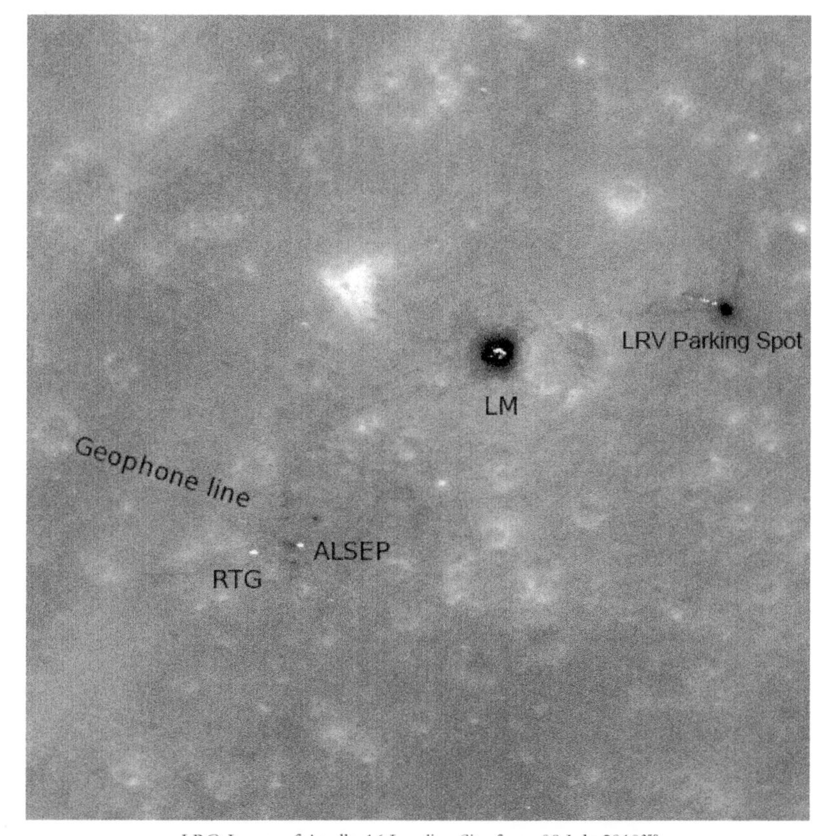

LRO Image of Apollo 16 Landing Site from 08 July 2010[359]

Did the LRV explode? Or perhaps it was hit by a small meteorite – and so was the LEM….

Of course, I am being facetious – and in doing so, I am expressing my scepticism of NASA's claims for these particular images. This is due to all the factors and evidence we have covered so far, and more that we will cover soon.

Again, most people will calmly state these LRO images are exactly as they should be and so you are free to agree with them, if it makes you feel better. For now, let us consider a comparison first made by Jarrah White in an episode of his Moonfaker Video Series.

Comparison of LRO / GeoEye Images

As Jarrah White has also observed, GeoEye imagery – here shown with a 1-meter per pixel image – seems considerably better and more detailed than LRO.[360] GeoEye is 423 miles above earth – travelling at 4 miles per second – 17,000 mph. LRO's close orbit was about 50 miles above the lunar surface – no atmosphere...

Here we can see a section of an image of Copacabana Beach, Rio de Janeiro, Brazil.[361] On the right, a 20 x 20 image segment where we can see more variation in the pixels than in the LRO image.

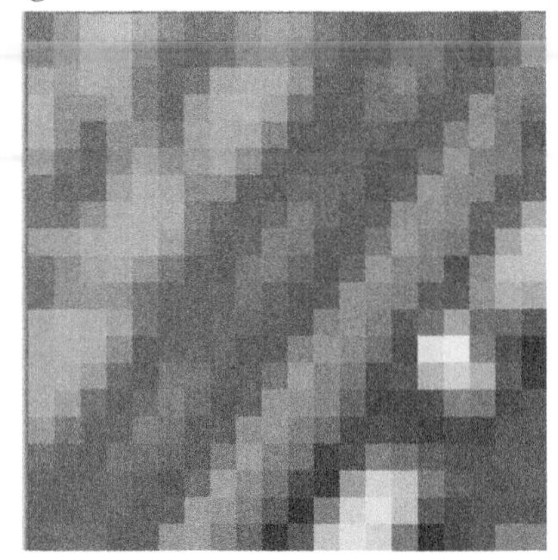

Try downloading and comparing some images yourself.[362] Jarrah White did this in 2011 and showed this comparison:

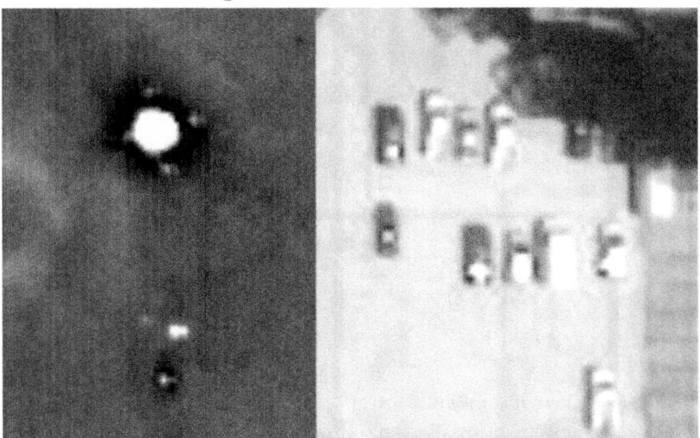

LRO/GeoEye Image comparison by Jarrah White.[360]

Conclusion

There are enough impossible elements in the Apollo photographs for me to say with a high degree of confidence that, at least as far as the photos that were allegedly taken on the surface, NASA have essentially faked the photographic record and, indeed, each "landing" was filmed/photographed in a studio environment. This is the only sensible conclusion I can draw. The reader is free to draw his or her own conclusion but do continue to consider the additional evidence in the next two chapters and give yourself a "fair crack of the whip" in understanding more aspects of the Apollo Hoax.

17. "It's a Hardware Problem..."

Back in the 1980s, when I was working with microelectronics, hardware and software engineers would sometimes try (normally in a light-hearted way) to blame each other for development problems. I remember one instance where I had been asked to write some software which generated a sine-wave oscillation pattern to test a servo motor control board. When we ran it and connected an oscilloscope, it looked like there was a weird hardware timing problem on the board. However, careful tests showed that it was actually caused by an error in the way I'd generated the sine-wave data for the test... So, it was a software problem...

I digress! Another area of the Apollo Hoax that is worthy of study is the stated specifications of some of the hardware. Now, I've had several debates over the years about these issues, but I am still comfortable that, mainly due to the unforgivingly hot environment, some of the hardware could not have functioned as claimed.

Most people, who think that the landings were real, will simply argue the heat issues were adequately dealt with in the design of the hardware. I disagree.

Things are Hotting Up!

To me, with prolonged deliberation, the lunar surface heat issue becomes the most obvious problem in the Apollo record. If one works out the incident heat on the suits and how that heat would be dissipated, it becomes very difficult to accept that the suits could have provided sufficient cooling. Again, Ralph Rene goes into considerable detail about this in chapter 10 of his book "NASA Mooned America."[279] To summarise, however, the moon does not have an atmosphere to speak of – it is basically a vacuum. Therefore, two of the cooling methods that work on earth – convection and conduction – do not work in a vacuum. This means, then, that the lunar surface and any objects in contact with it, can only lose or gain heat by radiation.

While we are thinking about this... How did the cameras and film survive *the heat* let alone the radiation? The cameras were only modified to allow them to be chest-mounted and the film was, apparently, ordinary Kodak Ektachrome, yet (apparently) it experienced no undesired fogging due to radiation... Nor did it experience heat damage.

Also, as I have mentioned many times in presentations I have given, consider your own experience of parking a car in a car park/parking lot on a hot day and going about your business. You return to your car and the first thing you would do is either wind the windows down or switch on any air conditioning you might have. Otherwise, when inside the car, the heat is uncomfortable or unbearable. For the reasons stated above (no convection), Apollo craft would not have the same options for cooling down the occupants – or the equipment, for that matter.

Navigation

Ralph Rene covers several different hardware-related issues in his book "NASA Mooned America."[279] In chapter 7, he also raises the important question of how they actually navigated their way to the moon. Apparently, Raytheon built the computer that was used for navigation.[363] However, we have to remember that the moon is over 200,000 miles away so a journey there would represent the furthest distance ever navigated by human beings. This means that the navigation required would have to be much more accurate than that used over shorter distances. How did they measure the required angles with sufficient precision?

Space Suits

Ralph Rene discusses the pressurisation of the suits and how that would, in all likelihood, have meant that the astronauts would not have had enough freedom of movement in the gloves/fingers to do what was shown in the videos and photos (Rene built a vacuum chamber and inserted a mock-up glove to test this). Other problems include radiation protection (we will mention a little more about this later) which seemed woefully inadequate.

Suit cooling would also have been problematic. Just by moving around, the astronauts would generate heat and this would need to be dissipated somehow. In chapter 12 of his book, Rene discusses this in terms of the volume of water that would be needed to transfer heat out of the suit by evaporation – and he shows that the volume of water they would be required to carry to keep themselves cool would be impossibly large in the lunar environment.

In a similar manner, but less savoury to consider, perhaps, is how they handled the urine and faeces over the course of the mission. Working in such confined spaces (i.e. considerably smaller than any of the space stations that have been in operation in earth orbit), how was this made acceptable. (Yes, I am quite sure this has "all been explained..." sorry for even broaching the question...)

The Lunar Roving Vehicle (LRV)

A question asked by Bart Sibrel in his "A Funny thing..." film, was "Why did the missions not take an optical telescope to the moon?" Various excuses may be given by apologists, but consider that a small telescope might weigh, say, 20kg. The LRV weighed about 200kg – but was that weight commensurate with the value of the scientific data that the LRV allowed the actornauts to gather? What did the rover allow the astronauts to really do? Perhaps it was just "theatre" or for the spectacle... At least one YouTuber has noticed that the chassis of the LRV seem to match that of a Willy's Jeep[364]. But perhaps the designers of LRV borrowed heavily from the design of a jeep[365] – so that, alone, proves nothing. Whatever you think of this idea, there are some additional potential hardware problems.

Again, the heat issue raises its head. According to the LRV Operations Handbook, care was needed with the rover battery.

LS006-002-2H
LUNAR ROVING VEHICLE
OPERATIONS HANDBOOK
APPENDIX A

3.1.0 Continued

Component	Maximum Operating Temperature Limit °F	Survival Upper Temperature Limit °F	Minimum Operating Temperature Limit °F	Minimum Survival Temperature Limit °F
*Battery	125	140	40	-15
DCE	159	180	0	-20
*Traction Drive	400	450	-25	-50
Wheel	250	250	-200	-250

Seen on Page A-30 of the LRV Operations Handbook

According to the above page, the "survival upper limit" for the LRV battery is 140F / 60C. The battery would be quite close to the Lunar Surface while the rover was operating. As we have seen, the Lunar surface in sunlight can be over 100C. If you watch the video about the LRV's construction[366], one of the engineers describes how they used a "paraffin wax coolant." The wax which sat around the battery and would melt and therefore provide cooling. They claim that when the Actornauts stopped the LRV, they could raise gold-coloured cooling panels up and then the wax would cool and re-solidify. This is highly questionable – as the cooling panels cannot use convection to transfer heat out of the system – only radiation. Shiny materials are not good heat radiators.

Another observation made by Jim Collier is that in the videos of the LRV "outings," the surface dust did not fly up high enough when considering the one-sixth gravity that the moon has. Also, if the videos are sped up by a factor of two, it looks like a vehicle driving across a desert more than it being on the surface of the moon! Similar observations can me made regarding all the footage of the astronauts moving around the lunar surface.

Other people have suggested that the Lunar Rover, even when folded, seems to have been too big to fit into the allocated space in the LEM adapter (the part of the Saturn V stack that the LEM was stored in).

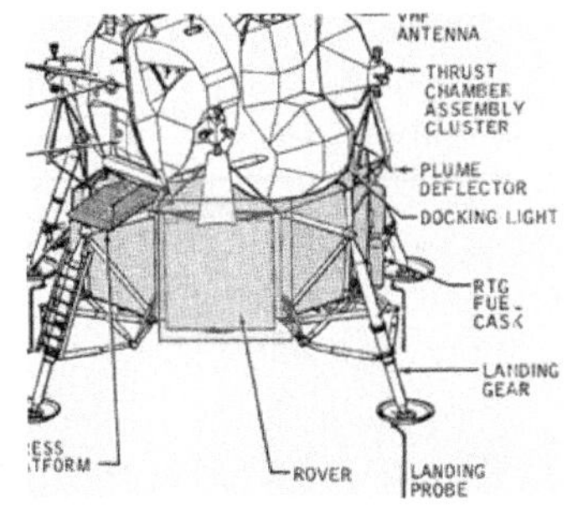

Left: Schematic showing where LRV was attached to LEM.[367]

Wouldn't the placing of the LEM make it lopsided and therefore even more difficult to land…?

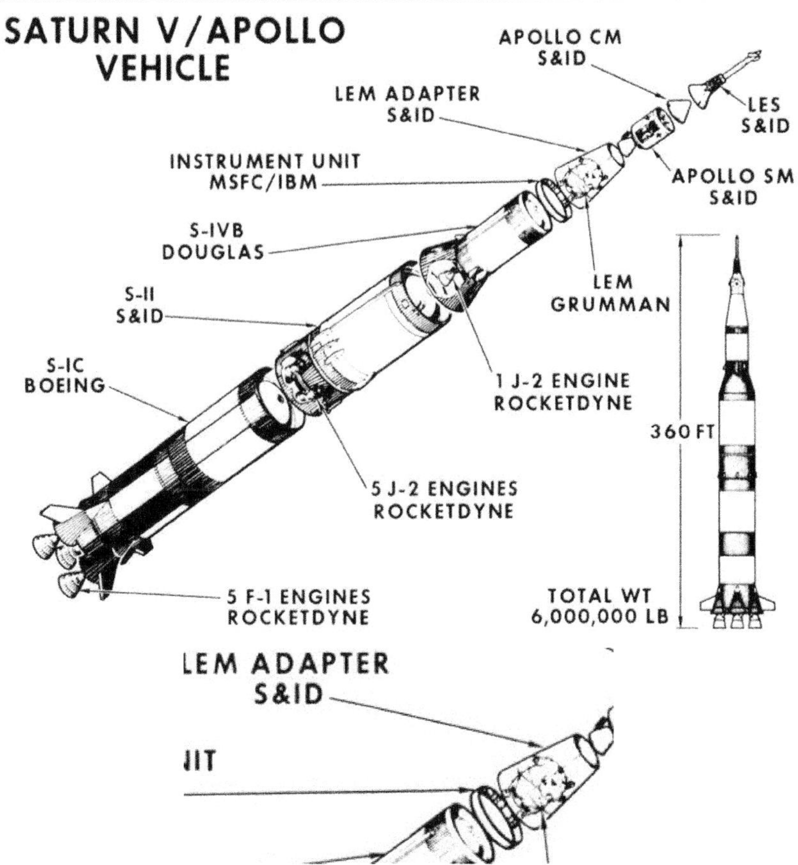

General Schematic of Apollo Saturn V Rocket[368] – Was there sufficient Space for the LRV?

Also, the LRV was attached to the *side* of the LEM - and a video shows how they removed it[369].

While we can see in the video that the idea of the folding LRV was plausible and *could* have fitted in and been unfolded successfully on the surface, what about the *mass/weight* of the LRV? The Lunar Roving Vehicle had a mass of 210 kg[370]. Whilst this mass was, of course, lighter on the moon than the earth, we have to consider if having the LRV on one side made landing the LEM more difficult on the later missions. Did they have counterbalance on the other side? Did they re-distribute the weight in the LEM? Having to deal with all this extra mass means that extra fuel would be needed both in the launch from earth and any manoeuvres during lunar landing. Is this realistic?

Considering the LEM itself, we can ask in engineering terms, is it reasonable to believe that a vehicle that (according to Astronaut Alan Bean) could not support its own weight in earth's gravity, worked perfectly six times in a row, on a different "planet"?

Cumbersome Space Suits and The Apollo 17 EVA

It is worth reconsidering the action of getting into and out of an Apollo Space suit. Indeed, there is an interesting video showing British Science Journalist (of the 1960s – 1980s) James Burke performing this activity. He needs two people to help complete some of the stages.[371]

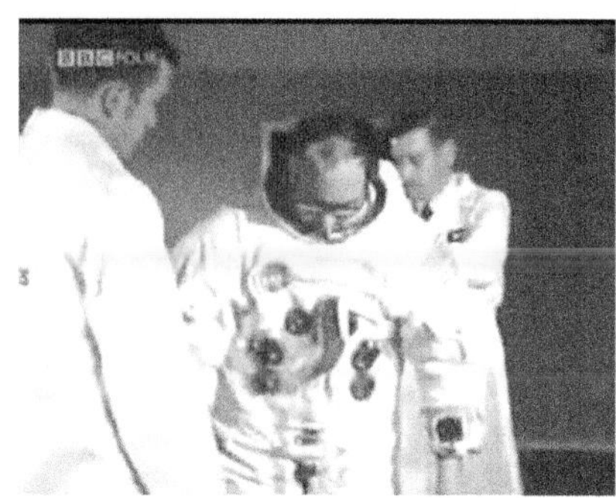

Two people help James Burke remove one of the space suits.[371]

With this in mind, let us consider an episode in the Apollo 17 mission where, *during the return to earth*, one of the astronauts *left the Command Module* to retrieve some items from a storage chamber outside!

All three astronauts were definitely in the Command Module at this point, after they had jettisoned the LEM (they didn't carry the LEM back to Earth). This is confirmed in the Apollo 17 Lunar Surface Journal (see the section on "Ron Evans Goes Spacewalking")[372]. The LSJ describes how they took 2 hours to prepare for the EVA, but inside the command module was very cramped, it being only about 12 feet in diameter. Have a look at James Burke standing up, in the only place possible, in the Command Module.[373] In the photos, the central couch has been removed to allow them to film the interior.

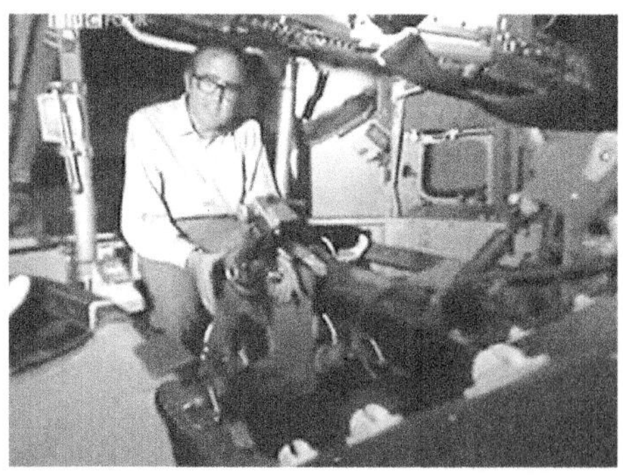

James Burke inside the Command Module (central couch removed)[373]

Would it be possible for three astronauts to put on then take off their space suits *and* backpacks in this space? Could they do this without breaking any switches on any of the panels or without puncturing or ripping their suits, or damaging one of the seals? You decide...

Radiation on the Moon

This topic always comes up – and again Ralph Rene goes into a lot of detail about this in chapter 15 of his book "NASA Mooned America."[279] Hence, I direct the interested reader there for detailed study. Here, let us just consider a few basic things – such as the published risks about radiation that can be experienced by "frequent flyers."[374] So, how much should the astronauts have been concerned in their week-long journeys *outside* of the earth's atmosphere, into cislunar space? Did their space craft and suits *really* give them adequate protection?

Also, radiation effects have been seen by the space shuttle crews which are still within the magnetosphere – only 125 miles above the earth. Bart Sibrel discussed this with Alan Bean in an interview in Astronauts Gone Wild, and asked him whether he saw flashes of light in his eyes (caused by cosmic rays). Bean responded saying such flashes were not seen on his mission because "they hadn't been discovered yet."[375] (Yes, he really said that. But he then later contradicts himself.) Alan Bean, in this interview at least, does not come across as being particularly well-informed about the issue.

Another oft-quoted problem is with travelling through the Van Allen belts – zones of radiation - outside the earth's magnetosphere. However, Dr Van Allen, who discovered them, is quoted as saying[376]:

> *The recent Fox TV show, which I saw, is an ingenious and entertaining assemblage of nonsense. The claim that radiation exposure during the Apollo missions would have been fatal to the astronauts is only one example of such nonsense.*

But again, it's difficult to get a clear answer on these issues. For example, in 2012, a video was posted on the NASA Johnson YouTube channel describing preparations for the launch of the Orion space capsule[377] (one of the proposed "reboots" of manned spaceflight, which we will discuss in chapter 18) The capsule is meant to be designed to be used to take people to Mars. The video contains two contentious quotes:

> *For these missions, Orion has to be one tough spacecraft withstanding high speeds, searing temperatures and extreme radiation. Before we can send astronauts into space on Orion, we have to test all of its systems, and there's only one way to know if we got it right; fly it in space.*

Later it states (emphasis added):

> *As we get further away from Earth, we'll pass through the Van Allen Belts, an area of dangerous radiation. Radiation like this can harm the guidance systems, onboard computers, or other electronics on Orion. Naturally, we have to pass through this danger zone twice, once up and once back. But Orion has protection, shielding will be put to the test as the vehicle cuts through the waves of radiation. Sensors aboard will record radiation levels for scientists to study.* ***We must solve these challenges before we send people through this region of Space.***

Why is this video making reference to a problem as being unsolved when it was supposedly solved in the Apollo missions? Why is it apparently harder to go to the moon now than it was in the 1960s? This very question brings us onto another interesting area of study - the legacy of Apollo.

The Apollo Legacy – is it Real?

In the video about the Orion module/craft, no reference is made to either the radiation shielding or heat shielding used by Apollo. Indeed, it seems rather peculiar that the Smithsonian Institution had to supply the Orion project team with an "unused" heatshield from 1966 that they could study, as reported in October 2008 by Science Daily[378] (and even that story is almost a decade old, as I write this).

On David Percy's "Aulis" website, a series of articles – published in 2014[379], 2015[380] and 2016[381], by a New Zealander using the Pseudonym Phil Kouts, Ph. D studies some of the NASA documentation relating to the Manned Space Programme "reboots" dubbed Constellation (CxP) in 2004 and the Orion Programme in 2010 (announced by Obama – which we will discuss below). Of particular note in Kouts' 2014[379] article is his Table 2, reproduced below.

Event Program	Announced	Problems & Critics	Unmanned Trial Entry, Interim Condition	Unmanned Trial Entry, Full Range	First Fly-By with Crew	Lunar Lander	Man on the Moon	No of years to completion
Apollo	1961	1963	1966	Never	1968	1969	1969	8 Claimed
CxP	2004	2009	2011	2017	Was not planned	2018	2020	15 Abandoned
Orion	2010	2011 onwards	2014	2018	2021	Not yet planned	Not yet planned	More than 20

This direct comparison, to me, makes it abundantly clear that NASA could not possibly have landed men on the Moon in 1969, as they claim. As I write this in 2018, the US manned space programme has had a 7-year-long hiatus and the private space programmes (as we shall see in chapter 19) have achieved nothing that hadn't already been achieved by the mid-1960s.

When I was writing this book, I came across a short video from 1986 – about the Cray 2 Super Computer in use at the NASA Ames Research Centre[382]. The video proudly stated:

> *The Cray-2 Supercomputer - "World's Most Powerful Computer"*
>
> *The Cray-2 will also play a key role in technology development for the NASA DoD national aerospace plane program leading to an entire new family of aerospace vehicles in the next century…*

Well, there is no new family of space vehicles this century, despite billions of dollars of white-world spending.

18. Apollo – Reboots, Revelations and Conclusions

Rebooting the Public Manned Space Programme

As we briefly mentioned in chapter 17, the USA seems to have attempted to "reboot" the manned space programme on two occasions. The first of the attempts took place[383] in Feb 2004, on the 100th Anniversary of Powered Flight, when warmonger George Bush announced the "President's Commission on Moon, Mars, and Beyond."[384] This resulted in a series of public hearings – where Pete Aldridge and General Tom Stafford (former Astronaut) talked of missions to Mars using Nuclear Thermal Propulsion and Ion Propulsion – with trip times as short as 60 days! Nuclear Electric Propulsion was also mentioned. What tweaked my interest was their discussion of "leveraging technology held by the Air Force" to "help" the space programme. In one exchange, General Les Lyles and General Tom Stafford agreed that "the [US] Air Force does have a lot of technology." However, nothing has come of this commission's report[385] and Obama "killed off" the Constellation programme in 2010[386].

Obama @ NASA - 15 April 2010

In a "second reboot" of the manned space programme, Obama backtracked on the proposed lunar landing[387] (which had apparently made far less progress in six years than Apollo ever did.)

Here is a segment of Obama's speech:[388]

> *"... we will ramp up robotic exploration of the solar system, including a probe of the Sun's atmosphere; new scouting missions to Mars and other destinations; and an advanced telescope to follow Hubble, allowing us to peer deeper into the universe than ever before.*
>
> *Now, I understand that some believe that we should attempt a return to the surface of the Moon first, as previously planned. But I just have to say pretty bluntly here: We've been there before. Buzz has been there. There's a lot more of space to explore, and a lot more to learn when we do. So, I believe it's more important to ramp up our capabilities to reach—and operate at—a series of increasingly demanding targets, while advancing our technological capabilities with each step forward.*
>
> *In developing this new vehicle, we will not only look at revising or modifying older models; we want to look at new designs, new materials, new technologies that will transform not just where we can go but what we can do when we get there."*

This very much sounded like they were simply going to re-develop old technology – and any development of "breakthrough" technologies, considered in the mid-1990s - such as those discussed in chapter 11, had long since been abandoned and forgotten.

Did Bill Clinton Almost "Blow the Whistle"?

Having mentioned a US president's words in relation to a moon-landing programme, let us now briefly cover some unusual remarks by another US President. It is interesting to wonder if Neil Armstrong's 1994 usage of the "truth's protective layers" phrase prompted a certain entry in the corrupt president Bill Clinton's autobiography (which is mentioned in the "Apollo Zero" Documentary) … On page 156 of Clinton's 2004 book "My Life," he states:

> *Just a month before, Apollo 11 astronauts Buzz Aldrin and Neil Armstrong had left their colleague, Michael Collins, aboard spaceship Columbia and walked on the Moon, beating by five months President Kennedy's goal of putting a man on the Moon before the decade was out. The old carpenter asked me if I really believed it happened. I said sure, I saw it on television. He disagreed; he said that he didn't believe it for a minute, that 'them television fellers' could make things look real that weren't. Back then, I thought he was a crank. During my eight years in Washington, I saw some things on TV that made me wonder if he wasn't ahead of his time.*

It would be nice to think that Clinton really was interested in revealing the truth about the Apollo Hoax, but anyone who watches the "Clinton Chronicles" videos[389] (and several similar ones) can only conclude that the Clintons are absolutely corrupt and contemptible. This can be said even before we start to research such scandals involving the Clintons such as "Tainted Blood."[390]

Why the Hoax has Not Been Revealed Sooner…

It is often said that, due to the political climate of the 1960s (i.e. the Cold War), if the Russians/Soviets had some kind of knowledge or proof that the Apollo landings were staged, they would have "blown the whistle on the whole scam" and showed how untrustworthy the leaders of Capitalist regimes were… However, it seems that the Russians/Soviets had run at least one "space hoax" of their own.

I distinctly remember watching a BBC Horizon programme in 1999, where it was revealed that Yuri Gagarin did not land in his capsule, following his "historic" first orbital space flight. This is also reported in an article from the same year:[391]

> *Disorientated, Gagarin failed to follow instructions from ground control. Though told not to eject too early, he did, and from an undetermined height. Deep in the countryside, tractor driver Jakob Lashenko was working in the fields when he was astonished to see a figure descending from the sky"*
>
> *Gagarin landed by parachute.*

Can we imagine him hurtling down in his cramped capsule… opening the door – inwards or outwards – then safely jumping out? I struggle to believe this one, too…

In David Percy's "What Happened on the Moon" documentary, there is a segment, approximately 8 minutes in length, which includes a number of important observations about the Gagarin mission.[391] One such observation includes a shot (reproduced below) which claims to show Gagarin in the space capsule, prior to launch, but this cannot be the case, because two shadows can be seen behind Gagarin's seat, indicating the presence of lighting rigs.

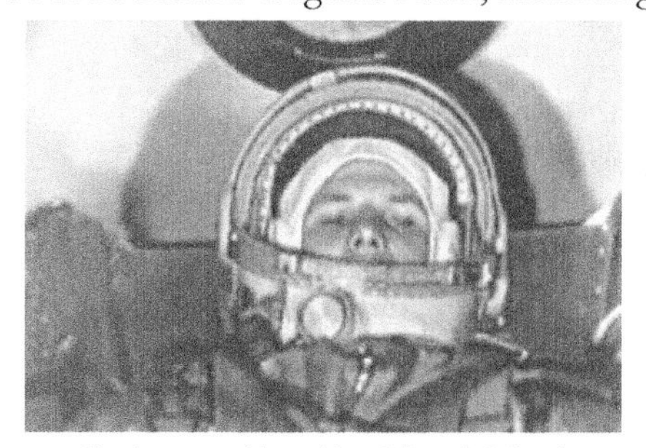

Shadows on either side of Gagarin's head

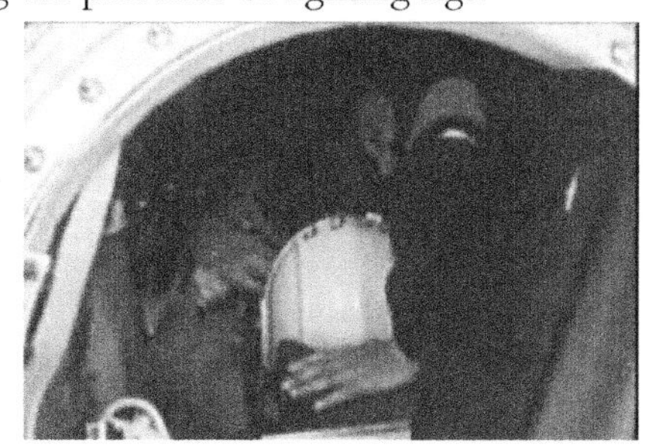

In the Capsule (no lighting seen)

The documentary and the Percy/Bennett book "Dark Moon" also discuss the experiences of a British Journalist, who, it seems, did not believe that Gagarin had flown into space in the way claimed:

> *Lord Bruce Gardyne[392] was, at that time, a journalist with the London Financial Times and until his death expressed grave doubts about the validity of the Gagarin flight.*
>
> *Throughout the spring of 1961, he'd been in the Soviet Union and was in a Moscow "alive" with rumours about this event. From April 6 until his departure on his way to the airport on April 11th, he was informed by his very well-connected minder that according to his sources, the flight had already taken place that morning. The source of this information had been none other than the chairman of the Soviet Committee of Sciences. Bruce Gardyne, travelling in Germany the following day, was then very surprised to hear the Gagarin flight being announced, live, over the East Berlin Airport tannoy system.*
>
> *Back in England, Jock Bruce Gardyne established that NASA initially denied all knowledge of the Gagarin flight on the 12th and this statement was later "corrected" by the White House. But were the Americans in fact fully aware that the Soviets did not launch a craft on that day - April the 12th?*

Was it the case that Gagarin was, as suggested by Percy and Bennett's research, removed from the capsule before it was launched, taken to an aircraft and flown to an area near the landing site and then he parachuted out? Definitely a "simpler scam" than Apollo – but enough to make the Russians "keep mum" about another global deception.

Other Questions Commonly Raised

I have not covered all the possible or even commonly-asked questions about the Apollo Hoax, and neither do I intend to – I leave it for the concerned and interested reader to do their own further research.

That said, I will try to cover some of the questions below. Be aware that lack of definite answers does not affect the other evidence we have covered here so far.

What about the role of facilities like the Parkes Observatory in Australia?

This was told in the 2000 film "The Dish" starring Sam Neill.[393] They relayed Apollo radio communications from space even during the first moon-landing itself! If the Apollo craft wasn't on the moon, there would be nothing to point the dish at, would there? The folks operating the dish would have known!

Well, it seems that before Apollo, *unmanned* craft had been sent to the moon – some with radio transmitters. It is therefore possible that a signal could have been relayed *from* the moon if a transmitter relay had been remotely landed there, or placed in orbit there. It is difficult to know how precisely such a dish would need to be pointed to determine whether a craft was just *in orbit* around the moon or whether it had *landed* there. Or, perhaps the Parkes Observatory itself became part of the hoax too... The TETR satellites were sent up as part of the early Apollo missions – these simulated radio communications coming from Space.[394] (These were part of what Bill Kaysing called the "Apollo Simulation Project," which he says was run from Nevada.) Did the TETR satellites remain in use throughout the hoaxed missions?

A similar situation applies to those who had advanced radio receivers and could apparently listen to ongoing conversations during the actual mission i.e. they could have just been receiving relayed signals.

What were all the people in Mission Control Actually Watching?

Perhaps they were watching a simulation – after all, the missions had to be simulated before being done "for real". How many people would be able to tell the difference? Each of them was only looking at a readout of one or two systems. Again, Bill Kaysing discusses "The Apollo Simulation" project in his "We Never Went..." book.

How could that many people – thousands who worked at NASA – be fooled?

As we have seen, getting people into orbit is, in of itself, difficult to do – requiring complex systems of rocketry and life-support. The bigger the cargo, the more difficult it becomes. Just to achieve this takes an enormous effort. Once the capsule is "out of sight," then there was, really, very little verifiable information from outside NASA to say where it was. Scientists who had no clue what was actually going on would not even begin to suspect a hoax. Most people at NASA would never have considered they were involved in something that was being hoaxed – they would never question what was on the screen in front of them. Even 50 years later, many people still have no idea about any of the evidence I have included in these chapters and so never question NASA's claims about the moon landings. Again, once a big enough consensus has been created that it becomes a "social norm," all the important hoaxing and cover up work has been done. One only has to study the events of 9/11 witnessed, close up, by tens of thousands of people, to know how sophisticated methods have been employed to create mass-deception.

Reasons for the Hoax

Some people ask "What *possible* reason could there be for *hoaxing* the Apollo Missions? You're *nuts* for suggesting that!" Whatever the reasons were or were not, the weight of evidence shows that large parts of Apollo were hoaxed. But let's suggest a few reasons and see if they are part of a "larger picture" – politically, economically and psychologically.

One (political) reason we can imagine is that those that were committed to supporting Kennedy and his vision would have needed a "get out of jail free" card when they realised that Gus Grissom was right – and they were not technically capable of landing on the moon. So, they went along with the hoax brigade.

There is another reason which is, I would say, implicitly assumed as being the main reason the missions were developed in the first place. For the bankers, and politicians alike, the "victorious" Apollo landing served to fuel the Cold War well. It "proved" that "capitalism was better than communism." It wouldn't matter whether they landed on the moon or not – they just had to convince enough people that they did land. (This is little different to the "string pullers" needing to convince enough people that the Gulf of Tonkin incident really did involve an attack by the Vietnamese[395], so that the Vietnam war could be initiated.)

Another reason could be that the Apollo missions made the USA "look good" – they were a fine public relations antidote to the horrors of the Vietnam War, for example. The publicity for the missions built people's confidence in technology, for example, and probably contributed to people's wider acceptance of it in that era.

Another less-considered reason was that the whole space programme was a way to "soak up" the brilliant minds in science and engineering who wanted to get involved in projects to do with space travel or space sciences etc. All this talent and skill was then absorbed into programmes which would ultimately achieve relatively little – only earth orbit for manned flight (which is all that has been repeated since). It distracts these people from thinking about much more advanced technology – such as we have already discussed – and will do so again in future chapters.

The majority of these gifted individuals would be held hostage to their beliefs – never suspecting that there was a larger, darker agenda being played out. Because they were so bright and gifted, they would then automatically reject any evidence shown to them that they had been deceived (through no fault of their own).

With the vast sums of money "sloshing" around, it was perhaps easier for some of it to be "siphoned off" into Black Programmes, which we have already touched on and we will do so again in later chapters.

Another reason for the hoax was that it was a great experiment in Social Engineering in the USA (and the world). Apollo was in progress as TV sets were being delivered and installed to households around the world. Could the folks running things successfully use "TeeVee Power" to make hundreds of millions of people believe an enormous lie?

On a completely different level, did those who planned the hoax realise that the "high" created by belief in the reality of the landings could, at some time in the future, be "played off" against the "low" created now when people discover it was a hoax. (I see this is kind of a "consciousness energy extraction" process.)

Apollo Conclusions

There are "mucky fingerprints" all over the Apollo record. A trail of fakery, possible murder, lies and deception seems to be apparent. If the astronauts went to the moon, it wasn't atop a Saturn 5 rocket. We don't have enough evidence to conclude whether the named astronauts or other astronauts went to the moon using more advanced technology, but I think the behaviour of most of the astronauts suggests they did not actually go to the moon – rather, they were somehow "made to believe" they did. We seem to have few photos or video sequences where the astronauts are clearly recognisable on the moon. (For example, you probably don't know Jack Schmitt![396]) It doesn't really matter what I write here, because the majority of people still need to defend the hoax, because like the official story of 9/11, uncovering the scale of the deception undermines our ideas about democracy, our "civilised society" and makes us deeply question "who is really in charge of things."

Apollo 17 Astronaut and Geologist Jack Schmitt[397]

19. Bigelow's and Amazon's - Rent-a-Rocket

In this short chapter, we will briefly review the status of commercial efforts to put a person into space.

Bigelow Aerospace

Robert Bigelow is an American Billionaire who started the company named above. From his (the company's) website, we read[398]:

> *Bigelow provides the daily strategic leadership at Bigelow Aerospace in its design, development, and testing of expandable habitat architectures where Bigelow Aerospace employs approximately 150 employees at its Las Vegas facility. Mr. Bigelow has successfully launched two subscale spacecraft called Genesis I & II into orbit as well as the Bigelow Expandable Activity Module (BEAM), which is attached to the Tranquility module of the International Space Station. Moreover, Mr. Bigelow serves as the program manager of the B330 spacecraft – Bigelow Aerospace's main habitation system for LEO and beyond LEO destinations.*

Bigelow, then, is not developing rockets or propulsion systems, so does not really concern us in this book. He is mainly concerned with developing some type of commercial space station – his Genesis and BEAM units represent a kind of inflatable or expandable module, which is easier to lift into orbit.

However, it is worth noting that Bigelow created NIDS – the National Institute for the Discovery of Science. In 2004, this "group" published a report entitled "NIDS Investigations of the Flying Triangle Enigma."[399] Soon after this report was published, Bigelow closed NIDS and started up Bigelow Aerospace. We will discuss Flying Triangles again in later chapters. It is also interesting to note that Bigelow has spoken of his keen interest in the UFO/ET issue on multiple occasions – and this has been mentioned in mainstream reports. This seemed to happen most prominently, in May 2017, when Bigelow was interviewed by CBS reporter Lara Logan.[400]

> *Lara Logan: Do you believe in aliens?*
>
> *Robert Bigelow: I'm absolutely convinced. That's all there is to it.*
>
> *Lara Logan: Do you also believe that UFOs have come to Earth?*
>
> *Robert Bigelow: There has been and is an existing presence, an ET presence. And I spent millions and millions and millions -- I probably spent more as an individual than anybody else in the United States has ever spent on this subject.*
>
> *Lara Logan: Is it risky for you to say in public that you believe in UFOs and aliens?*
>
> *Robert Bigelow: I don't give a damn. I don't care.*
>
> *Lara Logan: You don't worry that some people will say, "Did you hear that guy, he sounds like he's crazy"?*
>
> *Robert Bigelow: I don't care.*
>
> *Lara Logan: Why not?*

Robert Bigelow: It's not gonna make a difference. It's not gonna change reality of what I know.

Lara Logan: Do you imagine that in our space travels we will encounter other forms of intelligent life?

Robert Bigelow: You don't have to go anywhere.

Lara Logan: You can find it here? Where exactly?

Robert Bigelow: It's just like right under people's noses. Oh my gosh. Wow.

This explains an unusual graphic that has been painted on one of his buildings.

Google Street View showing Bigelow's[401] building with an unusual design on the top right.

Branson and Virgin Galactic

Richard Branson formed a company called Virgin Galactic in 1999 and has, since then, spent considerable sums of money to try and produce what is essentially a "space glider." The small craft is piggy-backed on a "normal" aircraft and taken up to as high as the aircraft can go. A small rocket engine on the glider then propels it to about 90,000 feet above the earth's surface. It is then meant to glide back down.

Branson originally projected he would be taking people into "space" as long ago as 2007, but even as I write this, 11 years later, no paying passengers have been taken into space by his company. Indeed, the project has been hampered by at least two fatalities during the development of the hardware.[402] In essence, then, Branson, after 19 years since his first announcement, has not yet achieved what NASA achieved on 5 May 1961 (when Alan Shepherd first made a sub-orbital flight[403]). I do not mean to be-little Branson's efforts here, but it seems to clearly illustrate the enormous technical difficulties that are apparent in working with chemical rocket propulsion for manned space travel.

Richard Branson in 2014 with a model of the glider.

Also, of note is that Branson's initiative is solely for "thrill-seekers" – it's not a scientific initiative and neither is it exploratory.

Elon Musk and Space X – a Re-usable Rocket

Elon Musk is a technology entrepreneur who became wealthy following his investment in ventures such as the PayPal electronic payments service, used initially in eBay transactions, and then even more widely.[404]

Elon Musk with the Space X "Dragon" Capsule.

Early in 2002, he founded the Space X company to try and develop reusable rockets for Space exploration. After 16 years, his company has not been able to develop a system which can carry a person into space.

In his April and May 2018 UK tour, Richard D Hall pointed out, during a series of presentations he made, that Elon Musk made some peculiar statements regarding space-related technologies. Hall showed a clip from an interview between Musk and Jaime Peraire, then head of MIT Aeronautics and Astronautics Department, on 24 October 2014. This interview was the final session of the AeroAstro 1914-2014 Centennial Symposium[405]. Near the 16-minute mark in this interview, the host asked a question about key technologies that would enable a manned or even a colonisation mission to Mars to be made successfully. Musk replied:

> *It would obviously be great to have some sort of fundamental new thing that has never existed before - and like, pushes the boundaries of physics... that would be great.*
>
> *But as far as, you know, the physics that we know today, I actually think we've got the basic ingredients... are there... If you do densified methyl ox rocket... Earth orbit refuelling... so you like load the spacecraft into orbit, you send a bunch of refueling missions to fill up the tanks and you have the Mars colonial fleet, essentially, that gets built up during the time between earth-mars synchronizations, which occur every 26 months... then the fleet all departs at the optimal transfer point...*
>
> *I think like we have...* ***we don't need anything ... we don't need any sort of thing that people don't already know about...*** *I believe we've got the building blocks...*

Hall pointed out how different the statement above would sound if we changed the word "people" to "the public." Thus, the statement would then read: "we don't need anything ... we don't need any sort of thing **that the public don't already know about...**" This would be highly suggestive that Musk has an awareness of some of the technology that this book discusses in later chapters.

Jeff Bezos and Blue Origin

Jeff Bezos is an entrepreneur and now one of the richest people in the world – because he was the founder of the world's biggest retailer – Amazon. An interesting article on Space.com notes his early interest in space travel.[406] In 2000, Bezos formed a company called Blue Origin, which he has continued to invest in. The company has been developing rockets and space vehicles since that time, but it kept a "low profile" for about 10 years. As with Elon Musk's "Space X" venture, Bezos has not yet got to the point where he can put a person in orbit. It seems, based on comments he made in public in 2016, that he realises how difficult it is to put a person into space. He took part in a discussion, held on 14 June 2016, at the Smithsonian Air and Space Museum's Lockheed Martin IMAX Theatre in Washington, DC. This discussion was posted on NASA's YouTube channel.[407]

Bezos sat next to Apollo 11 Actornaut Michael Collins. Around 15 minutes into the interview, he describes his thoughts on the Apollo Programme.

Major General Michael Collins (left), USAF (Ret.) and Jeff Bezos in June 2016

> *If you think back to the kind of the heyday of the 1960s and the Apollo program and all of that excitement...* ***my gut instinct on this is that we, as a civilization - we as humanity - pulled that moon landing way forward, out of sequence*** *from where it actually should have been. It was a gigantic effort, with what is, in many ways -* ***it should have been impossible...*** *and they pulled it off with, you know barely any computational power. They were still using slide rules. They couldn't numerically model in computers a lot of these important processes, like combustion inside a rocket engine - which is still hard today. But we can do it a little bit. They didn't have computational fluid dynamics to rely on. Everything had to be done in a wind tunnel - nothing could be done on a computer. So... I think the reason we've sort of taken a hiatus, maybe in part at least, is because ...* ***we pulled that forward to a time when it should have been impossible*** *and then once it was done, kind of had* ***to wait and let technology catch up you know****...*

My "gut instinct" is that Bezos is not in quite the same position as Elon Musk and he has not considered or not been told about the reality of reactionless propulsion technology (which can affect or control gravity), which is what we have been considering throughout this book.

20. Formerly Secret Space Programmes

In previous chapters, we have discussed elements of the public space programme. We can note that the last alleged (manned) Moon Landing took place in 1972 and the US Space Shuttle Programme ran for 30 years and ended in July 2011. The maximum distance the shuttle reached from the earth was about 360 miles (during STS-82).[408] Unmanned Space Probes have visited all planets in the Solar System.

Slightly less well known is that the USSR had a Shuttle Programme called "Buran" which only ever completed a single unmanned orbital flight in 1988.[409]

In 2015, photographer Ralph Mirebs published a superb set of photos, showing the fate of two of the Buran prototypes.[410]

Two Buran Shuttle prototypes, abandoned in a hangar at the Baikonur Cosmodrome in Kazakhstan. Photo by Ralph Mirebs.

The USAF's Space Shuttle - Vandenberg AFB

In 2004, or thereabouts, I heard a talk by William Hamilton, which was kindly sent to me by a friend in the USA. The talk was given in 1998 to a MUFON LA meeting where he mentioned a "Five-billion dollar launch facility" at Vandenberg AFB. I had never heard about this, but later found a photo showing part of this facility. (A cropped version of this photo is reproduced below).

Air Force Base Space Launch Complex 6 in February 1985.[411] The space shuttle Enterprise, mated to an external tank and solid rocket boosters, is resting on the launch mount.

A few more details about secret Space Shuttle Missions are given in an interesting article on the Air and Space magazine's website:[412]

> *Between 1982 and 1992, NASA launched 11 shuttle flights with classified payloads, honoring a deal that dated to 1969, when the National Reconnaissance Office—an organization so secret its name could not be published at the time—requested certain changes to the design of NASA's new space transportation system.*

The article also states:

> *"NRO requirements drove the shuttle design," says Parker Temple, a historian who served on the policy staff of the secretary of the Air Force and later with the NRO's office within the Central Intelligence Agency. The Air Force signed on to use the shuttle too, and in 1979 started building a launch pad at Vandenberg Air Force Base in northern California for reaching polar orbits.*

The article shows that NASA, the USAF and the National Reconnaissance Office (NRO) all worked together on the Shuttle Programme in the 1970s. As stated in the article, referenced above, the NRO was never talked about – I first heard of this agency in 2003 when I was listening to Steven Greer's Disclosure Project videos.

MOL and Almaz

A 2007 "Nova" PBS Documentary called "Astrospies"[413] tells us about the two secret US and USSR space programmes which were in development/usage in the 1960s and the 1970s.[414] On 10 December 1963, President Lyndon Johnson announced "the development of a Manned Orbiting Laboratory."[415] Crews were trained and hardware was designed and developed.[416] However, the MOL, which was based on hardware developed in the Gemini program, never went into orbit – as the NRO eventually decided they could achieve their objectives using unmanned satellites.

A 1967 conceptual drawing of the Gemini B re-entry capsule separating from the MOL at the end of a mission.[417]

The Russians had an equivalent programme (Almaz), which did get off the ground, although only after several years of delays. According to the PBS documentary, it only ran five missions and only two of those were deemed to be successful.[413] The Almaz programme ended in February 1977.

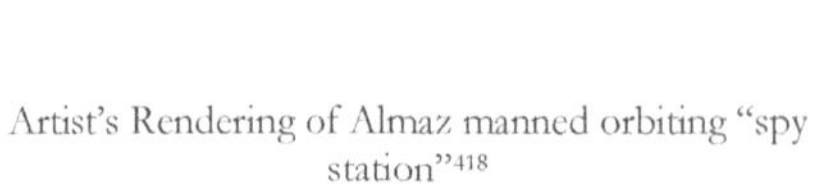

Artist's Rendering of Almaz manned orbiting "spy station"[418]

Surviving Almaz Module, kept near Moscow.[419]

Even in 2007, elements of the Almaz program were still classified.

X37-B Unmanned Space Plane

Although we are mainly concerned with manned space flight in this book, it is worth mentioning the X-37B project[420] – it is noteworthy that, almost from the outset, it has been "no secret that it is secret!" with landings having been reported in the press several times, since its first reported flight in 2009. An article on the Air and Space Magazine Website gives a few more details about the X37-B[421] and suggests the unit is for testing various technologies that are of interest to the USAF, but the craft only seems to operate in fairly low-earth orbit.

A crew tends to the X-37B after it landed at Vandenberg Air Force Base in October 2014, after 674 days in orbit. (Boeing)" from Air and Space Magazine[421].

Part 3

Rumours, Whistleblowers and Sightings

21. Regan, Rich and McKinnon

A number of rumoured secret space programmes have been talked about over the years, even going as far back as alleged landings on the Moon by the Vril Society in 1942.[422] From the research I have done over the years, a fair proportion of which I have written about in this book, I don't give the Nazi/German saucer and space programme stories much credence. The evidence seems sketchy at best. In this chapter, I will cover some of the more interesting rumours that are "out there." Wherever possible, I suggest you try and listen to the interviews/recordings of any whistleblowers mentioned, which you can find using the references and links I have included.

Ronald Regan and the "Secret Shuttles"

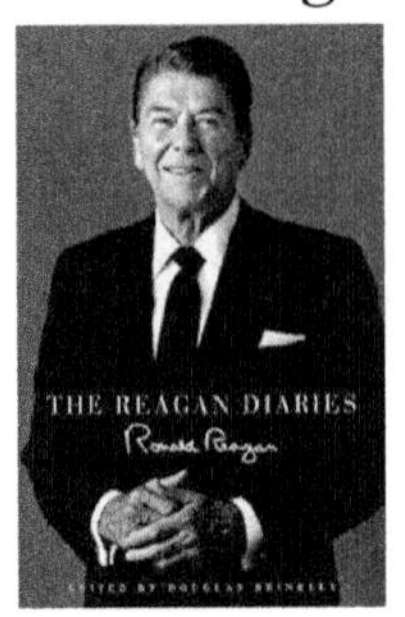

President Ronald Reagan alluded to a highly classified space fleet in the 11 June 1985 entry in his diaries. This is mentioned in the Reagan Diaries book, first published in 2007[423] where he apparently wrote[424]:

> *Lunch with 5 top space scientists. It was fascinating. Space truly is the last frontier and some of the developments there in astronomy etc. are like science fiction, except they are real. I learned that our shuttle capacity is such that we could orbit 300 people.*

A document in the Reagan Library shows, on Page 24, who these "5 top space scientists" were:[425]

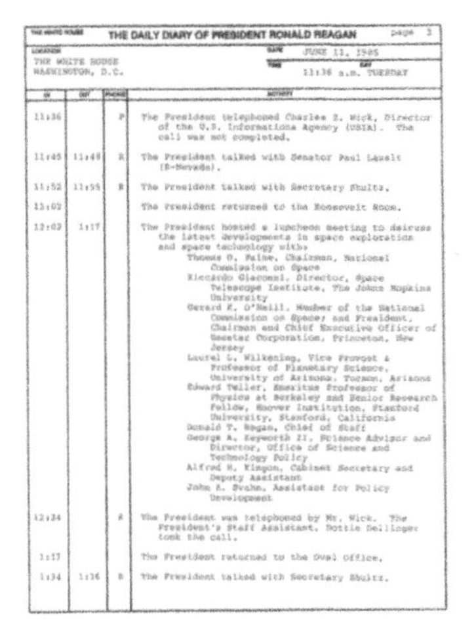

THE DAILY DIARY OF PRESIDENT RONALD REAGAN — page 3

LOCATION: THE WHITE HOUSE, WASHINGTON, D.C. — DATE: JUNE 11, 1985 — TIME: 11:36 a.m. — DAY: TUESDAY

IN	OUT	PHONE	ACTIVITY
11:36		P	The President telephoned Charles Z. Wick, Director of the U.S. Informations Agency (USIA). The call was not completed.
11:45	11:48	R	The President talked with Senator Paul Laxalt (R-Nevada).
11:52	11:55	R	The President talked with Secretary Shultz.
11:02			The President returned to the Roosevelt Room.
12:03	1:17		The President hosted a luncheon meeting to dsicuss the latest developments in space exploration and space technology with: Thomas O. Paine, Chairman, National Commission on Space; Riccardo Giacconi, Director, Space Telescope Institute, The Johns Hopkins University; Gerard K. O'Neill, Member of the National Commission on Space; and President, Chairman and Chief Executive Officer of Geostar Corporation, Princeton, New Jersey; Laurel L. Wilkening, Vice Provost & Professor of Planetary Science, University of Arizona, Tucson, Arizona; Edward Teller, Emeritus Professor of Physics at Berkeley and Senior Research Fellow, Hoover Institution, Stanford University, Stanford, California; Donald T. Regan, Chief of Staff; George A. Keyworth II, Science Advisor and Director, Office of Science and Technology Policy; Alfred H. Kingon, Cabinet Secretary and Deputy Assistant; John A. Svahn, Assistant for Policy Development
12:34		R	The President was telephoned by Mr. Wick. The President's Staff Assistant, Dottie Dellinger took the call.
1:17			The President returned to the Oval Office.
1:34	1:36	R	The President talked with Secretary Shultz.

The President hosted a luncheon meeting to dsicuss
the latest developments in space exploration
and space technology with:
Thomas O. Paine, Chairman, National Commission on Space
Riccardo Giacconi, Director, Space Telescope Institute, The Johns Hopkins University
Gerard K. O'Neill, Member of the National Commission on Space; and President, Chairman and Chief Executive Officer of Geostar Corporation, Princeton, New Jersey
Laurel L. Wilkening, Vice Provost & Professor of Planetary Science, University of Arizona, Tucson, Arizona
Edward Teller, Emeritus Professor of Physics at Berkeley and Senior Research Fellow, Hoover Institution, Stanford University, Stanford, California
Donald T. Regan, Chief of Staff
George A. Keyworth II, Science Advisor and Director, Office of Science and Technology Policy
Alfred H. Kingon, Cabinet Secretary and Deputy Assistant
John A. Svahn, Assistant for Policy Development

How interesting to see Edward Teller's name on the list! (See the discussion in chapter 10 of Bob Lazar's story).

Ben Rich "Taking ET Home" in 1993

One of the more interesting rumours revolves around fairly well documented remarks made in (and after) a lecture by Ben Rich former CEO of Lockheed's

Skunkworks from 1975 to 1991. You can watch an interesting interview between Ed Bradley and Ben Rich which aired on CBS in 1995[426], shortly after Rich's death. Although the interview seems to me to be something of a glorification of Lockheed's F-117 Stealth Bomber, it does give some insight into aspects of how black programmes are run.

Ben Rich with the Lockheed F-117 Stealth Bomber.[427]

As we alluded to in chapter 10, when we discussed Boyd Bushman's statements, Lockheed Skunkworks is said to be involved in developing even more advanced aircraft – with some form of antigravity or hybrid propulsion system.

An article by Mike Doughty posted on 30 June 2013 on the "UFO Disclosure" Website[427] discusses what Ben Rich said some 20 years earlier.

According to people who attended a lecture by Ben Rich about his 40 years at the Lockheed Skunk Works, given at an Alumni meeting of the School of Engineering at UCLA on 23 March 1993, Rich showed a slide of a Black Disc zipping off into space to end his talk. Rich then said:

> *We now have the technology to take ET home.*

After the talk Rich took questions. Someone asked where the technology came from. Rich mentioned that during research relating to the technology, they had found that "there was an error in the equations and that they had found the error and now knew how to travel to the stars – within one's lifetime."

Jan Harzan, Director of MUFON asked Rich some further questions and followed him out to the car park (parking lot!)[428] Harzan said that he had interest in this technology from a personal standpoint and he wanted to know how it worked. Rich asked Harzan the question, "How does ESP work?" Harzan replied, "I don't know … all points in space time are connected?" and Rich replied, "That's how it works."

I find this answer most interesting, because it ties in with what Wilbert Smith wrote in his 1950 Top Secret Memo regarding UFOs and "mental phenomena" that we discussed in chapter 4.

William Hamilton and a Colonel

As I wrote in an earlier book "Secrets of the Solar System," at a Los Angeles MUFON meeting on 17 June 1998, UFO researcher William Hamilton[429] discussed meeting with an anonymous Colonel who had a degree in Plasma Physics and he had worked as an aerospace engineer. He was also a consultant to NASA and to people at the Palos Verdes nuclear power plant, 50 miles west of Phoenix Arizona. The former Colonel claimed to have worked at several underground facilities in the USA (having ridden to them on a Mag-lev train). One entrance to the system was at the White Sands proving grounds. The Colonel told Hamilton that there was a base on the moon. Hamilton stated:

> *The other thing he said was there was a base on the Moon.*
>
> *I find that quite astounding. I mean I have heard astounding things. Not only from this Colonel, Okay, but also probably the highest rank was a retired three-star lieutenant general in the Air Force. And he knew another guy that I was in contact with and this other guy claims that he worked seven years on the US antigravity program. If a fraction of what he told me - and I've got a tape-recorded - is true - **whoa what a sham the space program is**. Because there's **another space program going on - a secret space program** and that's what I was attempting to find. In fact, you don't have to go too far to find that there is some secret space programs going on because they had their own shuttles and their own astronaut crews right out here at Vandenberg Air Force Base, unbeknownst to everybody. And they scuttled a five-billion-dollar facility - that was housed there, at one time, that was back in the 80s. **But the kind of space program he was talking about is using a spacecraft that can leave the bounds of our atmosphere and travel to the planet Mars at 0.8c – eight-tenths the speed of light...***
>
> *He claimed to work shoulder to shoulder with the race of human-like beings just like us - I forgot where he said they were from - if it really matters... But it was these people who are guiding him through the... he wasn't back engineering, so much, as he was trying to explain the physics of what the engineers were doing. He was investigating the physics of it - with the assistance of these beings - so he claimed. And he worked in underground facilities up at the Nevada test site, China Lake, Edwards Air Force Base and maybe other places.*

Karl Wolfe and a Moon Base

Another credible person who had talked about evidence of there being a Moon base was Sergeant Karl Wolfe, who was interviewed in September 2000 by Dr Steven Greer as part of Greer's Disclosure Project initiative.[430] I have quoted from Greer's Executive Briefing Document below[431].

Karl Wolfe was in the Air Force for 4½ years beginning in January 1964. He had a top secret crypto clearance and worked with the tactical air command at Langley AFB in Virginia. While working at a NSA facility he was shown photographs taken by the Lunar

Orbiter of the moon that showed detailed artificial structures. These photos were taken prior to the Apollo landing in 1969.

Again, we see the promotion of the Apollo Hoax by Greer, which we will mention again, later. But, let us now consider Wolfe's testimony.

> *So I was asked to go over to this facility on Langley Air Force Base, where the NSA was bringing in the information from the lunar orbiter ... As I walked in, there were people from other countries, a lot of foreign people from other countries in civilian clothes, with interpreters with them, with security badges hanging around their neck. ...*
>
> *We walked over to one side of the lab and he said, by the way, we've discovered a base on the backside of the moon. I said, whose? What do you mean, whose? He said, yes, we've discovered a base on the backside of the moon. And at that point I became frightened and I was a little terrified, thinking to myself that if anybody walks in the room now, I know we're in jeopardy, we're in trouble, because he shouldn't be giving me this information.*
>
> *I was fascinated by it, but I also knew that he was overstepping a boundary that he shouldn't. Then he pulled out one of these mosaics and showed this base on the moon, which had geometric shapes- there were towers, there were spherical buildings, there were very tall towers and things that looked somewhat like radar dishes but they were large structures... This fellow and I were the same rank, I think he was very distressed. He had the same pallor and demeanor as the scientists outside the room, they were just as concerned as he was. And he needed to discuss it with somebody... Some of the structures are half a mile in size. So they're huge structures. And they're all different sized structures in different photographs. Some of the shapes, as I said, were - some of the buildings were very tall, thin structures. I don't know how tall they were but they must be very tall. They were angular shots with shadows. There were spherical and domed buildings that were very large. They stood out very clearly, they were large objects… I didn't want to look at it any longer than that, because I felt that my life was in jeopardy. Do you understand what I'm saying? … I knew the young fellow who was sharing this was really, really overstepping his bounds at that point. I felt that he just needed somebody to talk to. He hadn't discussed it, couldn't discuss it, and he wasn't doing it for any ulterior motive other than the fact that I think he had the weight of this thing on him and it was distressing to him...*

This testimony is included, because Wolfe stated that he would swear an oath before congress that his story was true.

Gary McKinnon and Non-Terrestrial Officers

Gary McKinnon who might be called a "geek" or a "computer nerd," (not unlike me really!) became well-known in 2006 when he was threatened with extradition to stand trial in the US for "Computer Misuse" offences that he had committed in approximately 2002. At that time, he had successfully logged in to various networks in the USA and elsewhere, using a technique of looking for blank/empty passwords, having scanned various IP addresses. His motivation for this activity was that he was looking for information about free energy technologies.

He wasn't comfortable with being labelled a "hacker" – because all he had done was look for "open doors" in computer networks and "gone through" them. He reported that he had accessed NASA and the US Military's (Army, Airforce Navy and DoD) computer networks in various locations – using a piece of software called "Remotely Anywhere." He did this using dial-up communication (a 56K modem) and a Windows PC.

He was "caught" by UK Police and initially, before the US authorities got involved, was told that he would probably just have to do some community service as punishment for the crime he committed. However, after the US authorities got involved, he was threatened with extradition to the US, where he would stand trial and if found guilty there, he would face up to 70 years in jail! A campaign was launched to try to ensure that he was tried in the UK for the crime he admitted he was guilty of![432] The campaign lasted about 10 years when then UK Foreign Secretary, Theresa May, blocked the extradition. Following this, Gary's mother, Janis Sharp, wrote a book about the case.[433]

There are a number of interesting aspects to McKinnon's case, a few of which are relevant to the topics we have been covering in this book.

In 2006, Gary was interviewed in the UK by the BBC, Channel 4, Channel 5, ITV, and BBC World Service. In an interview for the BBC "Click" computing/IT news magazine programme[434], McKinnon stated[435]:

> *I was in search of suppressed technology, laughingly referred to as UFO technology. I think it's the biggest kept secret in the world because of its comic value, but it's a very important thing.*
>
> *Old-age pensioners can't pay their fuel bills, countries are invaded to award oil contracts to the West, and meanwhile secretive parts of the secret government are sitting on suppressed technology for free energy.*

McKinnon repeated his statements many times. For example, in an interview he gave to Kerry Cassidy in June 2006, who was, at the time, building up a website she called Project Camelot[436] (which has turned into a large mixed-bag of interviews and material), McKinnon repeated:

GM: I passionately believe that we should all have this technology. And not so much, obviously, if you could confirm the existence of extra-terrestrials and their contact with us, then that would be good. But to me it was more important to have this free energy system.

This point was talked about in the media, to some degree, in 2006, and so was the UFO/ET issue that he said he'd found information about. However, over time, these issues no longer formed part of the public discussion of the McKinnon case. The discussion became about political issues and Gary's health and the risks to it if he was extradited to the USA.

One thing I found interesting was that a fellow called Paul McNulty wrote McKinnon's indictment[437]. This was significant because McNulty also wrote the indictment for Zacarias Moussaoui[438] – the alleged "twentieth 9/11 hijacker."[439] One may presume that this was because McNulty had experience in dealing with cases of "terrorism" (or the requirement to invent terrorists or acts of terrorism to falsely accuse people of.)

Now, to return to the reason why McKinnon's case grabbed quite a bit of attention in the alternative knowledge community, we will discuss comments he made about finding a document relating to "non-terrestrial officers." Below, I will quote portions of Kerry Cassidy's interview with McKinnon (it seems that Gary, for whatever reason, did not discuss this in the later years of his case.)

KC: Right. So, can you tell me what else you found? Because I know you have some information in regard to Non-Terrestrial Officers. Is that right?

GM: Yeah. There was an Excel spreadsheet and the title was "Non-Terrestrial Officers," and it had names, ranks... it wasn't a long list; it didn't fill the whole screen, I don't think... 20, maybe 30 [names]... Definitely ranks, but nothing to say Army captain, or Navy captain, or US Air Force captain.

KC: Ah, so the designation wasn't there as far as which organization they worked for?

GM: Yeah. I mean, that was the title ***"Non-Terrestrial Officers"****, and obviously it's not little green men. So I was thinking: What force is this? And that phrase is nowhere to be found on the web or in official Army documentation or anything. And the other thing was a list of ship-to-ship and fleet-to-fleet transfers - and bear in mind fleet-to-fleet, that means multiple ships - movement of materials. And these ships weren't, you know, US Navy ships. Again, I don't remember any of the names, but I remember at the time looking and trying to match up the names, and there wasn't anything that matched.*

KC: So, now, this theoretically would have been pretty top secret information if indeed non-terrestrial is what it sounds to be, which is off-world, right?

GM: Yeah. I mean, I gleaned from that information... What I surmised is that an off-planet Space Marines is being formed. And if you actually look at DARPA, the Defense Advanced Research Projects Agency, literature at the moment and in the last few years, a lot of government and space command stuff is all about space dominance. It is really, you know, the final frontier. Yeah, so I think it's natural for them to want to control space and to be developing a space-going force in secret. But I think most likely using technology reverse-engineered from ETs....And also to be cloaked, otherwise how many other governments would see this going on.

> *KC: So, was this NASA? ...*
>
> *GM:... I remember thinking at the time that this must be NASA and the Navy, you know, or secret parts thereof. So it was either the Navy or NASA. I really think it was most likely the Navy. But I'm not entirely sure.*

A number of people suggested that what Gary had discovered was somebody's log or strategy file from a computer game – some type of trading game or "command and control" type game – such as "Homeworld," which was out in the early 2000s, as I recall.

Gary also described finding airbrushed or filtered or processed images in NASA's Building 8.

> *And having been all over other NASA installations already - I assumed the blank password scanning method will work the same at Johnson Space Centre - and it did. Once I was in there, I used various network commands to strip out the machines that were in Building 8. And I got on to those. And the very first one I was on literally had what she said. I can't remember if it was "Filtered" and "Raw", Processed" and "Unprocessed," but there were definitely folders whereby there was a transformation in the data taking place between one and the other.*
>
> *And what I saw, or was hoping to see, was what she was describing as a saucer, very definite imagery. And what instead I saw I assume was the Earth. This was in shades of grey. You had the Earth's hemisphere taking up about 2/3 of the screen and then halfway between the top of the hemisphere and the bottom of the picture there was a classic sort of cigar-shaped object, but with golf-ball domes, geodesic domes, above, below, and this side [gesturing to the right], and I assume the other side as well. It had very slightly flattened cigar ends. No seams. No rivets. No telemetry antennae or anything like that. It looked... it just had a feeling of not being man-made. There was none of the signs of human manufacturing.*

As several others have already commented, many years ago, what Gary describes sounds very much related to what former NASA Contractor Donna Hare experienced in 1970/1971 when she reported that she was told that some photos were airbrushed to remove UFOs.[440]

Meeting Gary in London, Tues 30 May 2006

Anthony Beckett, a friend who shares a common interest in the topics discussed in this and my other books arranged a meeting with Gary McKinnon on the date given above and so I was able to personally hear Gary give the same account of what he'd found. I have no doubt he had considerable expertise in the prevailing networking and other digital technologies that were in use then (and now). The meeting was arranged in order for former MI5 Officer David Shayler to offer Gary some advice on how to treat the media and other bodies he would have to interact with to prevent him from being extradited.

At the meeting, Gary confirmed that he had *confessed* to a crime, but he had not actually been prosecuted. This was actually one of the fundamental issues which caused Gary and his mother enormous angst over many years.[441]

Gary explained that he understood that his case had been "passed" to the US Office of Naval Intelligence (ONI) and that they were (illegally) in possession of his hard drive, which had been handed over to them by the UK police.

Since his case was dropped, he has given at least one interview - to Richard D Hall.[442] It is worth watching this interview to get more details about Gary's case and how it was presented in the media.

The Mars Records?

I was unsure whether to include a discussion of this case, here, because there is very limited evidence to support it. A long account of a man's clandestine posting to Mars was published by a lady called Stefanie Relfe in year 2000. Relfe has posted a detailed record of recalled experiences – which have been obtained from her husband Michael Relfe through kinesiology techniques. (This technique has been criticised because it uses a device similar to that used by the Scientologists. There are many references to Christ, Jesus etc at the start of the accounts.) Also, Relfe released some photos of the scars of the person who said he was involved. On the website Stefanie Relfe writes:

> *In December 1996 I began doing one-on-one Clearing sessions in Australia with an American man in his early 40s named Michael. I am an Australian. I had known Michael for a year prior to this. His life prior to 1995 had been spent in the USA.*
>
> *During these sessions we came to discover a vast number of astounding things that he had done, and which had been done to him, of which he had been totally unaware. These things included having worked on Mars for the US government, remote viewing, time travel, age recession by 20 years, being given false memories to cover his time on Mars, abductions and other experiences.*
>
> *Michael is a successful professional working in the information technology industry. He previously worked in the U.S. Navy. He is of sound mind and body, is highly intelligent and healthy. He has never taken recreational drugs or alcohol. Before the first clearing session, one would have considered that he had lead a perfectly normal life.*
>
> *The first clearing session was carried out because Michael had memories of seeing spaceships as a boy, and he wanted to determine whether these memories were real, or a dream, or if they were something else entirely. In fact, the sessions uncovered a whole range of suppressed memories.*

If you wish to study this case further, you can download the entire books, containing the accounts, free from their website[443].

The accounts are clear and the reason why the original account "came out" in the first place also seems plausible. These people are not asking for money or proclaiming how good they are at public speaking, for example. Neither have they embellished their claims over time, nor do they ask you to join a group - or do anything in particular, for that matter. This case has not received much attention – neither in mainstream media nor in the alternative knowledge community. Do I believe the whole story? Not necessarily, all I am saying is that if you were to pick one of the "rumoured secret space programme" stories to read in depth, this could be it. There are some parallels to the Andrew Basiago "Project Pegasus" story – discussed in chapter 28, which came out later. Is it the case that Basiago's story was "put out" as a muddle up for Michael Relfe's story and other evidence, such as I discussed in the chapters about Mars in "Secrets in the Solar System: Gatekeepers on Earth."[444]?

22. Mark McCandlish, the ARV / Fluxliner, Maslin Beach and The Dart

Mark McCandlish at the 2001 Disclosure Project Press Conference.[445]

Mark McCandlish is a gifted aerospace illustrator who worked for many US Defence Companies such as General Dynamics, Lockheed, Northrop, McDonald-Douglas, Boeing, and Rockwell International, mainly doing conceptual artwork. In his Disclosure Project witness testimony,[446] he spoke of a UFO sighting he had in 1967 at Westover Airforce Base Massachusetts. In the same testimony, he discussed an account told to him by a friend called Brad Sorensen. It is worth listening to or reading McCandlish's testimony[7], which contains a lot of details (some of which can be easily verified) and it does appear to tie in with aspects of what we have discussed earlier in this book. (i.e. as we shall see, some of the details he describes about the craft he "got wind of" seem to be characteristic of what you would find in an electrogravitic device).

The Airshow

McCandlish recounts that his friend Brad Sorensen (who he later became more reluctant to mention, for whatever reason) went to an "invite only" Airshow at Norton Air Force Base in Southern California, which took place on 12 November 1988. McCandlish had planned to go there with Sorensen as Sorensen wanted him to meet a contact with a view to them getting some more illustration work through this contact. However, due to McCandlish being asked to do some work at the last minute for the magazine Popular Science, over that same November weekend, he dropped out and apologised to Sorensen. A week after the Airshow, McCandlish called Sorensen up, to ask him how the Airshow was. Sorensen reported that the person who he met there (whom McCandlish thinks was probably Frank Carlucci – who was the US Defence Secretary from 1981 to 1983 under Reagan[447]) then authorised a flight up to Palmdale during one of the fly-bys at the Airshow. They visited the Lockheed Skunkworks there and Sorensen's companion said, "There are a lot of things in here that I didn't expect they were going to have on display - stuff you probably shouldn't be seeing. So, don't talk to anybody, don't ask any questions, just keep your mouth shut, smile and nod, but don't say anything - just enjoy the show. We're going to get out of here as soon as we can." McCandlish states in his Disclosure Testimony what Sorensen related to him about other things in the hangar they had entered:

> *Behind these curtains was another big area, and inside this area they had all the lights turned off; so, they go in and they turn the lights on, and here are three flying saucers floating off the floor - no cables suspended from the ceiling holding them up, no landing gear underneath - just floating, hovering above the floor. They had little exhibits with a videotape running, showing the smallest of the three vehicles sitting out in the desert, presumably over a dry lakebed - someplace like Area 51. It showed this vehicle making three little quick, hopping motions; then [it] accelerated straight up and out of sight, completely disappearing from view in just a couple of seconds - no sound, no sonic boom - nothing.*
>
> *They had a cut-away illustration… that showed what the internal components of this vehicle were, and they had some of the panels taken off so you could actually look in and see oxygen tanks and a little robotic arm that could extend out from the side of the vehicle for collecting samples and things. So, obviously, this is a vehicle that not only is capable of flying around through the atmosphere, but it's also capable of going out to space and collecting samples, and it's using a type of propulsion system that doesn't make any noise. As far as he could see, it had no moving parts and didn't have any exhaust gases or fuel to be expended - it was just there hovering… he told me how he had seen these three flying saucers in this hangar at Norton Air Force Base on November 12, 1988 - it was a Saturday. He said that the smallest was somewhat bell-shaped. They were all identical in shape and proportion, except that there were three different sizes. The smallest, at its widest part, was flat on the bottom, somewhat bell-shaped, and had a dome or a half of a sphere on top. The sides were sloping at about a 35-degree angle from pure vertical.*
>
> *The panels that were around the skirt had been removed, so he could see one of these big oxygen tanks inside. He was very specific in describing the oxygen tanks as being about 16 to 18 inches in diameter, about 6 feet long, and they were all radially-oriented, like the spokes of a wheel. This dome that was visible on the top was actually the upper half of a big sphere-shaped crew compartment that was in the middle of the vehicle, and around the middle of this vehicle was actually a large plastic casting that had this big set of copper coils in it. He said it was about 18 inches wide at the top, and about 8 to 9 inches thick. It had maybe 15 to 20 stacked layers of copper coils inside of it.*
>
> *On the inside of the crew compartment was a big column that ran down through the middle, and there were four ejection seats mounted back-to-back on the upper half of this column. Then, right in the middle of the column, was a large flywheel of some kind.*
>
> *Well, this craft was what they called the Alien Reproduction Vehicle; it was also nicknamed the Flux Liner. This antigravity propulsion system - this flying saucer - was one of three that were in this hangar at Norton Air Force Base…. The first one - the smallest, the one that was partially taken apart, the one that*

> *was shown in the video that was running in this hangar November 12, 1988 at Norton Air Force Base - was about 24 feet in diameter at its widest part, right at the base. The next biggest one was 60 feet in diameter at the base.*

As you can see, this is a very detailed account – and you can read even more detail about the craft in McCandlish's Disclosure Testimony. Of course, McCandlish, too realises that what he describes does sound as if it is the result of further research into the electrogravitics experiments of Thomas Townsend Brown, that we discussed in chapter 3.

ARV Schematic and Operation

Based on the details of Sorensen's account, McCandlish produced a very detailed schematic diagram of the craft, which has been distributed widely on the Web since 2001. A version is reproduced below, but most of the detail is unreadable, as the original is drawn on a large poster.

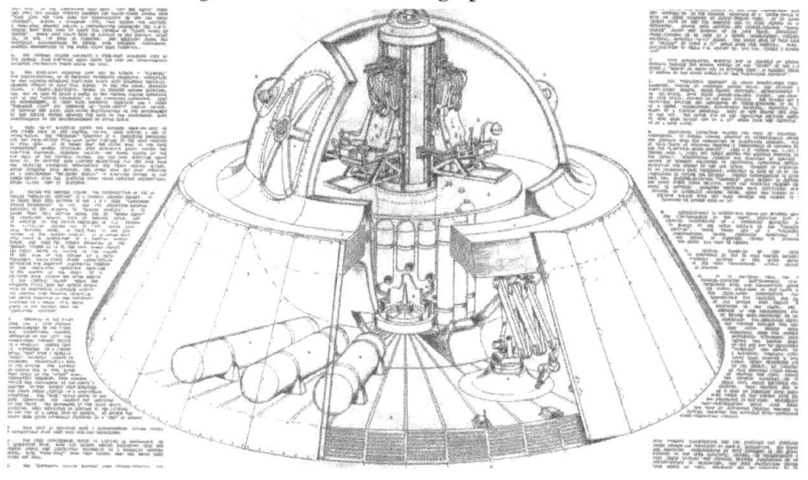

Schematic of ARV by Mark McCandlish[448]

In his testimony, McCandlish gives some intriguing detail:

> *Brad said that in this exhibit at Norton Air Force Base, a three star general said that these vehicles were capable of doing light speed or better. Oh, by the way, the largest of these vehicles was about 120 to 130 feet in diameter. I mean, that's massive when you think about it - it's just huge….*
>
> *Brad maintained that inside this big vacuum chamber in the central column that's inside everything else - inside the flywheel, inside the secondary coils of the* ***Tesla coil, inside the crew compartment - there is mercury vapor. Mercury vapor will conduct electricity, but it produces all kinds of ionic effects.*** *These little molecules of mercury become charged in unusual ways, and if you fire a tremendous amount of electricity through mercury vapor that's in a partial vacuum, there is something special, something unusual that happens in that process.*

Some years later, a 3D rendering was produced.

3D Rendering of ARV/Fluxliner.

McCandlish stated he has talked to Tom Bearden about the technology and Bearden suggested that, because they had this classified technology, NASA's space programmes had all been cut back and the "really advanced stuff" was kept exclusively for use by the Intelligence Agencies such as the NSA. Bearden even went so far as to claim these vehicles could travel through the solar system in a matter of minutes.

McCandlish also related that he had found other witnesses (who he names) who had, at one time or another, seen this craft, or something very similar. He also felt that an object that had been photographed (shown with his picture at the beginning of this chapter) was one of the ARV's in flight.

(McCandlish also stated that he has spoken to… surprise, surprise – Dr Harold Puthoff. At this point we can note that Puthoff also gave "testimony" in Greer's Disclosure project, though what is included in the Executive Briefing document is quite brief - less than 400 words – and it is really just a set of non-committal statements, not disclosing any of his direct experiences. He basically only acknowledges that he is interested in topics such as those discussed in this book and that there are anomalous phenomena reported all the time. He does not really disclose anything about black programmes.

Maslin Beach 1993 and ARV – A Connection?

In approximately 2002, some photos appeared on the internet which were taken by a World War II veteran, Eric Thomason on 10 March 1993, near Maslin Beach, South Australia. I was made aware of these photographs by David Cayton who, in turn, was made aware of them by Robert Hulse, who had posted them on the UFO Casebook forum[449]. Hulse even travelled to Australia in 2006 to meet Eric Thomason and get more information about the photographs. I discussed all these things in an interview I recorded with David

Cayton in December 2016.[450] Below, I include Eric Thomason's UFO report form, along with most of his sighting report (edited for brevity).

I have included the photos here because the shape of the craft is somewhat reminiscent of the ARV described by McCandlish – particularly once you see it with the legs retracted. However, the top part of the craft does not have a hemispherical dome.

U.F.O. REPORT FORM

This form has been designed to assist us in the interpretation of the phenomena observed by yourself. Your assistance in completing and returning it would be appreciated.

U.F.O. International Research
Australian Flying Saucer Research Society
Box 2004, G.P.O. Adelaide, S. Aust. 5001 Telephone: 272 3131

File #: ....
Conclusion: ....

SIGHTING INFORMATION

Sightee surname: THOMASON .... Christian name & initial: ERIC P.
Sightee address: 39, GULF PARADE MASLIN BEACH Post Code: 5170
Sightee age: 69 .. Occupation: RETIRED .....

SIGHTING DATA

DATE of sighting: 10th .. day of March .. 19 93 ....
LOCATION of Sighting: MASLIN BEACH (OCHRE POINT CLIFF) at 200feet
Number of witnesses if any: NONE. Day of week(Mon/Tue etc): WEDNESDAY .
SIGHTING STARTED —(hrs/mins-am:pm): 6 A.M. .. Duration (mins): APPROX. 2 MINS .
Height of cloud cover: 2,000 ft ..... Percentage of cloud: 10 PER CENT .
Was vehicle involved:(yes/no), if yes,how: .... NO ....
.... Type of vehicle: ....
Wind speed(kms): 20km/h Direction: WEST .... Blustery/constant ✓ etc.
.... 14°C (57°F) ... Temperature:(tick) hot...warm...cold. ✓
DID OBJECT LAND: NO ... Landscape: hills ✓ .. flat...sloping...undulating...
Ground cover: not known...stone...sand...grass ✓ .. small/large trees...
Direction object travelling when appeared EAST .. when disappeared: VERTICALLY

OBJECT/S INFORMATION

NUMBER of objects: TWO .. COLOUR: LIGHT GREY .. SHAPE: CIRCULAR ...
Apparent SIZE (circle): 30 m ...distant plane...moon...several moons...
Height above ground in degrees: 30° at X .. NOISE (yes/no)... NO ......
(tick) loud..whirring..swish..soft..whistle..screaming..bang..muffled..
Speed (tick): fast..slow..very fast..stationary..or any other movement..
SLOW .... Estimated DISTANCE: 2 Kms (first sighting)
Texture (tick: solid..foggy..light..black..many colours.... LIGHT GREY
Did object (tick): zig-zag..falling-leaf..or any other movement... NO ..
.... Could you see into object (yes/no): ..... YES ..
Anything noted inside: LIGHTS .. Were photographs taken (yes/no): ... YES
How did object disappear: .... VERTICALLY ....

PLEASE TURN OVER

Eric Thomason's Sighting Report

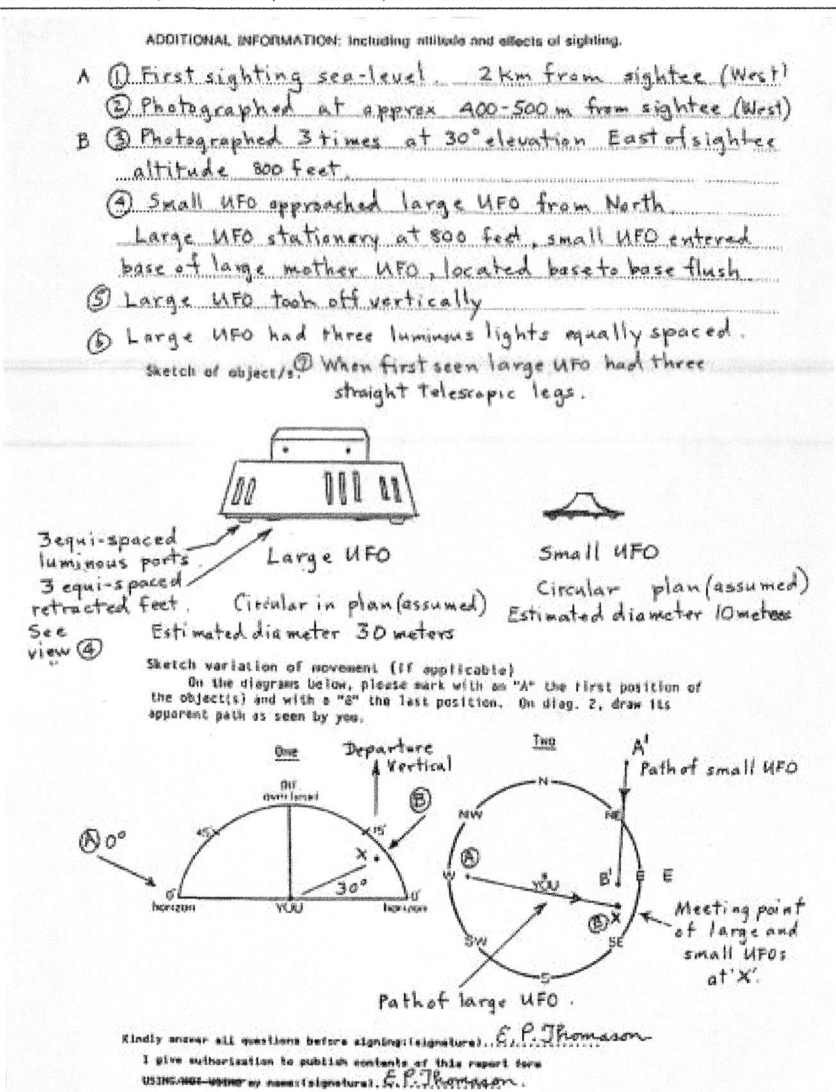

ADDITIONAL INFORMATION: including altitude and effects of sighting.

A ① First sighting sea-level. 2 Km from sightee (West)
② Photographed at approx 400-500 m from sightee (West)
B ③ Photographed 3 times at 30° elevation East of sightee altitude 800 feet.
④ Small UFO approached large UFO from North. Large UFO stationery at 800 feet, small UFO entered base of large mother UFO, located base to base flush.
⑤ Large UFO took off vertically
⑥ Large UFO had three luminous lights equally spaced.

Sketch of object/s. ⑦ When first seen large UFO had three straight telescopic legs.

Sketch variation of movement (if applicable)
On the diagrams below, please mark with an "A" the first position of the object(s) and with a "B" the last position. On diag. 2, draw its apparent path as seen by you.

Kindly answer all questions before signing:(signature)..E.P. Thomason
I give authorisation to publish contents of this report form USING/~~NOT USING~~ my name:(signature). E.P. Thomason.

Thomason explained in his account:

> *Early in 1993 residents of Maslin Beach, South Australia were notified that Maslin Old Quarry would be rehabilitated as soon as the winter rains ceased. Excavation of sand and clay had commenced in 1928 and by 1990 when work ceased 82,000,000 tons of material had been removed.*
>
> *I had lived opposite the quarry for eleven years in 1993 and frequently walked the dog through there. …*

Thomason then recounts how he had decided to try and take a photograph, so he could enter a World Solar Challenge Photographic Competition (sponsored by ABC-TV) in the "Southern Times – Messenger" weekly newspaper. He wanted to catch a few sunrises and the obvious position to take the photographs from was the top of the cliff on the north face of the quarry.

About 200 metres north of this point is Ochre Point, a favourite launching place for hang glider enthusiasts. The cliff along there is 200 feet above sea level and offers an unobstructed view to the east, south and west over the Gulf of St. Vincent. So for several mornings I set my alarm clock for 5.45 a.m., arose and made my way up to Ochre Point. The result of these efforts was disappointing. Low cloud on the hills obscured the rising sun every time. Maslin Beach is noted for beautiful coloured sunsets over the sea to the west but the sunrise over the land to the east was disappointing.

Wednesday 10 March 1993 was one of the mornings when I arose at 5.45 a.m. collected a warm coat, gloves, stick, camera and my faithful dog from the back garden and set forth for Ochre Point. The stick was in case we encountered any snakes in the long grass in the quarry. Dawn was just lightening the eastern sky as we climbed the hill.

I watched as the light brightened beyond the far hills but there was the usual cloud, it was no good for a snap. The wind was cold, it is always windy up there, so I turned and walked north toward the hang glider launching section of the cliff.

Looking out to sea there was nothing in sight, the sea was only slightly ruffled by the wind. Then I saw a disturbance far out, a white swirl of foam caused by a grey object which looked like the conning tower of a nuclear submarine.

The submarine vehicle rose fully out of the sea leaving a white swirl of foaming water. It started moving towards the shore, rising higher above the sea. It had three legs projecting straight down below it. I remembered that I still had the camera in my left hand with the lanyard round my wrist. I raised the camera, sighted the vehicle in the viewfinder, and snapped one shot.

The vehicle had rotated slowly and the three legs were retracting into the body. I suddenly realised that I was silhouetted against the sunrise in full view. I then went down to a ledge about fifteen feet below the cliff top. There I felt safer and less noticeable. The vehicle passed over slightly to the south of me and stopped over the middle of the quarry.

Then I saw another object coming from the north towards the large vehicle. I took another photograph but the sunrise was full in the viewfinder so I moved to the left to try to get out of the glare, and took another snap. The small vehicle came beneath the large one, hovered a few seconds then moved up into a circular recess in the base of the large vehicle. There were lights around the recess inside and there were also three large luminous lights beneath the large vehicle. I had taken a final photograph when the big vehicle rose straight up and vanished above me. I felt some water falling on and around me.

I crawled on all fours back up the slope to the cliff top and when I looked round there was no trace of the vehicle. There was a broken vapour trail very high up above the clouds. This could have been from a civil jet climbing away after taking off from Adelaide Airport 25 miles to the north. Or maybe an F111 or Hornet fighter from Edinburgh RAAF base forty miles to the north. Following the event, I decided to say nothing to anyone until the reel of film was developed. The three photos taken into the sun could have been overexposed due to sun being in the lens. There was also the possibility that someone else might have seen the event who would mention It. As far as I could see at that time there was no one on the beach and no boat fishing out at sea. Sometimes boats stayed out all night fishing off shore. Again, someone in the jet which left the vapour trail might have seen the vehicle below although its

> *grey colour would blend into the sea. When the film was developed I saw on the first picture that there were two other objects which I did not notice at the time of the event. A black dot in the far distance, half way between the vehicle and the right-hand side of the picture. Also, a peculiar luminous shape of unidentifiable form on the far left above the vehicle. The black dot could have been an aircraft many miles away or It was the second smaller vehicle coming in to rendezvous with the large mothership.*

The photos themselves have been cropped and contrast-enhanced to bring out some detail and they are definitely among the clearest I have seen. People posting on the forum, and elsewhere attacked these photos as "obvious fakes." The arguments often go like this:

> *Sceptic: How come there aren't any clear pictures of these UFOs?*
>
> *Evidence Analyst: Have you, for example, studied the Maslin Beach case – where we have a complete witness account detailing how and why the photos were taken?*
>
> *Sceptic: Maslin Beach? Those pictures are "too good to be true" – they are therefore almost certainly a hoax.*

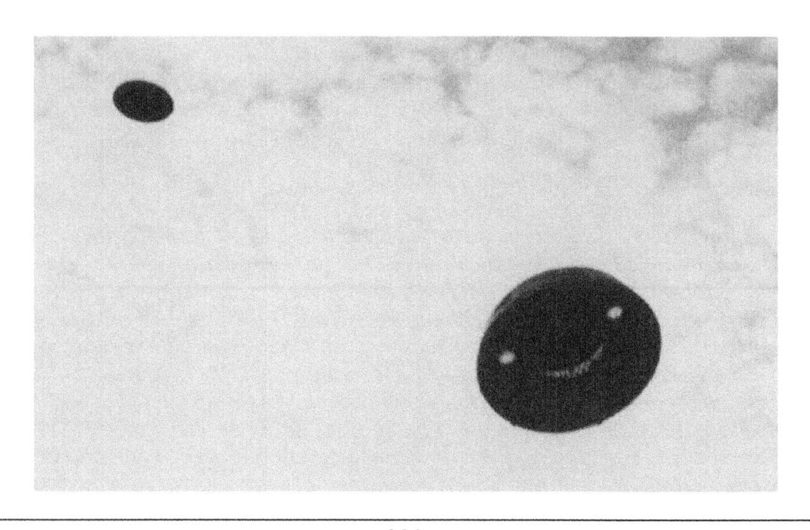

There are a number of details in these photos which I discussed in the interview with David Cayton. Chief among these are the light-coloured streaks seen in the last two photos. Of course, sceptics have said that this proves the photos are fakes – because "these light-coloured lines are clearly a length of fishing line to hold up the model." However, if you study the details of the case,[451] the likelihood of that being the correct explanation drops to zero. From Thomason's account, "far out" though this may sound, it appears that the streaks are some kind of guidance or "tractor" beam. But who would believe *that*?

Robert Hulse and David Cayton have both investigated and studied this case and they concur that the photos are real and Thomason did, indeed, witness a real physical craft, emerging from the sea, docking with a smaller craft and then departing.

Mark McCandlish – The Dart

Another very interesting account that Mark McCandlish has discussed on more than one occasion is that of a secret aircraft which crashed in East Germany in

1989. He discussed this in an interview with David Wilcock and Corey Goode. This interview, aired on 31 May 2017, was transcribed in a posting entitled "Cosmic Disclosure: Zero Point Energy And Advanced Propulsion Technology."[452] I have reproduced a couple of the images from this posting.

McCandlish described how a vehicle was retrieved by a special military team that was "off the books." He explained that the way this works is that a unit is formed, it will do maybe two missions before the unit's existence is put "on the books." Hence, any early missions are less likely to be acknowledged as ever having taken place.

McCandlish said that US Intelligence agents were operating in Germany at the end of the Cold War, as the Soviet Union was collapsing, he was given the story about the crash retrieval by a young man who was part of "an insertion" into East Germany to pick up some of the intelligence people that were thought to be at risk. McCandlish's contact was redirected to a crash site, which turned out to be 10 kilometres southwest of the city of Halle, in East Germany.

His contact described a "dart-shaped" vehicle, but when the vehicle flew through the air, it flew "with the blunt end forward." It had a cockpit which looked very similar to that of the F-117A stealth fighter.

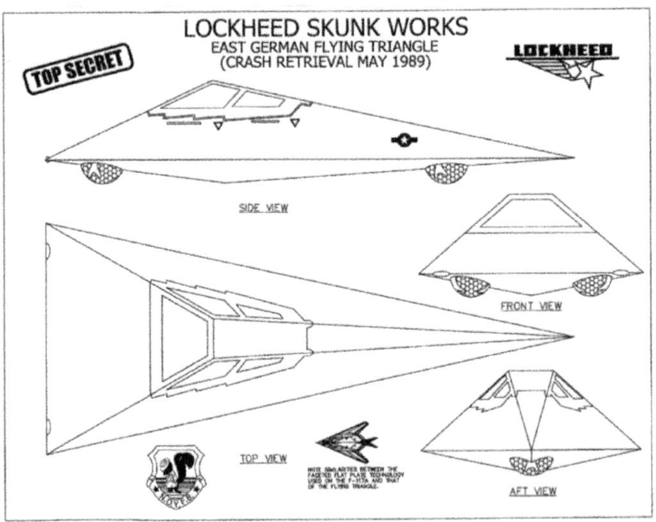

McCandlish reported that "from the side, the craft would look like a pyramid that had been stretched out on one of its three corners. And then under each of the front corners was about a 2½, or maybe about 24~30" diameter white sphere embedded up in the body of the vehicle at each corner in the front."

About two thirds of the way back, on the long point end, there was a third sphere. The spheres were held in place by a three-pronged clasp, almost like the setting of a pearl in a ring, that was silver. Also, it looked as though it could be articulated and pointed in different directions, perhaps for steering the vehicle.

McCandlish's contact reported that when he arrived, a salvage operation was already in progress, with components being removed. He learned the craft had been on a surveillance mission and crashed after clipping a large pine tree, which crushed one of the three spheres and caused a loss of control of the vehicle. It hit the ground tail (sharp end) first and this caused the fuselage to crack open and the sphere in the sharp end, broke loose and began floating off around the hill side, which is what he saw when they arrived.

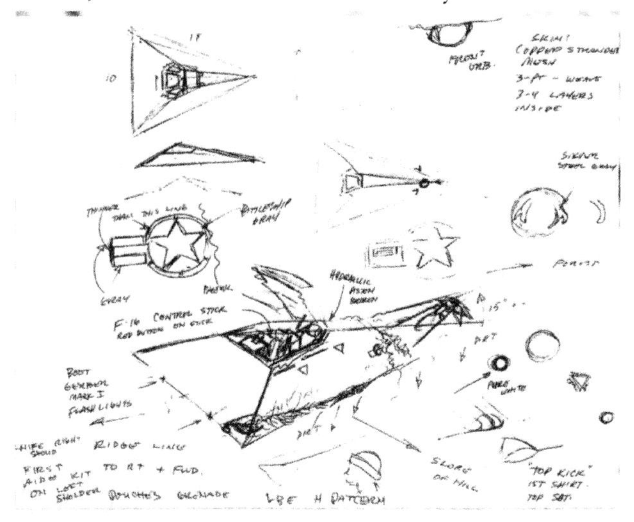

The sphere was intact and looked like a "white, ceramic pumpkin." A conduit protruded from the top and it was probably metal like a brushed aluminium, and there were a number of wires coming out of it. The sphere bobbed around and occasionally there was an electrical discharge between the wires. They had to re-capture this sphere in a large utility-yellow, aluminium carrying case. This was difficult due to the erratic movements of the sphere and the fact that as they tried to get close to the sphere, it would be repelled by the aluminium case. They managed to "capture" the sphere into the aluminium box, but as soon as they did, it "stood up on end like a gravestone." Then, his contact said that the container had to be airlifted by helicopter – on the end of a 100' long (at least) – synthetic lanyard. This was done because the magnetic fields produced by the sphere were so powerful, there was concern that it would interfere with the avionics on the helicopter. The contact also described how Navy SEALs, armed with M-16s rifles and grenade launchers surrounded the site and two Cobra helicopters, fully armed with the missiles orbited the site while the recovery operation was in progress. They collected as much debris as they could and carried it to a nearby vehicle. They put thermite grenades into what was left of the fuselage and burned it down to nothing.

Rendlesham Craft?

As Richard D Hall has observed (and perhaps others too), this "Dart" craft, according to the description, has some similarities to the craft which landed near the Bentwaters Base in Suffolk, UK - about eight or nine years before the East German crash - in 1980. It was sketched by Sgt Jim Penniston, when he went off-base, into the Forest to study what had landed. Based on his sketch and description, a sculpture/model has been created, which now resides in Rendlesham Forest.

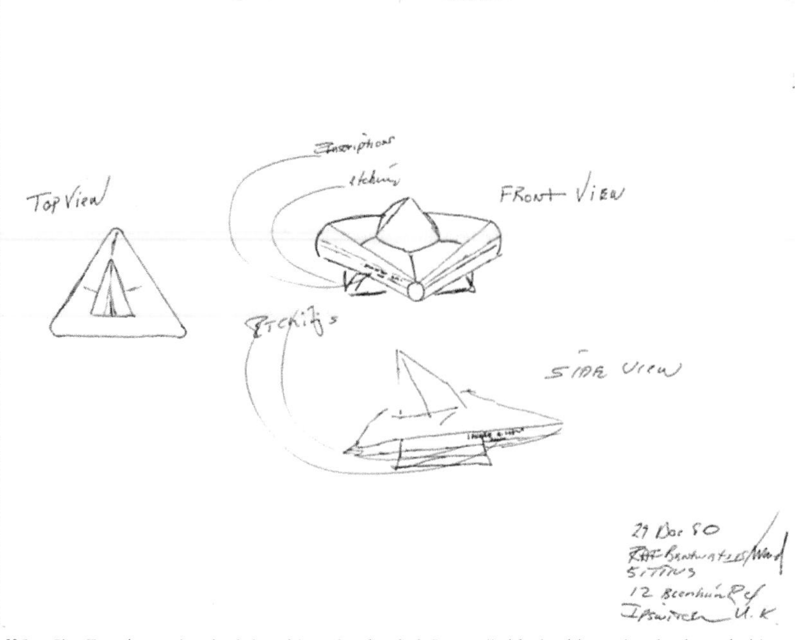

Staff Sgt. Jim Penniston sketched the object that landed. It was "a black, shiny, triangle-shaped object - 3 meters long by 2 meters high (9.8 feet by 7 feet)."

In 2014, a sculpture was put in the place where the craft landed.[453] Was it an earlier model of "The Dart" described by Mark McCandlish?

In his 2018 tour, and in our own conversations, Richard D Hall has pointed out the possibility that there is probably a desire to "link aliens/ET in" to sightings of advanced, secret aircraft/vehicles – especially when sightings have residual evidence and/or accounts can be corroborated among witnesses.

Conclusions

I find the accounts and detail given by Mark McCandlish to be some of the most compelling that I have come across in the the last 15 years. He has taken an interest in how the craft operate and his descriptions regarding the history of electrogravitics are, as far as I can tell, accurate.

One point to consider is that the information regarding the "Dart" craft is hosted on a website which is promoting Corey Goode. We will discuss Corey Goode in chapter 28.

Among the enormous body of UFO/UAP accounts, there are a number of sightings which include descriptions which match those of the ARV and the Dart, although this, of course, does not prove that these craft are owned by the US DoD or any of their allies. However, it looks to me like the electrogravitics propulsion systems *were* successfully developed in black programmes in the 1960s and the 1970s and were then deployed. In chapter 25, we will hear the story of a different craft, about which we know certain technical details. This, too, appears to have been developed and deployed.

23. David Adair and a "Symbiotic Engine"

David Adair in 1997[454]

In about 1999, David Adair came forward telling an amazing story about being involved in a military project which was investigating some unusual hardware.

Adair was a child prodigy, and using tools and materials in his father's "racing car workshop," built his first rocket when he was only 11 years old. By the age of 13, he had built and launched a better rocket, which achieved a launch speed of 3500 mph!

In 1971, at the age of 17, one of his rocket projects was recognized by the USAF and following this, he says he was asked to help someone in Area 51 analyse some unusual hardware.

12 Mount Vernon (O.) News Monday, April 19, 1971

David Adair, 17, wins Air Force certificate for rocket project

Centerburg High School's rocket expert David Adair, 17, received a U.S. Air Force award at the 10th annual Central Ohio Regional Science Fair Friday and Saturday at Otterbein College.

Adair's entry was a 10-foot-tall rocket.

He improvised parts from automobiles and other equipment to construct the engine located inside the slim red, white and blue rocket.

The certificate given the high school junior says his exhibit "has been selected as the most outstanding in the field of engineering sciences."

Officials of the Center of Science and Industry in Columbus asked Adair to exhibit his rocket during the month of June.

Adair has named his rocket Pitholem, a Greek word which means center of energy.

Only 50 of the top student projects in Central Ohio had been selected for display and judging.

Since then he spent 10 years in the US Navy, working on jet engines. He has worked as a Space Technology Transfer Consultant – where technology from the space program is adapted for use in commercial applications.

In the 1970s he was, according to his story, recognised as an expert because he actually designed a type of fusion rocket. He described this in a 1997 interview[455] (which, it appears, wasn't released until 1999, according to captions at the end of the video). Adair stated, in 1997:

> *The rockets kept getting bigger and faster till finally I was able to get funded by a federal grant by a congressman named John Ashbrook and without help I was able to go ahead and complete a larger rocket with an entirely new type of engine which was a fusion containment engine. Two primary engines were used there - solid propellant and liquid propellant...*
>
> *I manufactured a magnetic fusion containment engine which - its simple essence is to create a magnetic field capable of holding a thermal fusion reaction inside - like a chunk of Sun contained in a magnetic bottle. When you tap that power, it allows you to have a tremendous amount of thrust which is what makes the rocket go forward...*
>
> *Once the engine and the rocket was completed we obviously couldn't launch it where we launched smaller rockets, which [was] out in cow fields in Ohio, so we had to go to White Sands in Mexico. That was arranged by the congressman and a consultant at the time who was a retired general named Curtis LeMay[456]. He personally made the arrangements from Wright Patterson Air Force Base in Dayton Ohio. We rode down to White Sands on a C-141 Starlifter and when my rocket and myself got there, we started prepping it for a launch. At the time that we launched, which was June 20th 1971, a plane flew in the same day we did and out stepped these men in black suits and a pair of sunglasses and their DoD agents. So DoD was interested in the operation and they were alerted to it by the Air Force.*

Adair then related how he recognised some officials from NASA who came to witness the test. (In a later interview, Adair stated that one of the people that inspected his rocket that day was Dr Arthur Rudolph – who was mentioned in chapter 14. Rudolph was the chief architect of the Saturn 5 engines.) He was then instructed to direct the rocket to a particular place – a dry lake bed. The rocket was launched – at such high speed, it was "like watching a bullet leave a rifle." The test was a success. Adair then reported that they boarded a DC9 and flew to this dry Lake Bed – which was Groom Lake in Nevada. He said he thought they were going to see where his rocket had landed, which was at the end of the runway (he could see the parachutes), but something else happened.

> *We got in golf carts and rode to this Center hanger - which was a very low flat type hanger - but extremely light inside about the size of about four football fields. So we pull in there and stop in these little golf carts, we're just sat in there ... the hangar's empty... I thought "This is interesting," you know. And then in a few minutes, little yellow and red lights came on - flashing - and then... out of the floor come these rails with chains on... They came up around all the doorways and then the entire floor dropped out from under us... an elevator... we went down about 20 stories...*

He described that underground, there were many workshops and offices and "smaller, but still rather large" work-bays. They entered one of the work bays where something was set up, but with covers over it. The work bay had lighting running the whole length of it – like in a car/auto paint shop (where no shadows are cast, allowing unhindered inspection of the painted surface). The covers were removed and Adair recognised the device as an engine.

> *It's an engine that's about the size of a school bus and this engine is a much larger version of the thing I built. It's also an electric-magnetic fusion containment engine but it is sophisticated and it is powerful. If I were to compare mine to it, it would be like I had a "Model A" and this thing was a Ferrari. I mean there's just tremendous power difference... probably both were in the same engine breed. And this one had been damaged right in the center of the core.*

Adair describes some of the details relating to how he thought the engine operated, based on his own design. He wanted to inspect it close up and they said

> *"Sure, as a matter of fact you can climb on it if you want." So I got up to the engine and there was something really unusual about the engine - other than the design, the alloys... just a sheen color... looked like iridescent color... like when you hold a CD disc to the Sun and you see the rainbow... imagine those kind of colors covering this entire engine and glowing like that. Really pretty. The color of the engine was unlike any metal alloy that I was ever familiar with and when I walked to the engine there was a shadow on the engine. Now I just told you that rooms were built where you couldn't have shadows... When I stuck my arm out to it, there was a shadow in the engine, but right on the table there was no shadow, so it's not a light casting the shadow on the engine, the engine is picking up radiation off my body.*

He then realised if he backed away, the "shadow" disappeared. Later, he touched the engine, which he now saw, in places, had a translucent appearance. Swirling patterns went around his fingers. He inspected the engine more closely and realised that the underlying design was very different – there was no obvious wiring, which his engine had needed a lot of, to fire the particular accelerators and cyclotrons in the correct order. He said touching the engine was like "touching a frog's belly" and it felt wet, but it wasn't wet. He then began to suspect that "this may not be from around here." Studying the engine further, he felt the engine had been damaged by a breach in the plasma containment. He went on to relate what went wrong with the engine and he started asking questions about the firing system. However, they couldn't answer his question (note, at this point, the assumption seemed to be that the military personnel wanted Adair to think it was one of their engines, rather than it being from "elsewhere.") Adair then suggested to them that the engine had come out of a craft 300+ feet in size (based on the size of the engine). He asked "what they had done with the occupants" of this craft. Adair then said his hosts became uncomfortable and "it was time to leave." At this point, he had to touch the engine again as he climbed down, away from it. He reported:

> *I put my hands back on the engine. Something interesting [happened]... instead of the nice smooth wave lines, they all now look like small tornadoes coming off all my fingertips moving through the metal and I put my ... back on there they were but they were starting to calm down a little bit. And then I realized what this thing was ... a symbiotic engine. The engine is alive. It is an engine that's capable of... the reason I couldn't figure out the firing order is that the tubing that was cascading all over its body looked like the same pattern you would have from a brain stem of a neural synaptic... ordering firing*

order... so what's happening is when the pilots sit down in their seats, their thought waves drive the engine.

When I first touched the engine, I was curious and awed by it. The wave [patterns] were smooth off my fingers. When I was getting off down from it I was totally angry, the engine was not picking up radiation or heat waves - it was picking up mental thought energy. That engine is capable of a symbiotic relationship with the pilots... The pilots' brains are the wiring and the circuitry of the engine needs them in order to do all of its functions. In return the pilots know exactly what the engine feels and how the spacecraft feels.

(In the interview, Adair then pointed out that there were certain control systems in development which would use thought control, or "brainwave" control in some of their systems.)

Following Adair's inspection of the engine, he returned to the surface – back in the hangar. He claims he thought about the implications of what he had seen and what they were trying to do with this engine. He considered what would happen if they used his rocket technology to deliver nuclear warheads and destroy installations in the USSR. He considered that destruction of the USSR could be done much more quickly due to the higher speed that his rocket could travel. That is, the target would not have much warning time due to the speed of his rockets. He therefore decided he would take some "graphite grease" from the hangar door wheels. He was taken, with two sergeants went to where his rocket was.

*I walked over to the open hatch - open up the particle accelerators, [I] put the graphite grease - smeared on the doors inside - and closed 'em. When you turn on particle accelerators with fuel using deuterium, ask any physicist what happens when deuterium meets graphite. It's pretty bad – it's a major reaction – explosion. So I turned on the particle accelerators - no fuel involved - and I told the two sergeants "It's got a fuel leak - we got to run for it." So we get in the golf cart and take off. **I gave us 60 seconds on a time delay...** It exploded and blew a hole the size of a football field out there. There's nothing left. Now we have no rocket. Now they have nothing to work with. But they got this kid so, I was told I'm going nowhere for the rest of my life and it was scary.*

*They put me in a room and kept me there for about ten hours. And then a lot of noise on the outside... Then there was this big stogie chomping figure in the doorway ... his name is **Curtis LeMay.** So, he came there and got me took me all the way back home to Wright-Patterson. He took me back home to my parents. He told me that "they're not going to be through with you. They'll deal with you again." And he was right.*

Adair then related how, a year later, on 10 June 1972, he was drafted, but the way he describes this is a little unclear.

*I'm now property of the United States Government - it's called conscription... So after about three hours I'm back out. Then I was on a plane... I end up at Langley Virginia - home of the CIA and more fun activities. What they wanted was the engine to be rebuilt again and **we had to work out some compromises**. I ended up spending 10 years in the Navy. I never did build the engine again - that was the last thing **Curtis LeMay** told me to do "Never build another rocket as long as you live - you might make it out of here." And he was*

> *right so, um instead I agreed to build jet engines. So, I was a jet engine designer technician in the Navy ...*

Later in the 1997 interview, Adair is asked how he came up with the design of his rocket. He laughed in a way showing reluctance to give an answer, but then said a lot of his ideas came in dreams. He said he was working with Quantum Physics when he was 7 years old, but was limited in what he could achieve because he didn't have a calculator or a personal computer (they were not available to the average person in the late 1960s and early 1970s). He was working with advanced equations when he was in his teens and he says his Science teacher helped him. At this point, Adair claims he visited Ohio State University with his Science Teacher in July 1969 to discuss his equations. There he met "Stefan Hawkings" and he claims they worked "in parallel" because Adair needed to "contain" a fusion reaction explosion using a black hole. He also claimed in the interview:

> *When this engine turns itself on, eventually the vehicle is going to catch up with the speed of the exiting matter, out the tailpipe. How fast is it going? 186,000 miles a second – the speed of light. This engine's capable of ... light speed and we had it in 1971...*

Asked why he decided to tell his story after 20 years of silence, Adair stated he felt there had been a shift in public perception regarding the UFO/ET question. He mentioned the June 1997 report on the Roswell Incident, released by the USAF, which claimed that the reports of bodies were the result of test dummies being dropped in the area.[457] He also pointed out what a problem it is for the US Government that they have been lying for 50 years about the UFO/ET issue. He also stated:

> *The engine I saw did not come from the Roswell crash. The engine I saw was as big as the Roswell craft. So this is something even bigger - that's another one. So how many do they have? How much they got? You know, that's where the real problem is...*

Later in the interview, he reports the he saw symbols on the engine:

> *I was standing on top of the engine that that piece of metal came from. As a matter of fact, there's about 98 different emblems on this engine. I was born with a photographic memory and I got them all. So if you'd like to see them, here they are. And they're interesting the way they were grouped in clusters and they were on every single device - that was a separate device - and I realized what they were actually - not some great cosmic messages or anything. They're serial numbers for parts! That's how they figured out where their parts are. After looking at the emblems some of the emblems repeated themselves. There was one that had a very similar [appearance to the] mathematical symbol of Pi but it was a modified symbol of it - and it would show up every time there was a curved radius... Some of the emblems would be reversed on the other side of the engine, when the two parts were the same. Left and Right! I'm going "That's all it is!" No big great cosmic message – its serial numbers. Just like we do on jet engines in the Navy. We group out our numbers and but it would give you the alphabet of their language. So, I was able to retain that. Also I memorized how the engine was built. So, let's build another one.*

Also, in the same interview, Adair mentioned that he testified before the US Senate and Congress in April 1997, as part of an initiative by Dr Steven Greer, where they discussed the issue of aliens interacting with humans, as well as telling his story to the group of people, some of whom Adair said he recognised.

> *I was pretty impressed with it so uh I finally decided to jump in there. He had 12 other witnesses besides myself and we all testified to him. The next day we were supposed to talk to the press. The other 12 were allowed to talk to the press, but I was not allowed to talk to the press. Steven Greer asked me not to talk to the press and I think he was absolutely right at the time. ... So I think Dr. Greer was right for holding off - keeping me back a little bit. The reason being is that I was "too much" for the press to get into right then - because he's got more hearings more strategy... We talked about that later and we had a little bit of a misunderstanding at first, there, but we got it straightened out. But I agree with the way he's doing it, because there's more testimony and more things I can produce later.*

This section of the interview is peculiar, because it seems to contradict what Adair said later, when he met Greer at an event in LA on 4 Aug 2001 – a few weeks after Greer's Washington DC Disclosure Project Press Conference.[458] Adair reported in this other interview, in 2002, that at the LA event:

> *Once again Dr. Greer failed to take this chance to explain or apologize as to why I was treated so rudely at the Congressional Briefings in Washington, DC after he invited me.*

But, in the 1997 interview, he had apparently "straightened out" this misunderstanding.

Analysis/Commentary

As with Bob Lazar's story, Adair's story seems like a mixture of fact and fiction. Adair has not produced any pictures or documents to back up his story, but it appears he did build rockets in his teenage years. Did he fabricate the entire story about going to Groom Lake? Whilst I find aspects of his story tie up with other topics I've covered in this book – like, for example, the control of the craft by "mental phenomena," as mentioned by Smith, I find it difficult to accept that the engine would operate on the principle of hot nuclear fusion. It is quite reminiscent of Lazar's "gravity amplifiers" needing element 115 to operate.

I also find it difficult to accept Adair's story about how he put graphite paste in the fuel chamber of his rocket and set a "60-second delay" before it exploded. This sounds too much like a scene from a James Bond film (or similar).

It is similarly unlikely that he met Stephen Hawking ("Stefan Hawkings") in 1969 at Ohio State University, as it would have been quite difficult for Hawking to go there with the serious health issues he was experiencing by then. It is also odd that he got Hawking's name wrong, especially as he claimed to have a photographic memory. Similarly, the physics that Adair was discussing was conventional, mainstream physics – nothing to do with "engineering the aether" or zero-point field etc, as we have touched on in this book.

However, it's difficult to understand why Adair would make such a story up as he didn't write a book and didn't have a Website or YouTube channel or podcast to promote. (Since then, a DVD has been made about him and his story, however[459].)

It appears that people in the USAF did know of David Adair's work and abilities. In a 2014 interview, Jan Williams, daughter of Colonel Bailey Arthur Williams (USAF Ret.)[460] seemed to confirm that at least part of Adair's story was true. She had visited the top-secret offices of Strategic Air Command Headquarters on "family days" as a child, since her father was a senior ranking officer of Strategic Air Command. Williams confirmed that she had contacted David Adair so that she could question him about his relationship with her father. Williams said that after a careful review of Adair's story they became long-time friends. When asked if there was any way Adair could know the things he knew about Colonel Bailey Arthur Williams without having a personal relationship with him, Jay Williams emphatically replied "No."

Of concern, in my view, is Adair's description of visiting CIA Headquarters in Virginia. Add to that his mention of "having to make compromises." Adair has promoted the validity of the Apollo programme. Anyone as smart as Adair should have figured out they couldn't have got to the moon as claimed. Though he spoke out about the termination of the US Shuttle programme[461], he does not discuss any more specifics of the black technology that he claims he saw, except, perhaps in a rather muted way. Instead, he describes how the Apollo Saturn V rocket which "took men to the moon" could be re-built and upgraded with new technologies. I have never, for example, heard him talk about electrogravitics, as we have in this book. Rather, what he has talked about is the threat posed by certain countries. In a presentation to the Granada Forum on 21 Aug 1997 which had been given the title "The Other Space Programme,"[462] 17 minutes in, David Adair started talking about the 1991 Gulf War in a rather jovial manner and said:

> *Did anyone watch the Gulf War on TV? ... It's like one of the neatest technological Soap Operas going. All this technology we hadn't even tried yet... We had different pieces of it come together... believe me, we dumped everything we had at that guy...*

He then describes how a cruise missile can target specific buildings, but that was still a problem in Iraq because targets they wanted to hit were built between a hospital and a school, so the cruise missile could be used to destroy single buildings. He then enthuses about the GPS system being used for military applications. Of course, for someone who developed rockets as a child and was in the Navy for 10 years, having "rubbed shoulders with top military brass," they are probably going to have a more militaristic mind than, say, someone like me! Perhaps this is why the 1997 Granada Forum presentation he gave sounded more like a public relations presentation for the US DoD than anything else. It seems he had "forgotten" the implications of what he claimed he saw and was no longer concerned about the 50 years of secrecy he mentioned in his face-to-face interview.

Adair has appeared on programmes like "Coast to Coast," for example in February 2018. I don't have a transcript or recording of this broadcast, therefore I am relying on the show description, which said what Adair was talking about:[463]

> *If North Korea dropped an atomic weapon in the atmosphere over America, the EMP waves would cause transformers to detonate across the continent, he warned, which would have the effect of napalm. Shortly after that, every plane in the sky will crash down (about 500,000 people are in the air at any time), he continued, adding that all telecommunications will be out. A massive CME would be even worse, he cited, as the effects would likely be global rather than confined to a specific region. He also described what would happen at atomic plants during these events with meltdowns leading to widespread contamination. According to Adair, the Fukushima effects have been under-reported, and once when he swept a Geiger counter over a can of tuna at a grocery store, it detected radiation.*

This sounds like more fearmongering to me, slipping in the reference to North Korea. Why doesn't Adair realise the implications of the secrecy? I can only presume he is still trapped in the illusion of Moslem terrorists hijacking planes and crashing them into the WTC...

24. Years After "The Day After Roswell"

This chapter explores some of the statements made by US Army Col Philip J Corso (ret) (1915-1998) and whether any of those statements relate to black programmes or other information that might be related to a Secret Space Programme.

Colonel Philip Corso (Left) and Gen. Arthur Trudeau (right)

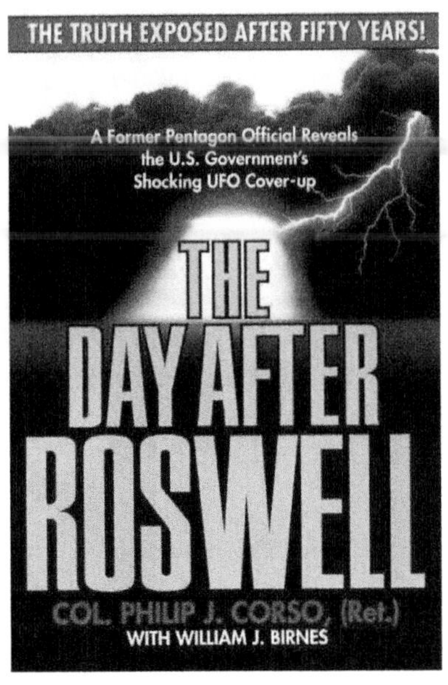

The 1997 Book by Corso and Bill Birnes

In 1997, a book called "The Day After Roswell," co-authored by Corso and Bill Birnes was released. This book created something of a stir and most researchers seem to agree that there are some important revelations in this book. In a posting by veteran researcher Stanton T Friedman[464], first civilian investigator of the Roswell case, Friedman writes:

> *...I checked with the Eisenhower Library. There certainly seems to be no doubt that Corso was with the NSC about 1953-1956, and that he served under General Arthur Trudeau in the Pentagon 1961-63. He was also very active in trying to obtain the release of POWs, some of whom he claimed had been taken away by the Soviets from the far east, while the US government did nothing. So, he is a real person with a solid military background.*

The book was criticised by many for its sensationalism and I had read, some years ago, that Corso himself was not happy with how the book turned out. In researching for this chapter, I found that some of his original manuscripts were released in a more "raw" form in 2011, via the Internet Archive.[465] The page where the file containing the manuscript is posted has an accompanying description about 1100 words in length. This description states:

> *Philip J. Corso, Sr. died in July of 1998 at the age of 83. Thirteen years after his death, we will finally get to see his manuscript, which has never released in*

> *the US [the only foreign edition was published in Italy by Pendragon in 2003]. People interested in UFOs will be compelled to read this book, because too much of the information in the bestseller "The Day After Roswell" does not stand up to what Corso really wrote in his diary. Philip Corso Jr., a man devoted to fulfil his father's legacy, described the manuscript in his lecture at the 2007 IUFOC in Laughlin.*

Corso's son, Philip Jnr states:

> *Dad went through "The Day After Roswell" and was yellow lining everything that he didn't say and didn't like. He got half way through it and quit because it was impossible. He had almost every page blacked out. This book [The Day After Roswell], my Dad was really very upset about it he did not proofread it.*

The "Dawn of a New Age" is mainly typed, but the very last page is handwritten. It contains some of Corso's original notes and several sketches. The description of Corso's file of notes contains the following statements (although the attribution is not clear):

> *Executing General Arthur Trudeau's orders at the Pentagon, Corso handled technical data and some mysterious components derived from the alien craft that crashed in Roswell, New Mexico, in early July 1947. Corso had also seen medical reports of the autopsies performed on the bodies retrieved at the incident site.*
>
> *According to Corso, they determined that the craft was a biological spaceship, functioning in conjunction with a crew of EBE's (Extraterrestrial Biological Entities). They were biological robots created through advanced genetic engineering, clones designed to withstand the extreme conditions of space travel. These EBE's were able to drive their starship through a particular neural interface, whereby they could connect with the craft, becoming an almost integrated part of it.*
>
> *Corso says he was responsible for handing parcels of alien technology to several industrial contractors connected to the US military. Their development would lead to discoveries such as night vision devices, Lasers, fiber optics, transistors, and more.*

The description notes that:

> *Unfortunately, we don't have a corroborative statement by anybody who knew what Corso was really doing in the early 60s when he was head of the Foreign Technology Research and Development department for the Army at the Pentagon.*

However, the page on the internet archive[465] also includes the text of an affidavit signed by Corso in May 1998, a few weeks before his death. A group called Citizens Against UFO Secrecy (CAUS), headed by Attorney Peter A. Gersten, started a lawsuit against the "Department of Army" and they used this affidavit. In said affidavit, Corso re-affirmed previous statements that he saw alien bodies in July, 1947 and had the opportunity to read their autopsy reports in 1961. He stated:

> *That on or about July 6, 1947, while stationed at Fort Riley, Kansas, I personally observed a four-foot non-human creature with bizarre-looking four-fingered hands, thin legs and feet, and an oversized incandescent light bulb-shaped head. The eye sockets were oversized and almond shaped and*

> *pointed down to its tiny nose. The creature's skull was overgrown to the point where all its facial features were arranged frontally, occupying only a small circle on the lower part of the head.*
>
> *There were no eyebrows or any indications of facial hair. The creature had only a tiny flat slit for a mouth and it was completely closed, resembling more of a crease or indentation between the nose and the bottom of the chinless skull than a fully functioning orifice.*
>
> *That in 1961, I came into possession of what I refer to as the "Roswell File." This file contained field reports, medical autopsy reports and technological debris from the crash an extraterrestrial vehicle in Roswell, New Mexico in 1947.*
>
> *That I have personally read the medical autopsy reports which refer to the autopsy of the previously described creature that I saw in 1947 at Fort Riley, Kansas.*
>
> *That said autopsy reports indicated the autopsy was performed at Walter Reed Hospital, which was under the authority of the defendant at the time of the autopsy.*
>
> *That said autopsy report referred to the creature as an "extraterrestrial biological entity."*

I will now highlight some parts of Corso's file of notes.

"Dawn of a New Age"

This collection of notes is described by Corso's son, thus:

> *"Dawn of a New Age" - the original notes of U.S. Army Colonel Philip J. Corso- are available through our website here. We can now get to the core of the classified activities and research on the Roswell files examined by Corso at the Pentagon between 1961 and 1963.*

I read through Corso's manuscript and so below, I discuss what I found to be some of the most interesting parts. I read a PDF version and I will refer to the page number of that PDF file. There are some numbers handwritten on the pages too, but these are not sequential from start to finish.

Foreign Technology Division

On page 30 of the PDF file, Corso writes:

> *In Army R&D, I had the title of Chief of Foreign Technology Division. I received all reports on foreign technology, which included foreign beings and events - those not of this world. I often headed teams of U.S. scientists, German scientist (Von Braun's group) and U.S. technicians. In the foreign technology division, I set up what I called "My nut file" - every wild eyed inventor or theorist that came or wrote to us was listened to, the promising ones were explored. Most were "nutty," but the effort was worth it because every so often we uncovered a gem. I was always the team chief and made all decisions. No one except the General knew of my clearances.*

The rest of the notes do not contain an enormous amount of detail about what came into his purview, and he does not state that he could not reveal details because of a security classification, or some kind of similar restriction.

Knowledge of the Roswell Incident

On PDF Page 141, for example, Corso writes:

> *ROSWELL AND COVER UP*
>
> *In the books and other articles about the Roswell incident, there seems to be quite a bit known of the craft and its elements. Pieces of the super strong thin metallic material still exist outside the U.S. Government but nothing is known of the beings that were aboard. The cover-up was thorough but there has evidently been contact in other areas and people are beginning to demand release of government information on ET contacts and confirmed UFO sightings.*

This seems to be at odds with what he writes elsewhere in the same document, where he describes something about the aliens' appearance. On pages 127 – 129, he writes about some of the US military's apparent knowledge of UFO activity, but he does not write about this as if he was directly involved. He mentions Space Command's Space Surveillance Centre being located deep inside Cheyenne Mountain, near Colorado Springs. He states:

> *The Center operates a Space Detection and Tracking System, a world-wide network of radars telescopes, cameras and radio receiving equipment.*

But this isn't really a secret, as such.[466] More interesting, perhaps is his next description:

> *The U.S. Naval Space Surveillance System, headquartered in Dahlgren, Virginia, operates an "electronic fence" stretching three thousand miles across the southern U. S. and extending a thousand miles off each coast. The "fence" is a man-made energy field reaching out nearly fifteen thousand miles into space.*

Corso also mentions INSCOM, who worked with John Hutchison in the 1980s:

> *INSCOM, the Army's Intelligence and Security Command, headed by Major General Albert Stubblebine, provided some funding to the UFO Working Group.*

But again, Corso does not give any additional information about these things.

One problem is that Corso refers to the Apollo missions several times, assuming they were "real" – which we now know is not the case.

Also, he stated in an interview (mentioned later) that this matter "wasn't his full-time job." As he said himself, he wasn't a scientist and, whilst he was highly intelligent, he was a soldier and he apparently knew how to assert himself when it mattered.

Stanton T. Friedman argues[464] that much of what Corso wrote about could have been gleaned from sources already in the Public Domain – such as books already published about the Roswell case and the alleged MJ-12 group. Similarly, Corso writes about Project Horizon – a 1959 proposal for a military Moon base.[467] Friedman stated that the document about this had already been available in the Army Archives for many years before Corso's book came out. Media reports state the document was declassified in 2014[468], however. Corso does not really give any information about his sources.

Corso claims that the following technologies were utilised after their initial discovery inside the crashed saucer at Roswell.

- Fibre Optics
- Transistor
- Chip
- Night Vision
- Craft can be controlled by the brain/thought.

Though he accurately describes the nature of these technologies and gives a few details of how they appeared when first discovered in the crashed saucer, he is quite vague about how these were then later transferred into US industries.

Corso was not a fan of civilian control of research - he seemed to favour the development of the Military Industrial Complex. On PDF page 8 he writes:

> *Now that our base of operation was firm and strong, we decided to marshal a preponderance of strength against which even the civilian control and left leaning politicians could not prevail. The partners that we sought and gathered were U.S. industry, a strong laboratory support and many leading universities.*

Also, on PDF page 8, he refers to flying Saucers built by the Germans:

> *Another important phase was German scientific developments, Army R&D organized operation "paper clip" and brought over Von Braun and his group. Many German scientific documents, even flying saucer successes of the Germans were gathered and brought to the United States.*

This theme is repeated on page 157, where he describes his encounter with a disk that materialised in the desert:

> *Finally, I saw something simmering like a heat wave, I could see through the simmering waves. Suddenly it materialized. It looked like a metal object on the order of a saucer, like your scientist Victor Schauberger had built for Germany. Seconds ticked away and abruptly it disappeared.*

On page 11, Corso writes about "metals with molecular alignment":

> *Reports from our laboratories on a piece of metal from Roswell showed the molecules/atoms had been aligned. This alignment created an incredible hardness, whereby it could not be scratched, bent, dented and repelled cosmic action and radiation, although it was paper thin. Since CRD thought this discovery could be bigger than Los Alamos, a special team was organized to investigate all possibilities. We found many precious gems and crystals had similar properties. A different alignment would allow radiation to pass through or make the element transparent. Up until 1963 or even 1993 we have not been able to solve this problem. It would make possible space vehicles of great strength, but light as a feather.*

Might Corso be talking about Quasicrystals (see chapter 25)? Or was he talking about the "memory metal" that was witnessed by a number of people after the Roswell crash.[469]

On PDF page 61, Corso writes about three artefacts that were allegedly in his possession:

> *I sat in my office one day and cleared my desk. I placed three items on it - a small piece of a greyish type metal, a faded photograph and a small charred piece of silicon with an integrated circuit. To the casual observer it would appear that I had scoured through a trash bin. However, I thought, these three innocuous bits of seemingly trash could change the future of the world as we knew it.*

It is presumed that the photograph he had was of one of the aliens, as that is what is shown in a sketch on page 166. Again, few specific details are given.

"Mental Phenomena"

One of the more interesting areas that Corso writes about in his document relates to the descriptions about the Roswell craft being controlled by the brain or by thoughts. For example, on PDF page 81 (marked 80), he writes:

> *We could transmit sound with an electric current and through a fiber optic, why not brain wave. In 1960-61 we started laser experiments with this in mind since EBES had such a beam. They also had a head band which could intensify brain waves possibly for mental telepathy.*

This ties up with what WB Smith wrote in his 1950 memo that we discussed in chapter 4 (Smith is mentioned once or twice in Corso's document). Smith is mentioned in a 1997 or 1998 Corso interview with James Fox.[470] At about 18 mins into that interview, Corso states:

> *Wilbert Smith is really the man that first put in my mind when nuclear binding... coming apart... on things and ... really he was a genius and the Canadian government really treated him bad[ly]. I was supposed to go to his laboratory visiting because the general told him "Smith, and the Colonel have a lot to talk... I'll let the Colonel come up and visit you at your lab at Mayborough[?] on Lake Ontario." Well I put it off, put it off. [In] 1962 I decided to go up. I called up and then they told me Mr. Smith had died of cancer. I never really got to go to his laboratory, but he gave us a piece of metal that he took from a flying saucer. We exchanged and he brought ours back later."*

In his document, Corso mentions that instinct is something that comes "from another dimension." On page 83 he writes:

> *Perhaps a brain or thought wave integrated in an electromagnetic wave could overcome gravity and eliminate the spacetime curve and travel straight to the missile, Science says mass against mass creates gravity. Could the insertion of a thought wave move into the fourth dimension and create an artificial gravity so that man could travel in space. A gravity magnetic field controlled by man himself (mass against zero). My analysis of the Roswell mystery leads me to believe that the above is a reality.*

On page 112, he refers to research into a "thought wave intensifier."

Nature of Craft and Aliens

Corso writes, in several places about the nature of the aliens that were aboard the Roswell craft. For example, on page 85 he writes:

> *From data available we could assume that the UFO's were reconnaissance craft and could quickly return to a mother ship to receive the necessities of life.*

> *The other was that they had solved gravity-time dimensional travel and could instantaneously return to their base.*

Again, there is no reference to where he got this data. He then postulates, as have other authors, whether the EBE's are clones, but adds no useful information:

> *If the extra-terrestrials were clones, how did they reconstitute or restructure themselves? Were they expendable and only used to perform a mission and then replaced?*
>
> *The former question seemed to be more logical and better fitted our thinking and our approach. They were delicate creatures and faded quickly, but still they were humanoids, I remember the chemistry classes of my younger days, when chemical reactions were described in plus or minus, or electrical patterns. I had many conferences with project officers and the German scientists. We invariably agreed the cells and even organs could be replaced by cloning and sustained by electro-magnetic means, applied to the basic structure.*

Hostile Aliens

He considered that the craft and the aliens were real and were hostile enough that action needed to be taken to protect ourselves on earth. He also makes something of a case for why he thought this. He had concluded that aliens were involved in abducting human beings and they were also responsible for animal mutilations, for example. Where he writes about the mutilations, all the details seem accurate, though he does refer to other sources, such as Linda Moulton Howe's "Alien Harvest" book. He does not refer, though, to cases that he himself investigated.

On page 93 he states:

> *They have violated our air space with impunity and even landed on our territory. Whether intentional or not, they have performed hostile acts. Our citizens have been abducted and killed. They have endangered our space vehicles and the occupants; scouted our sensitive bases and probed at will. They have mutilated animals on our territory and removed organs, possibly for study, for ulterior motives. The above are acts of war which we would not tolerate from any worldly source. It also appears they do not tolerate any such acts on our part on their bases.*
>
> *We in the military, long ago concluded they have voluntarily given us nothing. What we gained from their presence we obtained by accident and moved forward enough to challenge them with their own technology.*

NASA and Von Braun and General Douglas McArthur

In several places, Corso does refer to specific statements by other people that he may have known. For example, on page 91/92, he states:

> *During the Korean War, I was on Gen. Douglas McArthur's staff. He was a brilliant individual and did not scare before the enemy. His decisions and directives were clear, precise and well thought out. In 1955 he made an astonishing statement, "The Nations of the world will have to unite, for the next*

> *war will be an interplanetary war. The nations of the earth must someday make a common front against attack by people from other planets."*

He had a dim view of NASA and on page 93 states:

> *NASA should stop the impressions of having no strategy and stop irregular and disconnected strategy planning and publicize advanced concepts and technology just as the tremendous push NASA gave the computer ship and software section. Above all, they should reveal the truth of their findings on Moon and Mars gravity and other objects (UFO's construction, atmosphere, etc.) found and seen on the Moon and Mars. There is ample evidence of alien activity on the Moon, 122 photographs from NASA's science data center were examined and analyzed by photographic experts with startling findings.*

However, it appears that, again, at least some of this information came from other researchers, not his own direct contact with NASA officials, as he continues:

> *The authors of "Alien Activity of the Moon" Fred Steckling, himself, was a trained observer in astronomical studies. Moongate, the NASA military cover-up book contains 20 color photographs of U.S. astronauts on the moon.*
>
> *Various treaties demilitarizing the Moon, have been signed by the U.S., USSR and other countries of this world, but I wonder if the aliens recognize these treaties.*

On page 96, he includes a statement by Werner Von Braun

> *In 1959 Dr Werner Von Braun, another great space pioneer, made an intriguing statement, reported in Germany. Referring to the deflection from orbit of the U.S. Juno 2 Rocket, he stated: "We find ourselves faced by powers which are far stronger than we had hitherto assumed, and whose base is at present unknown to us. More, I cannot say at present. We are now engaged in entering into closer contact with those powers and in six or nine months' time it may be possible to speak with more precision on the matter." (Neues Europa, 1 Jan 1959.)*

This may have come from Timothy Good's "Above Top Secret" book. The story seems to be corroborated by Clark McClelland.[471]

Energy and Propulsion

On PDF page 85 Corso gives some interesting details about the army's own "compact nuclear reactor":

> *The Army had overseen Los Alamos and perfection of the atom bomb* We also had built smaller nuclear power plants which could be transported on barges. (This led to a reactor 6' x 6' weighing 6 tons, is self-operating, requires no attending, will produce electricity for 20 years and is safe.)*

He also refers to more conventional nuclear reactor technology:

> *Nuclear propulsion for ships and submarines is well known. We had built an atom bomb the size of a football and a 280 mm artillery shell, Radio-isotopes have powered dozens of space missions. Plutonium-230 was used in the interplanetary space craft to Mars, Jupiter, Saturn, Neptune and Uranus.*

The last sentence is probably a reference to the Viking, Voyager and later probes.

On page 92 he writes:

> *UFOs produce high tension electric charges and strong magnetic fields. Strong electrical and magnetic effects, affect the electrical systems on our space craft, often with serious and devastating results. Therefore, we must produce radiation or EMI hardened sensors, integrated circuits and other electrical equipment which cannot be harmed or distorted by electromagnetic forces. We will then be able to compete against their artificial fields of gravity.*

On page 110 he states:

> *SUPER PROPULSION SYSTEMS: We have advanced in rocket propulsion, nuclear and antigravity systems.*

Again, there are no references as to where this information came from. On page 112, he lists some related areas of research that were being undertaken by the army such as "Plasma research" and "Electromagnetic and antigravity propulsion devices."

Phobos

I was surprised to read Corso mentioning Phobos, the larger of Mars' two Moons. I wrote a chapter about Phobos in my previous book "Secrets in the Solar System: Gatekeepers on Earth."[444] On page 110 Corso writes:

> *TAMPERING WITH EARTH'S ENVIRONMENT: We have tampered with their environment. It appears that they have bases on the Moon and Mars and we have scouted the areas.* ***If they inhabit large ships like Phobos, we have sent radio and radar signals, which are a form of electromagnetic action****.*

On the next page he writes

> *RECOMMENDATION: In war, when in doubt about the enemy's intentions, strength etc, you must probe or force his hand. Since, there is so much doubt in the UFO-EBE area we should force them to react to our efforts.*
>
> *We should intensify "Star Wars" efforts and fire at their UFO's or mother ship.*
>
> *We should send a* ***nuclear armed probe at Phobos*** *on Mars.*
>
> *We should send an electromagnetic probe at the moon. De Gaussers as used in the Philadelphia experiment.*

Again, no references are given and neither is any further information given about Phobos. The original "Day After Roswell" book does not mention Phobos either, so this really stood out for me. It is quite possible that Corso read about Phobos in another book or magazine – such as the 1992 article by Don Ecker,[472] or even Scklovsky's "Extraterrestrial Life"[473] book.

Sceptics

Corso's attitude to sceptics seems ambivalent or even contradictory. On the one hand, he blames them for creating a "climate of disbelief and ridicule" with regard to the UFO/ET phenomenon, which he feels has hampered disclosure of important truths. However, on page 155, he thanks sceptics for helping to provide cover for their secret research operations:

In arriving at this conclusion we offer our thanks to the "debunkers". No one searched for or even suspected our approach. The "debunkers" through their shrill outcries and energy created the atmosphere that UFOs didn't exist. Our government and the liberals, adopted this policy and we quietly went along. If the UFOs, by virtue of national policy didn't exist, Roswell never happened and as a policy reality there were no aliens or hardware, or any other by-product, Therefore, why suspect anything or even think of any type of investigation.

We also wish to thank the "sceptics". The more they blared out, "show me something tangible," and it never appeared, the safer was our approach. Also, I wish to thank the "sceptics," because they made it possible not to share our discoveries from outer space with the scientific community since they would never have kept it quiet. We knew that many were Soviet agents.

Corso's Own Encounters

On page 156, Corso describes seeing a crashed disc in 1957, from the air. This was apparently when he was involved in missile tests and he was flying out to observe a missile that had been "lost" or gone in the wrong direction.

On page 157 Corso describes seeing a disk materialise in the desert when he was out on a patrol. Between pages 159 and 169, he describes an event that took place while he was in command of the U.S. Army Missile firing range at Red Canyon. He was out on another patrol when an encounter with a being that communicated with him telepathically and then departed. Corso stated that the being's craft had been brought down by radar. The being asked Corso to delay the next "run" of the radar equipment so it could take off safely.

The date of this incident is not clear, nor is it made clear if there were other witnesses.

Conclusions

Having read Corso's document, I can say that overall, there isn't a lot of detail (or at least, not as much detail as I would like to see). Additionally, there is a reasonable amount of repetition. There are different sections which seem to overlap, and it becomes easier to see why Bill Birnes needed to heavily re-work Corso's notes in writing the "Day After Roswell." That is, the book reads better and has better structure and repetition is removed etc.

I was left with the feeling that Corso really was involved with some of the work that he said he was, but he only had a partial grasp of what he was dealing with. He only had a "need to know" a small part of what was really going on. I think he probably did have access to a file of information about the Roswell incident, more or less as he describes. Perhaps he talked to enough people to realise that artefacts were recovered, and he even found out a little more about them during the course of his work. The mention of Wilbert Smith is interesting – especially if it is true that Corso had planned to go and see Smith.

I suspect that, after he retired, he became more curious about what he had really been involved with and so began to do more research, eventually deciding to speak out. However, he didn't really have all that much classified "Roswell-related" knowledge of his own, though he worked with people that had a lot more knowledge about "what was going on" than he himself did.

So, I can't glean that much information from Corso's material that might inform us about aspects of the Secret Space Programme, though one might suggest that Project Horizon (regarding a military Moonbase) would potentially have been (or be) part of a Secret Space Programme.

25. Master Sergeant Ed Fouché and the TR-3B

Introduction and Biography

I first came across Edgar Fouché in 2004 – or at least, I thought it was 2004, but I later discovered I had come across him about 5 years earlier in 1999! He was featured in a brief segment of a UK documentary called "Riddle Of The Skies".[474] Ed only appeared on the screen for a few minutes and, partly because of the way the information was presented in this documentary. I hadn't really "taken in" the huge significance of what he was disclosing to the world at that time. Edgar Fouché was disclosing information regarding what he had seen in a number of secret projects – or "black programmes."

As I relate below and in later chapters, a sequence of events that began in 2004 led me to come into contact with Ed Fouché and we became good friends.

Ed was born in 1948 in Americius, Georgia. In the 1960s, having completed his High School education Ed was drafted from college during the Vietnam conflict. He joined the USAF in 1967 and Ed was selected for the US Air Force Pararescue field and then, following an injury, retrained as an electronics and cryptographic specialist. His continually growing skills and knowledge in these and other areas led him to gain a top secret "crypto" clearance. He then spent about 20 years in the US military and another decade working for defence contractors, working across many levels of the military. All these factors led him, in the mid-late 1970s, to work, for some short periods, in top secret facilities in the Nevada Test Site – known as Area 51. This brought him into contact with some advanced digital technology – which, as part of his work, he held in his hands.

He had gained many awards for the high standards of his work. His reputation meant that his skills were often sought after by the people high up in his chain of command. People came to know and trust him (as did I). Over time, various people gave him their own accounts of incredible technologies and programmes that they had witnessed or worked on or knew about.

In the early 1990s, with the help of a few close friends, even though they knew that they would be putting themselves at risk of reprisals, the decision was made that they would disclose to the world what they knew – that the US government had been covering up the development of advanced, potentially world-changing technologies.

Ed collected together many documents to back up what he was presenting. Some documents were extremely sensitive and therefore had names or dates blacked out or deleted.

After years of careful research, interviewing, writing and preparation, Ed made his first disclosures in 1998, at a number of public presentations – including MUFON and IUFOC events in the USA.[475] Ed had also written down much of the information in the form of a book, also published in 1998, which was co-authored by well-known author Brad Steiger. The book was called "Alien Rapture – The Chosen"[476] and it used Ed's life as a "canvas" on which was "painted" the information he was disclosing. Due to the legal advice he received, most of the characters and scenarios in the "painting" are presented as fictional, but essentially, they are based on real people and true events. The only thing in his book that was fiction was the 'Alien Agenda.'

As well as being an author, Ed was also an artist and he even wrote some poetry, I understand. Ed's life should really be the subject of a separate book…

Ed Fouché passed away on 11 May 2017, having been in and out of hospital several times, whilst remaining in good spirits and well enough to talk to me and others he knew.

I recommend you take time to view Ed's 1998 presentation,[475] which contains nearly all of the essential information I will refer to in this chapter. I also recommend you try to get hold of a copy of "Alien Rapture – Unabridged" – as there is a lot of detailed information in this book[477].

Getting to Know Ed Fouché

In late 2013, Ed Fouché contacted me, following a dispute he had had with his friend Jeremy Rhys (who uses/used a pseudonym of "Alien Scientist"). I discuss more details about this in chapter 30.

In January 2014, Ed asked me to record some interviews with him, to help set the record straight about what he had done and said. You can listen to these interviews on YouTube, or examine copies of the documents that he released to the public over the years.[478]

In the five hours of interviews, and in other conversations I had with Ed, I learned what an amazing life he had led. Ed survived injuries that would be fatal in most cases (this also happened to Dr Judy Wood). He was stabbed in a fight. On a dive mission, something happened and he was found floating on the surface of the water – but his air tanks were full. In 1973, he was involved in a serious road accident where he severed his aorta – this kind of injury would be fatal in most cases. He took the Chrysler car company to court because the car he was driving had a faulty seat belt, which he argued, made his injury worse.[479] Though the initial judgement went against him, he won the appeal.

When I got to know Ed, he was suffering from COPD, diabetes and other problems. Not surprisingly, he continued to have ongoing health problems, but he remained very active in many areas – and in speaking to him, you may not have even realised he had such health conditions. (Listen to our interviews!)

I got to know a little bit about his family – he was a grandfather and as he explained in our interviews, when he was being attacked on websites and forums after his "disappearance" in about 2000, he decided he had to try to address the lies that were being told about him, because he had a family who might read these lies and "not know what to think."

Ed Fouché's Career – A Summary

Ed had a distinguished career, which you can get an impression of if you look at this document released to David Hilton (whom we will discuss later) in an FOIA (see below). People tried to say, of course, that Ed didn't even work in the air force and had "no proof" he did what he said he did. As he said himself, he had 900 documents which proved otherwise – and he produced 200 of these in 1998, because he knew that people would need to see them to understand that he was "the real deal."

As mentioned above, Ed joined the USAF and was stationed around the USA and abroad. When he was stationed at Canon AFB in New Mexico, Ed did some maintenance work on some of the aircraft equipment, which was ingenious, and this was appreciated by the "folks higher up."

In another role, he worked on systems to enable planes to send back friendly identification codes to scanners. During this work, Ed played a sort of "prank" and used the test codes that he had memorised to pretend that he had cracked these codes – this gained attention up to the level of wing commander. Ed told me that the NSA took notice due to the "crypto" implications.

During the Vietnam conflict, he was assigned to special projects at Kadena AFB Okinawa; Udorn AFB Thailand; Ben Hoi AFB Vietnam, and he spent time at many other South East Asian military bases.

He was considered an Air Force expert, with experience in working with classified electronics counter-measures test equipment, crypto-logical test equipment (owned by the National Security Agency), and Automatic Test Equipment. He worked with many of the leading military aircraft and electronics manufacturers in the US and DoD contractors and therefore participated as a key member in design, development, production, and Flight-Operational-Test and Evaluation. He worked in classified aircraft development programs, with state-of-the-art avionic, electronic countermeasures, satellite communications, crypto-logical and support equipment.

During his military career, he was "handpicked" for many of the Air Force's newest fighter and bomber development programs – some of which were secret. He received over 4,000 hours of technical training from the military and government, of which about half was classified training. He worked at Edwards AFB on the F-15 – worked under Franz Dietrich. (The F-15 could accelerate straight up, as it had more thrust than weight.)

He had a "Q clearance" I and this enabled him to work on the F-117 at Edwards AFB, when it was top secret. (Ed explained that when others talk of "levels of clearance above Top Secret," it is bogus.)

INFORMATION RELEASABLE UNDER THE FREEDOM OF INFORMATION ACT

NAME:
Edgar Albert Fouche

BRANCH OF SERVICE AND SERIAL/SERVICE NUMBER(S):
U.S. Air Force

DATES OF SERVICE:
May 12, 1967 to August 31, 1987

DUTY STATUS:
Retired

RANK/GRADE:
Master Sergeant

SALARY:
N/A

SOURCE OF COMMISSION:
N/A

PROMOTION SEQUENCE NUMBER:
N/A

ASSIGNMENTS AND GEOGRAPHICAL LOCATIONS:
7 Bombardment Wg (H), OL OOKE (SAC) Kelly AFB, TX; 380TH Avionics Maint Sq (SAC), Plattsburgh AFB, NY; 6100TH Logistics Support Sq (PACAF), Kadena Air Base, Japan; 57th Component Repair Squadron, Nellis AFB NV (TAC); 57TH Maint Sq Nellis AFB, NV; 6515 Test (Support) Squadron, Edwards AFB, CA; 6515 Avionics Maint Sq, Edwards AFB, CA; 366 Avionics Maint Sq Mt. Home AFB, ID;347 Avionics Maint Sq,Mt. Home AFB, ID; 27TH Fld Maint Sq, Cannon AFB, NM

MILITARY EDUCATION:
SEE ATTACHED

DECORATIONS AND AWARDS:
AF Good Conduct Medal w/ 5 OLC,AF Longevity Service Award Ribbon w/ 4 OLC, National Defense Service Medal,Small Arms Expert Marksmanship Ribbon,Vietnam Service Medal,AF Overseas Long Tour Ribbon,AF Commendation Medal,Republic of Vietnam Gallantry Cross with Palm,AF Outstanding Unit Award,Presidential Unit Award,AF Organizational Excellence Award, Meritorious Service Medal w/ 1 OLC

TRANSCRIPT OF COURT-MARTIAL TRIAL:
Not In File

PHOTOGRAPH:
N/A

PLACE OF ENTRY:
Dallas, TX

PLACE OF SEPARATION:
Kelly AFB, TX

FOR DECEASED VETERAN ONLY

PLACE OF BIRTH
N/A

DATE OF DEATH
N/A

LOCATION OF DEATH
N/A

PLACE OF BURIAL
N/A

NOTE: N/A denotes information is not available in the veteran's records

NATIONAL ARCHIVES AND RECORDS ADMINISTRATION — NA FORM 13164 (Rev. 02-02)

Ed spent twenty years working directly for the US Air Force and DoD Agencies, and retired in 1987 with diabetes, back and heart problems. He then worked for another eight years as a Defence Contractor Manager. His last position for the Air Force was as a Strategic Air Command Liaison.

In around 1993, he was told he would live approximately 1 year – so he "didn't care if the government came after him." However, he also wanted to pursue his own projects and developments.

Area 51/Nellis AFB Work

Perhaps of greatest interest to readers of this and similar books was Ed's description of his work in Area 51 – the Nevada Test Site – or whatever name you give it.

Ed had experience working in Electronic Counter Measures (ECM) and with satellite communications electronics. In about 1979, Ed's friend wanted Ed to work in Area 51 – but Ed was enjoying his current work and didn't want to work in the desert area, so declined the opportunity then. This friend must've suggested Ed for a job in Area 51, because he was asked to report to his commanding officer and told he "was needed on another job," but he was not given any details. It was then due to a combination of coincidences and technical experience and certification that Ed was requested to be temporarily assigned to a place, which had no name. He was one of the few personnel at Nellis who had a Top Secret clearance with Crypto access. He was certified to work on Mode 4 IFF, (an aircraft system which responded to classified, encrypted codes.) In August 1979, he was told by his commander that he was to report to an office on the base, and that he didn't have a clue where he was going or what he was going to be working on. Ed's commander was unhappy to have been left in the dark.

Ed left one Monday morning long before sunrise. It was 4:30 AM when he boarded a dark blue Air Force bus with all of the windows blacked out. There were armed guards on the bus. No talking was permitted, and it was a "dusty, uncomfortable journey."

Once he arrived, they had to walk from the bus to the work area. They were forced to wear "welders goggles" which blurred vision, but he could see over them slightly due to the goggles not fitting his large head too well. Ed explained that when wearing the goggles, he could see a short distance in front of him, but anything further away was blurred.

Ed worked in Area 51 for 10 days. His military records show he was re-assigned for these periods, but not the location that he was re-assigned to. Ed learned that he'd been assigned to "temporary duty." Ed was "covering" for someone who had had a bereavement and was given leave. Ed explained that he had done some repairs to equipment, but the problems he had tried to fix recurred, so he had to stay there longer than was initially anticipated.

Ed has recounted how on the third day on the job at Groom Lake, he had to remove a "Direct Orbital Code (or Communications) Link" module from a multi-bay piece of satellite communications equipment to inspect/service it. This is when he found the circuit board had VLSI electronics – and a 1 GHz processor. The paper inspection stamp on the chip was dated 1975. (The 1GHz Pentium chip was not seen in the "white world" until 2000).

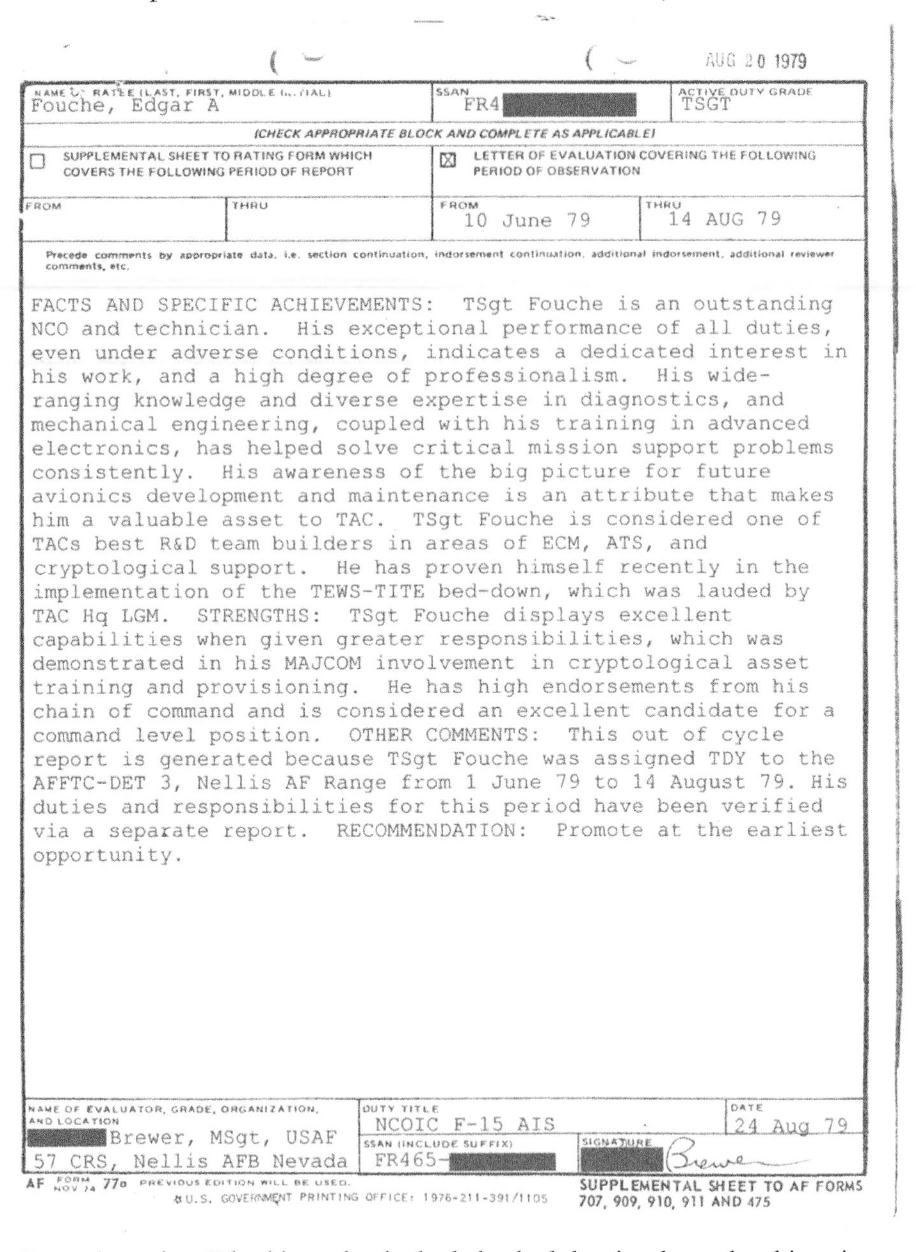

AUG 20 1979

NAME OF RATEE (LAST, FIRST, MIDDLE INITIAL): Fouche, Edgar A
SSAN: FR4[redacted]
ACTIVE DUTY GRADE: TSGT

(CHECK APPROPRIATE BLOCK AND COMPLETE AS APPLICABLE)

☐ SUPPLEMENTAL SHEET TO RATING FORM WHICH COVERS THE FOLLOWING PERIOD OF REPORT — FROM: THRU:

☒ LETTER OF EVALUATION COVERING THE FOLLOWING PERIOD OF OBSERVATION — FROM: 10 June 79 THRU: 14 AUG 79

Precede comments by appropriate data, i.e. section continuation, indorsement continuation, additional indorsement, additional reviewer comments, etc.

FACTS AND SPECIFIC ACHIEVEMENTS: TSgt Fouche is an outstanding NCO and technician. His exceptional performance of all duties, even under adverse conditions, indicates a dedicated interest in his work, and a high degree of professionalism. His wide-ranging knowledge and diverse expertise in diagnostics, and mechanical engineering, coupled with his training in advanced electronics, has helped solve critical mission support problems consistently. His awareness of the big picture for future avionics development and maintenance is an attribute that makes him a valuable asset to TAC. TSgt Fouche is considered one of TACs best R&D team builders in areas of ECM, ATS, and cryptological support. He has proven himself recently in the implementation of the TEWS-TITE bed-down, which was lauded by TAC Hq LGM. STRENGTHS: TSgt Fouche displays excellent capabilities when given greater responsibilities, which was demonstrated in his MAJCOM involvement in cryptological asset training and provisioning. He has high endorsements from his chain of command and is considered an excellent candidate for a command level position. OTHER COMMENTS: This out of cycle report is generated because TSgt Fouche was assigned TDY to the AFFTC-DET 3, Nellis AF Range from 1 June 79 to 14 August 79. His duties and responsibilities for this period have been verified via a separate report. RECOMMENDATION: Promote at the earliest opportunity.

NAME OF EVALUATOR, GRADE, ORGANIZATION, AND LOCATION: [redacted] Brewer, MSgt, USAF 57 CRS, Nellis AFB Nevada
DUTY TITLE: NCOIC F-15 AIS
DATE: 24 Aug 79
SSAN (INCLUDE SUFFIX): FR465-[redacted]
SIGNATURE: Brewer

AF FORM NOV 74 77a PREVIOUS EDITION WILL BE USED.
U.S. GOVERNMENT PRINTING OFFICE: 1976-211-391/1105
SUPPLEMENTAL SHEET TO AF FORMS 707, 909, 910, 911 AND 475

In our interview Ed told me that he had checked the signals on the chip using a portable oscilloscope and a frequency counter to determine this figure. Ed was also looking at other test points on the board and he told me (in Part 4 of our interview[480]):

> *I'm thinking "orbital communications link" and then I'm going like - okay it's gotta be just satellite communications. And of course, you know we didn't have a space shuttle back then [it was launched into space a couple of years later]. But the thing was modulated... the frequency this very high frequency...I think it's like 800 megahertz to 1gigahertz, but the modulated frequency... but the modulation of the transmitted signal was audio - human voice audio – and I'm sitting there you know... I really just sat back and looking through the schematics and I'm doing some test points here and there... and I'm goin' like "Who the hell are they talking to up there?"*
>
> *I'd never seen any type of ... I had worked on satellite communications gear, but it was all telemetry ... but I had never worked on satellite communication that was encrypted ... that was also modulated to audio frequencies –which is voice.*

This seemed to confirm that black programmes have technology at least 20 years ahead of the white world.

During his time working on classified programmes (I am not sure exactly when), Ed overheard talk of exotic technologies including "Quasicrystals". This is an important piece of evidence which Ed mentioned in his 1998 IUFOC presentation – and we will cover this in more detail later in this chapter.

Edgar Fouché – Knowledge of Advanced Aircraft

Edgar Fouché spoke to a number of people involved in black projects – including people who piloted the SR-71 (Blackbird).

In his 1998 presentation, Edgar Fouché describes a number of advanced aircraft. For example, even though the full specifications of the 30-odd year old SR-71 were still classified in 1998, it was known that it could fly at 85,000 feet at speeds up to Mach 3.3. Even then, the USAF reportedly had the SR-75 which can travel at up to 120,000 feet and Mach 5. It could fly from Nevada to Russia in three hours! Ed said this was part of Project Aurora.

Ed described the SR-75 as a "launcher" for the SR-74, which can reach a height of 151 miles (which is in space...). He said the SR-75 can also launch a 2-ton satellite payload. (Ed described the Space Shuttle as an antique.) He also described two orbital vehicles – the "Space Orbital Nuclear – Service Intercept Vehicle" (SON-SIV), code named Locust.

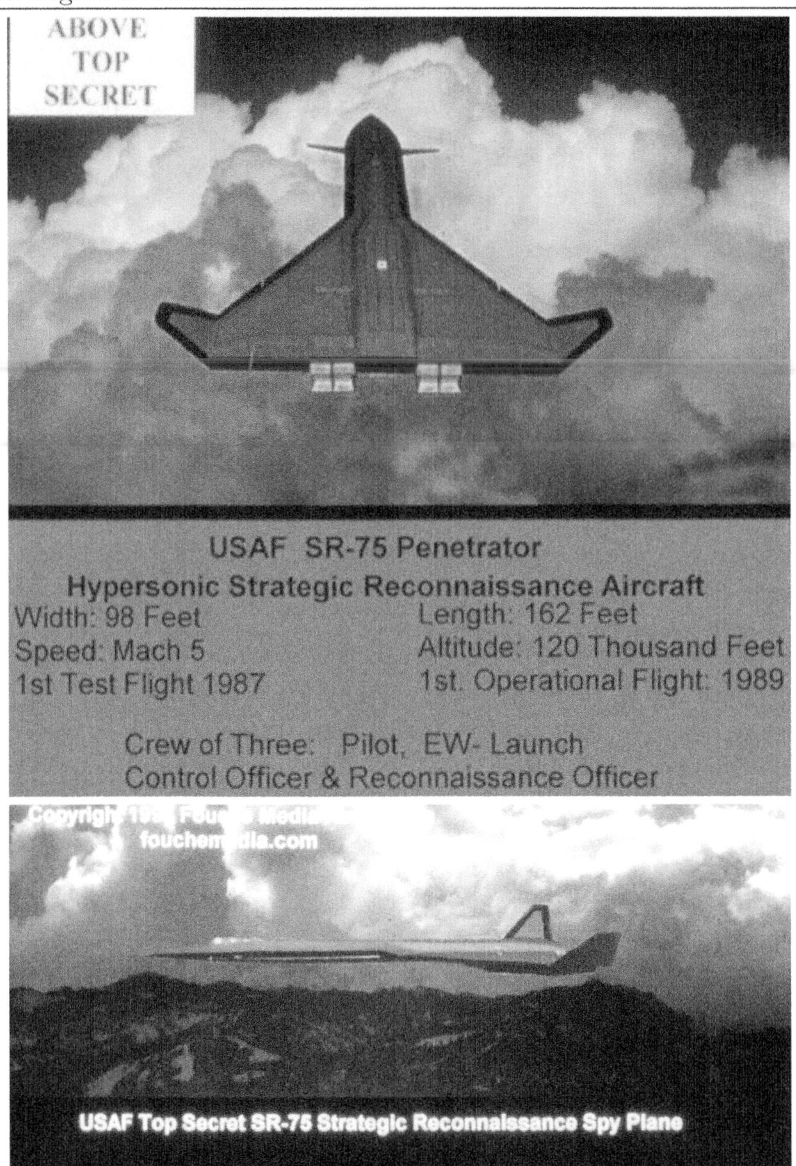

SR75 -Slides from Ed Fouché's 1998 presentation.

TR-3B – Flying Triangle

In his 1998 presentation, Ed Fouché described how he had seen photographs of a triangular craft called the TR3B, taken by someone who was giving Mission Support on a C-130. The TR-3B was thought to be part of the Aurora (Advanced Aircraft Development) programme. (Ed would later describe his own sightings of something that looked very much like this craft.)

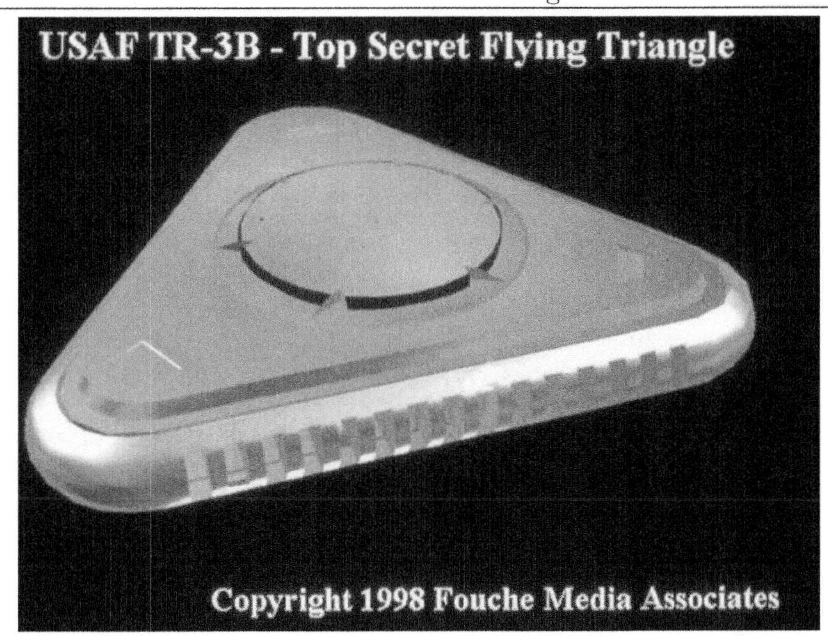

Ed Fouché's 3D-rendering of the TR-3B

One of Fouché's associates stated he saw the TR-3B at Papoose (part of Area 51). It hovered silently in the same position, for some 10 minutes, before gently settling vertically to the tarmac. At times a corona of silver-blue light glowed around the circumference of the massive TR-3B. The operational model is 600 feet across.

Fouché stated that at least 3 of the billion dollar plus TR-3Bs were flying by 1994. He said that the craft can travel at up to Mach 9 in any direction (due to reduced G-forces). He also said that the TR-3B's outer coating is reactive and changes with radar stimulation and can change reflectiveness, radar absorptiveness, and colour. Ed said that Quasicrystals are used in the vehicle's skin. Ed, at other times, said that the TR-3B could be considered as "a platform, not an aircraft." He did not know the full details of its operation. He had pointed out that the prototype dates back to the 1970s – so, we can ask, what have they got now? Ed also made a comment that "everything that NASA does is 'deception'." Of course, this is a slight exaggeration, perhaps – but as we saw in earlier chapters, this statement is true, in many important respects.

Quasicrystals

Ed stated he overheard conversations, while working in Area 51 in 1979, about Quasicrystals[481] which were an enabling technology for advanced communications and propulsion. Quasicrystals were discovered (in the white world) in 1982 and for many years, were not regarded as a "thing" worthy of study. Most scientists that studied crystals thought Quasicrystals could not exist. They, including Linus Pauling, a famous chemist, said that such properties of crystals were impossible.[482]

When Ed Fouché spoke of Quasicrystals in his 1998 presentation, hardly anyone knew what they were. (Even in 2004, when I first came across Ed's material, I had never heard of Quasicrystals – and I studied chemistry to A-Level – up to the end of secondary/high school). It was in 2011 (13 years after Ed's disclosures), that Daniel Schectman won the Nobel Prize in Chemistry for his work on… Quasicrystals[483].

In April 2015, as I continued to research the topics covered here, I stumbled across an interesting lecture, given at Imperial College London. The title was "The Schrödinger lecture 2012 - Metamaterials: new horizons in electromagnetism from 2012." In this lecture, Professor Sir John Pendry explained some of the theoretical (and two practical) uses of metamaterials in Optics. The lecture explains Snell's law – (the law of refraction). It introduces the idea of a metamaterial which has a *negative refractive index*.[484] Below, I include one of the slides from a section of the lecture on "Transformation Optics." The lecture implies that there is a possibility of using the crystals to enable a form of image projection.

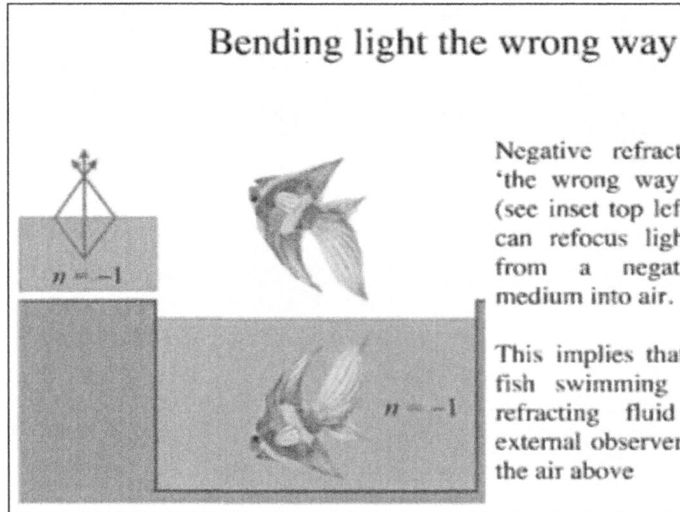

Professor Sir John Pendry's slide on using "negative refraction" in a meta material.

MFD and Mercury and the Vaimanika Shastra

Going back to Ed's 1998 presentation, he described that the TR-3B's propulsion system employed a method known as Magnetic Field Disruption (MFD), which had been reverse engineered. The drive system contained a Mercury based Plasma, pressurised at 250,000 atmospheres. This was held at a temperature of 150 Kelvin (-123°C) and spun at 50,000 rpm. This reduces the weight of the centre of the craft by 89%. This also increased manoeuvrability by 89% as well. Fouché stated that MFD research had started as early as 1955. Ed has made it clear that he did not work on the TR-3B and received information about how it worked from various contacts. He has kept the names of these contacts confidential, to protect them.

Ed also stated that the TR-3B uses Nuclear Thermal Hydrogen/Oxygen Engines[485] for manoeuvring (a technology mentioned by Fouché years before the 2004 NASA Moon to Mars briefings). It was, in fact, this type of technology that veteran UFO researcher Stanton Friedman worked on – so why isn't he more interested in Ed Fouché's disclosures[486]?

Now, earlier in this book, we discussed how any ongoing antigravity research seemed to "go underground" in approximately 1956. We also discussed the field of electrogravitics. However, we did not go into much detail about the use of magnetic fields, but obviously magnetic and electric fields go "hand in hand," so to speak. It is likely that the magnetic field in the MFD drive is created from a rotating electric field.

We also heard previously that some type of mercury compound was allegedly used in the Nazi Bell device. Although I came to the conclusion the Bell *wasn't* specifically an antigravity device (though it may have exhibited some type of levitation or weight loss), the mercury used in it may have facilitated the creation of a powerful magnetic or electric field. Mercury is, of course, the only metal which is liquid at room temperature.

Another interesting, though anecdotal, consideration is the apparent use of Mercury in the propulsion of the Vimaana Craft, mentioned in the Hindu Vedic texts. However, this information comes from a document called the Vaimanika Shastra, which was written out by dictation in 1902 during sessions channelled through Subbaraya Shastry, a mystic from Aneka.[487] Are we therefore to dismiss the details that are described in this document?

Rotating Plasma

Ed Fouché seemed to be fairly certain about the details of the technology which powered the TR-3B, so we can ask, was the idea sensible or feasible? Well, we can certainly find at least one scientific paper about "Generating a magnetic field by a rotating plasma"[488] – although it seems that this is based on theoretical work, not experimental work. Further research does reveal that it has, at least, been considered as a technique in the white world.[489] However, the physics involved is highly complex and therefore beyond the scope of this book.

In 2014, I discussed Ed Fouché's disclosures with Richard D Hall[490] and Richard later included a segment on Rotating Plasma in one of his lectures. This led me to a paper which shows a graph of electron density (in a plasma) against temperature. To generate a strong magnetic field, a high electron density is required. As Richard D Hall pointed out, it can be difficult to generate a high electron density in a wire or metal conductor. You generate a high electron density by increasing the current. However, wire or metal of a certain thickness, can only carry a current of a certain amperage before becoming too hot – due to electrical resistance in the wire or metal. (And yes, a superconducting wire will do better.)

If we now study a bit of astrophysics, we can get some figures regarding electron density in a plasma. In a paper called "Advances in Numerical

Modeling of Astrophysical and Space Plasmas" by Anthony L. Peratt at good old Los Alamos National Laboratory, New Mexico, USA[491], on Page 5, we find the graph shown below.

This graph basically vindicates the idea that Ed Fouché talked about. However, in considering this graph we do have to bear in mind the following[492] (for the physicists reading this!)

> *Practicing astrophysicists routinely refer to temperatures in units of eV or keV, even though this is wrong, because temperature is not dimensionally equivalent to energy. Nevertheless, they still do it, with the Boltzmann constant being implicitly included in the conversion.*

We therefore shouldn't convert between electron volts and Kelvin – but astrophysicists do, so we will too!

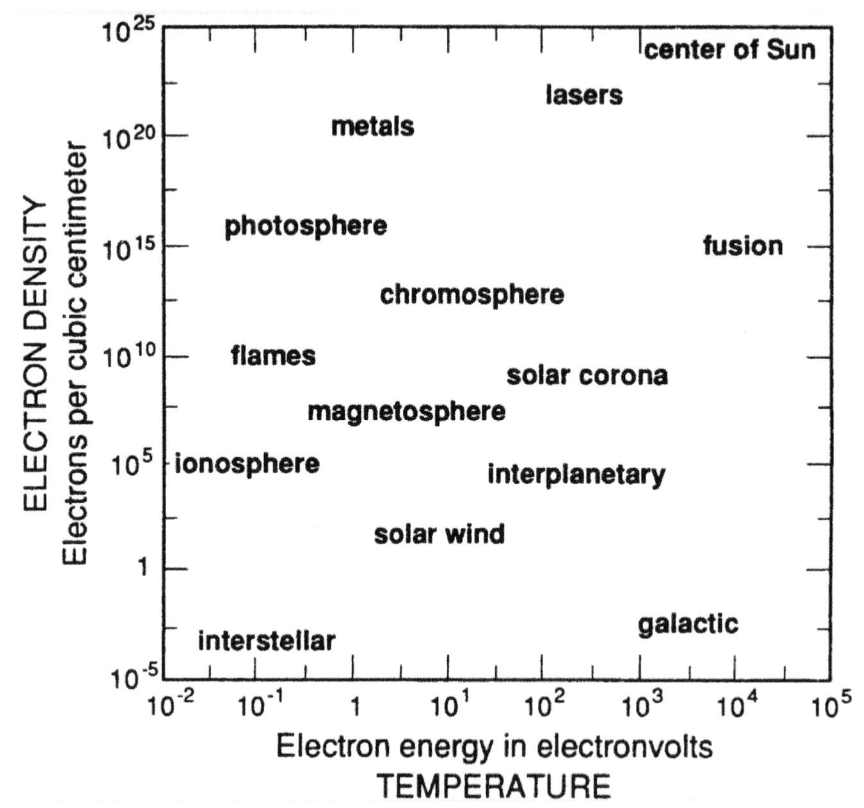

Above: The remarkable range of temperatures and densities of plasmas is illustrated by this chart. In comparison, solids, liquids, and gases exist over a very small range of temperatures and pressures. In "solid" metals, the electrons that carry an electric current exist as a plasma within the more rigid crystal structure.

In converting the energy in electron volts eV[493], we use a constant:

$$T\ (\text{eV}) = 8.625 \times 10^{-5}\ T\ (\text{Kelvin})$$

To get electrons in a plasma even closer together – i.e. increase the density, the plasma can be kept under high pressure. We can say that Magnetic field strength is governed by charge

q (amount of electrons) & **v** (speed) of electrons

In a 100ft diameter circular plasma container, where plasma is rotated at 50,000 rpm, it can be calculated that the magnetic field would be 87 million (8.7×10^6) times stronger than in a typical wire. Thus, the magnetic field strength would be 1000s or millions of times bigger than anything produced elsewhere in the white world…

Hence, in terms of physics, we see that a cooled, compressed Mercury Plasma would probably be a "good thing" for creating a very strong magnetic field.

Ed Fouché's Sightings of the Flying Triangle (TR-3B?)

In part 6 of our 2014 interview[494], Ed described to me how he had had three separate flying triangle sightings in the 1970s, with the third being the most significant. Here's his account.

> *Jerry Elrich worked with me and for me at Edwards Air Force base from 1973 to 1976 and I believe it was around 1975. It was the middle of the night. We'd climbed up the metal staircase and gone to the top of the [Fairchild] hangar and we're up there smoking… (We had access to many areas – we'd get out of the house - I mean he had babies and I had babies and sometimes would just go for a drive and go down to the [work] shop - to see who was working and what they were working on… You know just mess around - hanging out like buddies do.*
>
> *We were on top of this hangar and Jerry points to the sky, almost above us and it's very dark - but the stars were crystal clear. It wasn't overcast. He says, "You see those three stars up there?" At first, I didn't know what he's talking about. They were fairly small I guess ... but he says "They're coming closer..." I said "No... baloney!" and I finally saw what he was looking at and they were coming closer and then .... 3 stars in the shape of a perfect triangle got closer. It darkened out one of the stars behind it and Jerry got all excited and he says "Man that's something solid up there! You see that star blink out?" I said "Yeah..."*
>
> *It kept getting bigger and bigger and bigger - to where it would be like - being a half a mile from the airport... You see a 757 take off except it looked, like, four times bigger and it stopped at some point... or seemed to stop and then move for a fraction of a second, and then just went off and was over the horizon in, like, one second... just [a] massive burst of speed and it went north... It was definitely triangular, from what we could see - it was a perfect triangle… If you go north or northeast of Edward's, you end up in central Nevada. So, either southern Nevada or central Nevada and the only [place] a vehicle would be going to would be either Nellis Air Force base or Groom Airbase.*
>
> *So that was the first time and we told everybody about it. A few other people said "Yeah we've seen some strange triangles" or whatever. We didn't think much of it after that. I mean we didn't obsess about it until we got to Nellis and had another experience.*

Ed and three or four friends had another sighting of a flying triangle in Nevada in 1978.

> *Jerry Elrich and Brian and Ray and me had Dirt Bike's - cross-country bikes… and we would go all the time any chance we got we'd go on our motorcycles and usually take our 22 pistols or 22 rifles in and we'd shoot snakes or*

Jackrabbits, whatever. We're up in the high desert - north of Nellis Air Force Base - which now would be called part of the Nellis range. Now, Area 51 is also a subsection of the Nellis range.

So, we're riding our motorcycles and whoever was in front stopped and pointed. I guess a little bit south of us - and all of a sudden, the hill was covered with this massive shadow and then we all looked up to see what was causing the shadow and there was this triangle that... I mean I thought and Jerry thought it was like more than 400 feet across, but it happened so fast, it could have been anywhere from 200 to 600 feet. There was no way of knowing but we first noticed this massive shadow and we looked up and it was - this thing was low enough that you could probably have shot it with a BB gun. It was a very overcast day, and then before we knew it, within a couple of seconds it was gone. Somebody said they heard some kind of noise... I felt like I heard some kind of hum... It had no discernible high noise like a jet or a helicopter or anything like that in it...

It was moving at such a speed it sure as hell couldn't have been a balloon. It was so massive and there was no vertical stabilizer or horizontal stabilizer... there was no fuselage. There were no engines that you could see. It couldn't have been a conventional aircraft of any kind.

The final sighting of the triangular craft that Ed told me about was the most significant, because this occurred while he was on temporary duty (TDY) in Area 51 in 1979.

I was outside the hanger... I [am] thinking it was evening. It was a nice day, it was hot but still fairly nice. I was smoking a cigarette and had the goggles on, but they used to kind of make me a little bit dizzy, because I couldn't wear my glasses with them. So, I got in the habit of kind of pulling them down just a hair, so I could just tilt my head forward and see over the top of them with my baseball cap on - my squadron hat. Nobody seemed to notice that I could see over top of... At that point, they recognized my face - after being up there a few times, so I don't think they paid any notice to me.

But I can see at the end of this long hanger, on the tarmac or flight line whatever you want to call it, there was this massive vehicle and I could see the front tip and... it was curved. It was almost like the flat end of the vehicle was parallel to the hangar, except that from where I'm standing, I could see about a third of it.

It was set up about ten feet off the ground, it was several stories [about 30 feet] high - the body [was] - and it was massive. The body of a 757 is like 20 feet tall ... Something triangular [like that], you can store a massive amount of equipment on something that big.

I was literally amazed - it was like Captain Kirk had landed his ship on the ground. It didn't make any sense to me how this thing could fly because... I couldn't see any discernible engines. There were ducts - duct-work variable vectored veins or something - some of that coming out of the bottom that were noticeable. It was a greyish black... In the bright sunlight, it almost had a kind of a bluish undertone. But it was basically very dark grey. I mean it wasn't a shiny finish on it. But I immediately thought back to the sighting that Jerry and I had on the Fairchild hangar at Edwards and then of course when a number of us sighted this triangular shape vehicle in the desert, on our motorcycles, so I was just pretty much stunned. I kept looking around me because I was afraid someone's going to come up and club me on the head or something... Gawking at this thing...

Well I mean you know if you think you see a triangular vehicle in the sky and then a few years later, you see a huge, massive triangular shadow and a vehicle above you scooting across, and then you see something similar on the ground, it has to be a triangle - because one of the angles is what you can see. I mean it wasn't (then) a usual shape of any type of known aerospace vehicle.

USAF Top Secret Nuclear Powered Flying Triangle - The TR-3B

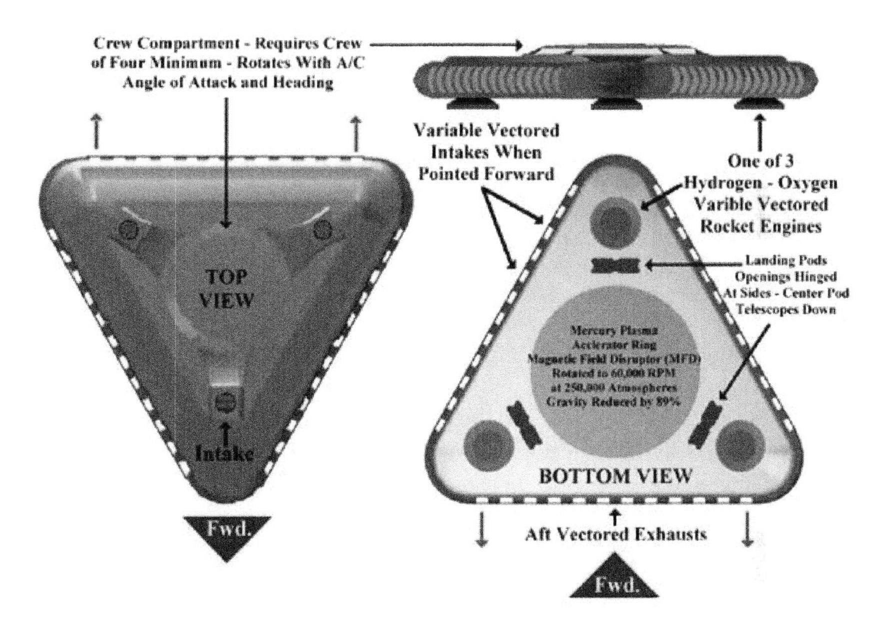

Ed Fouché's schematic of the TR-3B, based on his sighting and what he was told by other contacts.

In our interview, Ed also talked about the reactions to sightings of unusual aircraft in the Nevada area.

Thing is, like I said if you lived in Las Vegas, or damn near anywhere in Nevada, there were rumors - weird rumors about everything. [Some people] call it the conspiracy state because there's so much weird stuff going on and you'd hear stories about these lenticular circular shaped saucers or ... these cylinder- shaped vehicles or triangular shaped vehicles... You'd hear about them - you pretty much discounted most of it, because I mean, there were conspiracy theories A to Z.

As I said to Ed at the time, "It helps to keep the secrets if you've got 150 different theories and stories as to what these things are and where they come from. Then, obviously, nobody knows which is the correct one." He responded:

Well, first of all the newspapers and radio for the most part wouldn't talk about it, because of all the military and government agencies in Nevada. And they

> *were the biggest state employer - other than the casino industry. They just didn't talk about it. There's so many sightings - unidentified flying objects of different shapes - so often, that it was just common knowledge. Everybody had seen something. If you'd lived there a year, you'd seen several "somethings" in the sky that didn't make any sense. It was like "oh we had a mosquito a storm" or "we had a lot of mosquitoes tonight." I mean it was so mundane and so often the news media just never mentioned it. That was until George Knapp and John Lear and Bob Lazar fabricated their stories.*

One thing that we discussed in our interviews was Lazar's case and Ed Fouché stated emphatically that Lazar had made up his story and put out false information, particularly regarding the S4 lab/facility. Ed was particularly critical of Lazar's lack of documentation which would have proved, more comprehensively, where he worked.

Crashed Saucers, FTD and MJ-12

In part 15 of our interview, I asked Ed about the Roswell incident and where the materials and bodies went. He replied:

> *From my investigations over the last couple of decades it definitely appears like Wright-Patterson Air Force Base in Dayton Ohio and Carswell Air Force Base Fort Worth Texas which is now closed - is where the majority of the materials from the crashed saucers went. However, there were certain elements - either they were radioactive or they were unknown fluids or gases or traces of fluids or gases ... that was taken to Los Alamos - because they had the scientific equipment - in order to analyse those elements and materials.*

He also talked about the "FTD" in the Airforce. The Foreign Technology Division and MJ-12 are mentioned in Ed's "Alien Rapture" book. In our interview, Ed told me:

> *The foreign technology division is mentioned even in classified documents - some related to the majestic 12 or MJ-12 group - about the Foreign Technology Division at Wright-Patterson Air Force Base - or FTD - and then if you look at the MJ-12 charter I received - it's in the book, it calls aliens and UFOs "foreign artifacts." So, I think not only did the foreign technology division analyse and reverse-engineer our enemies' weapon systems and aeronautical systems, but at some point, they started analysing and reverse engineering alien technology.*

This, of course, ties in closely with what Colonel Corso wrote in his own book, discussed in chapter 24.

Underground Bases

In our interviews, Ed stated that he worked at several facilities that had underground levels. When he was at Groom Lake, he worked in a Defence Advanced Research Centre (DARC) facility which had 10 stories underground.

At another time, he was working at the Jet Propulsion Laboratory (JPL) facilities at the Dryden Flight Research Centre, which was at Edwards AFB. It too had large underground tunnels. Some of these tunnels had 30-foot-wide entrances and Ed explained one or more of the tunnels he was in had other tunnels branching off. Ed did not know how long the tunnels were, nor did he know where they went.

Ed also said in another interview that he knew someone who worked at Nellis who spoke of there being tunnel boring machines there that were "in operation 24/7."

Ed Fouché and the NSA

Ed told me that the NSA was far bigger and more powerful than most people realised. He gave an example.

> *NSA has ...100 acres of underground computers in different facilities - including Fort Meade and the Fanex[?] facility and other facilities. The NSA is such a key mover most people don't realize that when Truman created... the CIA and the FBI, he also created the National Security Agency. The CIA was supposed to investigate overseas people and people in America that communicated with people overseas, if they seemed to be a threat. The National Security Agency was for internal threats and tracking people and information and to make sure classified material wasn't revealed - or certain Americans weren't plotting against America on American soil. But over the years, it's obvious what's come out in testimony, that the CIA's meddled, quite often, in America - as opposed to their charter, and the NSA has meddled in affairs around the world - as opposed to their charter, and their original intention. So the NSA has, without a doubt, had the most advanced computers - going all the way back to the 60s. Like I said, they have acres of underground computers and supercomputers ... it's incredible they would delay corporations releasing breakthrough technology... they would hire the best scientists and engineers and material scientists available - to work for the NSA to develop and keep the NSA's technology 20 or 30 years ahead of the rest of what is commercially available.*

While on this subject, Ed also described what he thought was the reason for the existence of the cover up regarding UFO's and aliens.

> *There were several studies done- in fact multiple studies - by the RAND Corporation and the Brookings Institute and other studies, that basically [said something similar to the] Jack Nicholson quote "You can't handle the truth." In other words, their conclusion was that civilized people - especially Americans because of our infrastructure and constitution - a lot of people believe was based on Christian faith, that it would destroy our faith and therefore destroy our society. So that was the number one reason - they didn't think we can handle the truth and covered it up. But over the ensuing years and decades since then, there's been uncountable atrocities - probably murders, people disappearing, people locked up in prisons or institutes. I mean, there's been so many crimes committed, as part of this cover-up - keeping the information from us, there's actually hardly anybody that was in charge of committing these crimes and unconstitutional acts it wants to be held accountable...*

The TR-3B – Part of a Secret Space Programme?

Though we perhaps assume that the TR-3B could fly in space, we don't have any evidence of this. Craft resembling the description of the TR-3B have most often been seen quite close to the ground (as we shall see in chapter 26). However, based on what Ed says about the propulsion system, and what witnesses have seen, there may be other "versions" which are used in space – who knows…

Ed said that he had heard that they did achieve "a full antigravity" propulsion system and that there *is* a space fleet, according to what he has been told. Ed thought the TR-3B (600 feet across) as a space platform is/was big enough to supply a fleet. He said "you can store an awful lot of stuff in that!"

Ed Fouché – Notes about the 1998 Disclosure

Around the time the Fouché/Steiger "Alien Rapture" book was first published, Ed did several interviews. For example, I already mentioned he gave an interview for Channel 4 when he was at the International UFO Congress. "Riddle of the Skies"[495] (it was also Shown on Discovery Channel). Ed told me that that he spoke to the documentary makers for several hours and showed them many of his documents – and they recorded all this. However, the final interview, as you will see in the programme, is only about 4 minutes in length.

Ed was scheduled to appear with Brad Steiger on the Art Bell show, but Bell cancelled for unknown reasons. Bell apparently said "he wasn't interested." Ed felt that Art Bell did not want people with real credentials to appear on his programme. When Ed and I discussed this, I suggested that this was because the "Coast to Coast" programmes were more for entertainment than anything else.

The Alien Rapture Book

As mentioned earlier in this chapter, Ed created most of this book, essentially, as a vehicle for disclosure. As he was an unknown author, it seemed like a good idea to "team up" with a known author. Ed was advised by lawyers ("JAGS") to write the book as fiction.

Of course, this means that many will disregard Ed as just a story teller and, as I am referencing Ed Fouché in this book, they will feel more comfortable writing off the entire contents of this book as being irrelevant because the author of it is prone to believing fictional accounts and wild fantasies…

Technology

From the way parts of "Alien Rapture" are written, however, one should consider that Ed *did* have a lot of detailed knowledge. For example, in chapter 4 of Alien Rapture, Ed described a particular surveillance satellite:

> *The solar powered TRW Stationary High Orbital Reconnaissance Electro-Optical Satellite (SHOREOS) was a computer linked telescopic system that utilized solid-state CCD array sensors to scan and to provide real-time infrared and visible wavelength images of objects of less than two inches—even through heavy cloud cover and darkness. Real-time data was provided by a Burst Laser Encrypted Compressed Digital Signal Transmitter (BLEC-DST). The ground-based computers at the Nellis and Groom facilities received surveillance, acquisition, tracking, and assessment data that tracked movement by type, speed, direction, size, mass, and infrared signature for the*

> *western United States. What made the SHOREOS unique was that the data transmittals could not be intercepted by anyone outside of Area 51 because of the Laser transmitter and other unique provisions of the system. SHOREOS allowed the operator to differentiate between a rabbit and a coyote, to read tail numbers off aircraft, and to identify license plate numbers on ground vehicles. The surveillance system's only weakness lay in areas experiencing severe weather-related electrical disturbance. Even that single flaw was a fluke or FLUC, something of a bad joke inside the NSA special operations arena. NSA operatives had switched the Hubble's 94-inch light-gathering mirror at NASA with one that had failed the manufacturer's quality control, along with a space fuzzy logic Unix computer (FLUC). They needed it for the SHOREOS satellite; the mirror was perfect, but the FLUC was flawed.*

There are a number of passages like this in the book which describe technology which seems much in advance of things that were described anywhere in the white world in 1998. Of course, to back all of this up, Ed produced documents to illustrate that he worked in areas where he *would* as part of his work hear about developments such as the one described above.

Another passage reads:

> *"Under the Department of Defense's supervision, Los Alamos works with Livermore on nuclear weapons and nuclear propulsion. Los Alamos also works with the materials associated with containment of the prototype magnetic field disrupter and the down-sized reactors. Livermore developed the plasma containment and superconductors."*

Was some of the MFD development work for the TR-3B done at Los Alamos?

Assassin

Also discussed is how assassinations are sometimes carried out – and he discusses one assassin called "Crawfoot" who is based on a real person and regarded himself as a true patriot. The person that the character Crawfoot was based on felt assassinations really needed to be done – it was just a job to him. He actually asked for a draught of the Alien Rapture book. He wanted Ed to use his real name, so his relatives would know that he did what he did in the name of patriotism…

Majestic Twelve

The book also describes a fictional situation in Washington DC - a meeting of some members of MJ-12 watches an organised march, where protesters are demanding the release of UFO information. Reference is made to the fact that from about 1994, there was more open discussion in the media of formerly secret agencies and black budget programmes. Ed writes how one of the MJ-12 leaders is not happy about any of these things. Ed describes how:

> *Until the end of the Cold War, NSA and its inner-workings had been the best kept secret in the government. It was the premier intelligence agency in the world, and yet the average citizen had no idea that it was larger than the F.B.I. and the C.I.A. combined in personnel, resources, budget, and, most of all, power.*

In the story, an outline is given of how MJ-12 came into being. Also discussed are details about MJ12, such as Eisenhower being "read in" to some of the programme. Another scene is described where MJ-12 members watch a fictitious NBC special "Exploring the Unknown" which includes a compelling video clip of a TR-3B – flying, then hovering - and then being "visited" by a larger extra-terrestrial craft. In the story, the clip was sourced from a fictional, long time investigator of UFOs called Blake Webster. He is characterised as "an arch enemy of MJ-12." Again, in the story, MJ-12 members are incensed and agree to get someone to make a statement that the video clip is an obvious hoax and thereby start a disinformation campaign.

> *General Stewart also rose to the challenge. "I'll fax our folks within the civilian UFO groups to start spreading the rumor that Blake Webster is a C.I.A. agent. They love to spread that shit about each other. They're always denouncing each other in those groups."*

In the story, the UFO film is then declared "a fraud" on the news.

The Sumerians

In the story, the lead character, Joe Green learns about the Sumerians and the civilisation that started there. With another character called Carlton Farr, he discusses the Sumerian civilisation and the belief that there were 12 planets – including one between Mars and Jupiter. The Joe Green character learns that ancient scrolls which discussed early history were obtained by the CIA. The Farr character remarks:

> *"The two greatest conspiracies conducted by church and state have to do with the suppression of all evidence of the ancient 'gods' and their involvement in the creation of humankind...and the fact that these so-called gods have returned."*
>
> *"And that planet existed, at one time in ancient prehistory, between Mars and Jupiter, where the asteroid belt now lies. And it is very probable that Mars served as a temporary home base for the aliens before they visited Earth."*

Aliens

Ed has stated on many occasions that he has not seen an alien, but talk of them cropped up many times in conversations he had with his contacts. In the Alien Rapture book, contact with the Aliens is discussed - with people going mad, becoming withdrawn and even developing psychic abilities. The fictitious alien race in the book is called the Jexhovah. The idea that the Jexhovah interfered with the genetics of Neanderthal man is covered – and that it is an ongoing programme of genetic modification. It is said that "seed androids" were also involved in this programme, earlier in history. Ed also writes, in the story:

> *These abductions are another matter. They are performed by an android that seems intent on a kind of genetic manipulation that involves the stealing of sperm, eggs, or fetuses from us for some kind of quality control process.*

The Secret Autopsy Report

The discovery of this report is woven into the story. For example, the characters find the report on a floppy disk they got from the Carlton Farr character. The text of this Alien Autopsy report from 1947 is included in the book. What is interesting is that this tied in with documents received independently by Jamie Shandera, Bill Moore and Stanton Friedman. It appears that Ed's contact gave him a re-typed copy of "attachment D" – which was the autopsy report. However, Ed pointed out that he knew the Ray Santilli video/film that was released in 1996 was a fake/remake. It was created and released, according to Ed, because the original film had "gone missing" so they needed to create disinformation about the film, in case it ever got released into the public domain in the future.

Other MJ-12 documents are also mentioned in the book, but discussion of these is outside the scope of this book.

Conclusion

Ed Fouche has probably provided more information about black programmes that were running in the USAF in the 1970s than anyone else. Of course, as he himself openly states, much of the information is second-hand. However, important parts of it are first-hand and he has provided ample documentation to prove he worked where he said he worked. As I mentioned earlier in this chapter, everything that Ed said that I have been able to verify makes sense. He was being truthful in everything he said, as far as I could tell. Perhaps this is why he needed to be "managed" and also attacked, when he became active again in about 2010. We will discuss these issues in chapter 30.

26. Sightings of "Secret Space Programme" Hardware?

In this chapter we will briefly discuss a few significant sightings, but of course YouTube and other searches will reveal many, many sightings – some with video. In many cases, the authenticity is questionable, but here, I have tried to focus on several cases of credible sightings. In chapter 28, we will discuss the circulation of disinformation, including at least one fake Flying Triangle Video.

Flying Triangles

Flying triangles, with a very similar general appearance to what Edgar Fouché described, have been sighted around the world since at least the 1970s.

In 1996, Omar Fowler of Derby, UK, published his independent study of these sightings.[496] His study, published before Ed Fouché's disclosures, contained this summary:

> ***Size:*** *The size of FT's appears to be very large, variously described as "ginormous," "the size of three Jumbo Jets" "the size of a football field, " etc.*
>
> ***Sound**: The FT will normally appear to be completely silent, although witnesses in close proximity have heard a low hum or drone. There are also reports of FT's making a "roaring noise" and "a penetrating booming sound" (Southport 23/2/96)*
>
> ***Speed of Travel**: They have been reported travelling as slowly as 30mph. They may also be seen in a stationary position, as low as 300ft. The FT's are capable of rapid acceleration from a stationary position to several hundred miles an hour in a few seconds.*
>
> ***Mode of Travel:*** *They usually 'fly' point forward, although there have been cases of them moving 'flat' side forward... They have been observed moving with a 'jerky' (stop and go) motion.*
>
> ***Electromagnetic Effects:*** *Magnetic disturbance may be noticed if a F T is in the vicinity, e.g. TV pictures may waver and the electrics may be affected on motor vehicles (lights cut off, etc.). A compass needle will flicker.*
>
> ***Associated Phenomena:*** *Balls of red and white 'star like' lights have been seen associated with FT's. They may detach from the FT and fly off independently.*
>
> ***FT Departure:*** *A flash of light resembling sheet lightning, may also be seen in the sky shortly after a FT's departure (Derby 15/2/95).*

Later in this chapter, we will briefly discuss the Belgian sightings in 1989 and 1990.

There was also a significant case of a flying triangle being sighted near why I live in 1997 – the year following publication of Omar Fowler's study. This is discussed in the next section.

Sheffield / Howden Moor Incident, Northern England.

This took place on 24 March 1997 and was thoroughly investigated by Max Burns[497] – who lived very near the area that the sighting took place. It appears the object was chased by Tornado jets, which flew from RAF Coningsby. After many months of investigation, Max Burns established a "sightings timeline" for the event. It appears that one of the Tornado jets crashed on Howden Moor in the UK Peak district, west of the city of Sheffield. (You will easily find debunkers such as Dr David Clarke and Andy Roberts attacking Burns' investigation.)

One witness was Emma Maidenhead who was interviewed by Miles Johnston (audios of the interview are available online). Emma heard two Jets fly over her home and "this huge thing" flew over the house. She said "It certainly was not a plane." She then said that "six jets were over the house - in three pairs." Mrs Maidenhead said that, "It was strange that the Jets were so low at that time of night." It was one of the nights that Comet Hale Bopp was in the night sky. (The same month when the Phoenix Lights were sighted in the USA.)

Time	Event
19.40	Bryan Haslam saw a triangle shaped object hovering
20.45	Tornado Jets witnessed leaving RAF Coningsby
21.45	Mr Brassington saw jets and light aircraft over Dronfield
21.45	Both sides of a local football match saw jets over Dronfield
21.45	Emma witnessed the UFO, jets, light aircraft, & helicopters
21.50	Mr. Rhodes witnessed a glowing orange object
21.52	The first seismic anomaly is recorded over the Peaks
21.55	Husband and wife both saw the triangle being pursued by jets
21.55	The UFO is tracked on radar for 10 minutes
21.55	Mrs. Dronfield saw a cigar shaped object glowing orange
21.55	The Gamekeeper & wife heard loud explosion & saw orange glow
21.55	Mr. Morton & mother saw an orange glow & smoke plumes
22.00	Sharon Aldridge & Joanne hear a very loud whirring sound
22.02	Sharon Aldridge & Joanne are circled by, an unmarked helicopter
22.06	The second seismic anomaly is registered
22.30	P.C. Alan Jarvis thought he saw Hale Bopp moving across the night sky
23.00	Mr. Dagenhart, encountered a man at Ladybower smelling of fuel
23.30	Dan Grayson had the triangle insight for 15 minutes & helicopter

The Star City Final

SHEFFIELD Tuesday, March 25, 1997 28p ★ YORKSHIRE'S BEST-READ NEWSPAPER

Relax

TUESDAY TRENDS

I'M MUCH STRONGER NOW

Moors plane crash riddle

Rescuers in massive search for wreckage

By Paul Whitehouse and David Clarke

EMERGENCY services from four counties were today involved in a massive operation to solve an X-Files style air crash riddle in South Yorkshire.

The operation was launched after a suspected air crash and explosion were reported on Peak district moorland near Sheffield.

Police treated the reports seriously because callers reporting the incident were so specific - even though air traffic authorities had no official reports of missing aircraft.

A young couple at Bolsterstone raised the alarm after seeing what they thought was a plane flying low over the moors at 10.15 last night, followed by a red flash and smoke.

Others in the area - including a group watching the Hale-Bopp comet through a telescope - are said to have seen an object trailing smoke and a strange red glow on the moors.

Minutes later gamekeeper Mike Ellison and his wife Barbara heard "one hell of an explosion" which shook their home near the Strines Inn.

Mr Ellison searched the moors without success, but a

■ Turn to Page 2

9 times a winner

NO wonder she's got a winning smile . . . actress Juliette Binoche has just picked up the Oscar for best supporting actress in The English Patient, bringing the British film's haul of trophies at Hollywood's biggest night of the year to an amazing NINE awards

FULL STORY: P10/11

NEWS BRIEFS

Naz backing booze battle

Boxing champ Prince Naseem Hamed is backing The Star's campaign against under-age boozers.

PAGE 3

Hospital date

SHEFFIELD's £2m new accident and emergency unit at the Northern General Hospital becomes the city's only centre for emergency patients. We look at the implications.

PAGE 6

Kick him out

THE latest victim of a Sheffield sex pest has demanded his eviction.

PAGE 7

ACTION DESK P14 WEATHER P22 STARS P24 BREAKTIME

Michael Schratt and Flying Triangles

Michael Schratt, who we mentioned in chapter 10, has done extensive research into the flying triangles[498] and talked about them in many of his presentations – which are available on YouTube[499] – although he doesn't seem to have his own website. Michael has done detailed CAD drawings and renderings and visualisations, based on the descriptions given by witnesses. His work has been very important, although he does seem somewhat reserved when talking about certain cases.

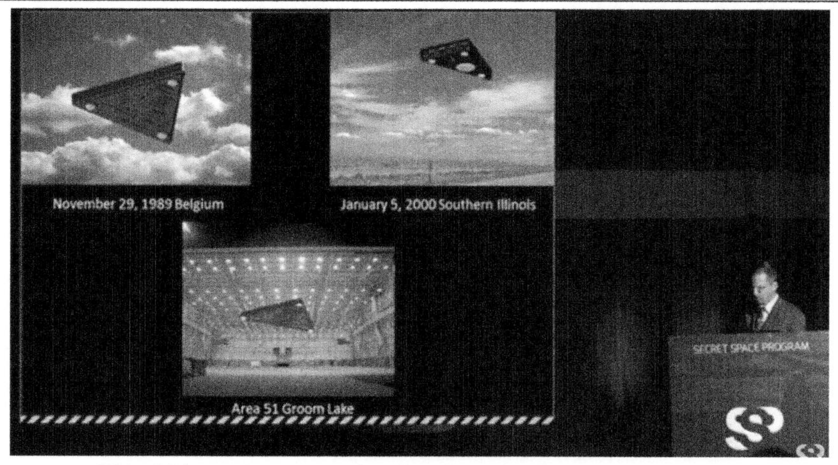

Michael Schratt – 2014 Lecture San Mateo Secret Space Programme Conference

Richard D Hall and "Almost Identified Flying Objects"

In 2013, chartered engineer, film maker, researcher and investigator Richard D Hall produced a new documentary called "Almost Identified Flying Objects."[500]

Hall's film mentions some of the information we have covered earlier in this volume and includes a segment of an interview with Military Intuitive Communicator Dan Sherman[501]. Hall proposes that some of the craft we are seeing could be the result of collaboration between humans and aliens. This has been claimed by several witnesses such as the one that William Hamilton mentioned (and we covered in chapter 21) and Capt. Bill Uhouse, for example, who was one of Steven Greer's Disclosure Project witnesses.[502]

Hall's film references the research of Michael Schratt (mentioned earlier) and also includes reference to a story published in the UK Independent Newspaper from March 1997, which refers to the crash of an advanced spy plane in 1994 at Boscombe Down, UK.[503]

Hall collected various UFO accounts and also noted that several people described they had seen (among other things) a large black triangle "the size of a football pitch." Hall and Schratt have also noted the 1981 Hudson Valley UFO flap[504], where many witnesses reported seeing triangular shaped UFOs.

Hall's film also covers the 1990 Belgian UFO triangle sightings in some detail and includes a clip from a hard-to-obtain interview with Professor Emile Schweicher.

Image from Richplanet TV – note the correct spelling is Professor Emile Schweicher.

Halls states (and this affair has been written up extensively by other researchers)

> *In 1990 the Belgian Air Force scrambled F-16 fighters to try and intercept them, but were outmanoeuvred every time. After days of activity the Belgian Air Force had no explanation. ... Professor Emile Schweicher ruled out any form of radar anomaly and concluded the radar systems were tracking real objects.*
>
> *Incredibly, in 1991 Schweicher was asked by the British Ministry of Defence to come over to the UK to give British military chiefs a briefing on the movements of this unknown craft and explain the radar data of the event. We know this because Schweicher himself told researcher Harry Harris when Harris managed to get an interview with him in his hotel whilst in the UK.*

Schweicher then states, in the clip that:

> *What I found by calculations starting from the data recorded by the pilot, that the UFO made, at one point of his trajectory, a very strange motion and during that very strange motion the normal acceleration was, let's say, about 30 G's. The trajectory was, firstly, the UFO was flying almost level and then suddenly made a 90-degree motion. At that point where he made the 90 degree turn, that point, that normal acceleration was about 30G's - climbed almost vertically and then descended obliquely to the ground during the ... descent the UFO was flying supersonic...*
>
> *I would say that with the present technology, that would not be possible.*

As Hall notes, the UK MoD have never publicly admitted they requested a meeting with Schweicher in the UK. The rest of Hall's film contains edited interviews with FT witnesses from around the UK and he notes the consistency of the characteristics they describe. Hall also discusses the possibility that some vehicles "hide" by submerging themselves in reservoirs – he discusses several witness accounts relating to this. He also notes there are no accounts which suggest a triangle has landed at a UK air base. Overall, this adds more weight to an argument that the Flying Triangles are in fairly widespread use.

NIDS Report on Flying Triangles

As we mentioned in chapter 19, in 2004, Robert Bigelow's National Institute for Discovery Science published a report entitled "NIDS Investigations of the Flying Triangle Enigma."[399] This report attracted enough attention to warrant a posting on the popular Space News Website Space.com.[505] This posting stated:

> *A key NIDS conclusion is that the actions of these triangular craft do not conform to previous patterns of covert deployment of unacknowledged aircraft. Furthermore, "neither the agenda nor the origin of the Flying Triangles is currently known."*

The NIDS report includes this discussion:

> *The years 1990-2004 have seen an intense wave of Flying Triangle aircraft, as measured by three separate databases. The major finding in this report is that the behavior of the Flying Triangles, as related by hundreds of eyewitnesses, does not appear consistent with the covert deployment of an advanced DoD aircraft. Rather, it is consistent with*
>
> *(a) the routine and open deployment of an (unacknowledged) advanced DoD aircraft or*
>
> *(b) the routine and open deployment of an aircraft owned and operated by non-DoD personnel.*
>
> *The implications of the latter possibility are disturbing, especially during the post 9/11 era when the United States airspace is extremely heavily guarded and monitored. In support of option (a), there is much greater need for surveillance in the United States in the post 9/11 era and it is certainly conceivable that deployment of low altitude surveillance platforms is routine and open. The data from the NIDS, MUFON and NUFORC (see below) Triangle databases show an increase in sightings of Triangles after 1997, so if option (a) is correct, large-scale deployment of the Triangles was initiated over the United States, possibly towards the late 1990s. Whether the deployment of the Flying Triangles is exclusively confined to population centers is unknown, since the majority of the NIDS reports come from cities and from Interstate Highways where people are clustered. NIDS does have some reports of Flying Triangles from remote areas.*

The conclusions imply, perhaps, that FT's were being deployed sporadically in the 1970s and 1980s and, perhaps, once they had monitored the reaction to this and decided what to do, they continued to deploy more craft.

However, we can disregard the comments relating to the "post 9/11 era," because whomever wrote the report did not know what really happened on 9/11 (although they can be forgiven for this, because the report was written four years before the known truth about the technology that destroyed the WTC was fully disclosed.)

"Night Vision" UFOs

There are a number of interesting videos online, now, showing what appear to be aerial objects filmed through third or fourth generation night vision equipment. In some of the videos, fast moving objects can be seen rapidly changing direction.

Oakmont PA, USA

I found a video published in 2011 on a channel called "seeingUFOsPA[506]" which contains such objects. The video, with a title "Fastwalkers" - Many High-Flying UAP[507] was filmed on October 10, 2010 near Oakmont, PA., at a summer house, using an ATN brand generation 3A Nightvision Scope with a 5X magnification lens attached to Sony HD camera, providing extra zoom. The videographer freely admits that some of the objects could be bats, birds or insects, but I don't think this explains all the objects seen in this video.

Freemont, CA, USA

Another video was filmed on 5 December 2008, over Fremont, California[508]. A caption at the beginning of the video reads: "A Big Thanks To ded995064 For Sharing His Video With This Channel -- knightskross." According to the video description, it was "filmed with ATN PVS-7 third generation night vision goggles, a Sony IR/UV modified camcorder (good for daytime sightings) and a Sony HD camcorder (recent night vision footage)." The video poster notes "Any camcorder will do, but you will need a third generation night vision, device if you want the detail/quality.)

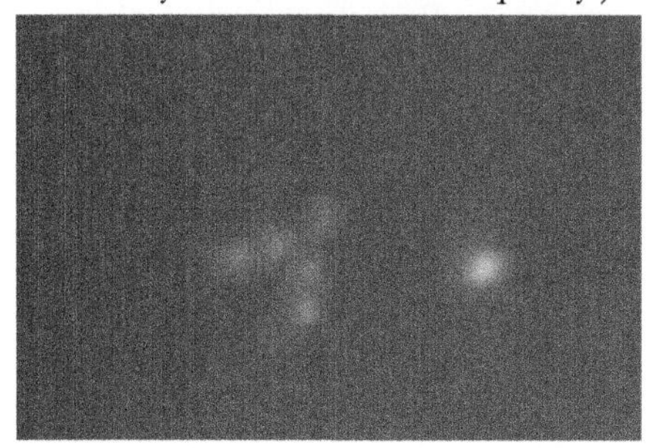

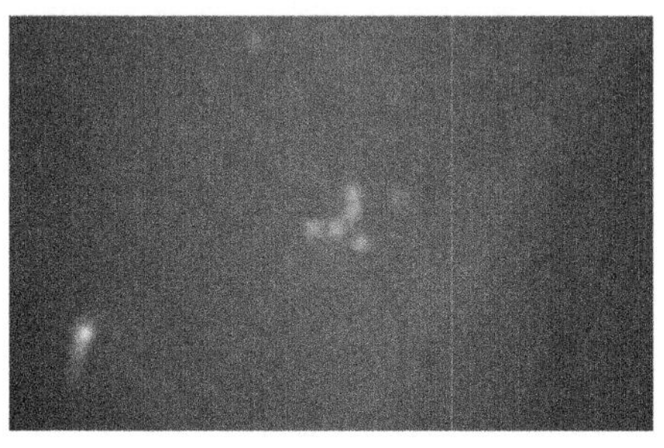

Stills from an extraordinary Night vision video clip.

The video is only 1 minute in length and the trajectory seems realistic and the object goes in and out of focus and it appears that wires get in the way of the view. Similarly, it appears that aircraft contrails are still present in the sky. Of course, there is a possibility that it could be a small drone, but as this video was filmed in 2008, this is far less likely, as drones were far less common then. Are we looking at a video of a variant of the TR-3B, going into space?

Ed Grimsley – San Diego, CA

Ed Grimsley[509] received some attention for videos he recorded starting about the same time as the one described above – i.e. about 2008. A particular video he shot on 30 April 2011[510] seems to show a UFO pause and reverse direction. This is worth watching. Other videos of his are less compelling and it is not always easy to see what he claims is there.

Space Shuttle Mission STS-48

There are at least three space shuttle missions from the 1990s which yielded some highly interesting evidence. A full study of these videos is beyond the scope of this book, but I can recommend a video study by David Sereda, made in 1999, which he called "Evidence - The Case For Nasa UFOs".[511] This video shows and analyses a number of interesting sequences and also includes comments by Dr Edgar Mitchell, which we will cover in chapter 27.

For now, I will focus on STS-48 and an analysis of a section of video supplied by Don Ratch – an independent investigator who recorded the original from NASA Select – a cable TV channel. The video may have also been recorded separately by Martyn Stubbs.[512] This clip - recorded on September 15, 1991, between 20:30 and 20:45 UT - shows an object flying towards the earth, on the night side, and then suddenly changing direction, just before a faint streak is shown rapidly moving towards the vicinity of the object. This is illustrated in this helpful timelapse image, produced in 1999 by Richard C Hoagland.

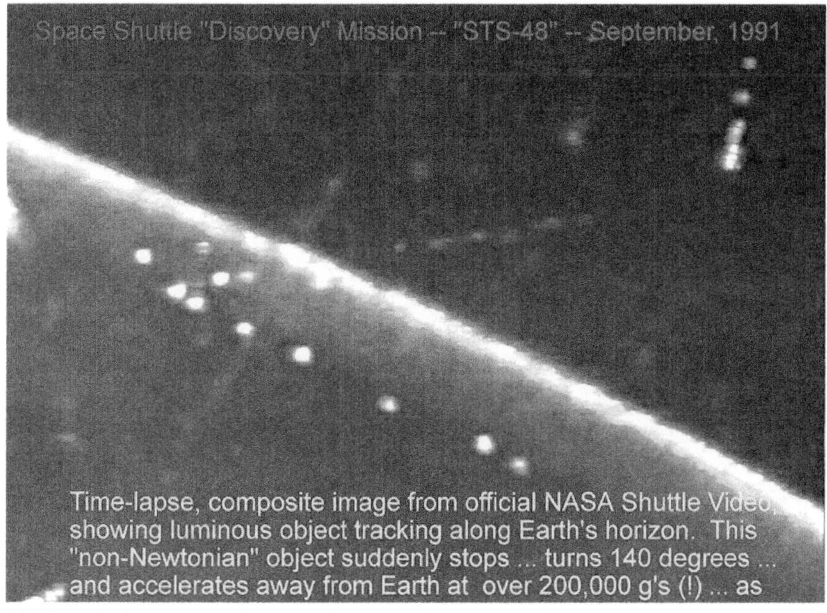

Space Shuttle "Discovery" Mission -- "STS-48" -- September, 1991
Time-lapse, composite image from official NASA Shuttle Video, showing luminous object tracking along Earth's horizon. This "non-Newtonian" object suddenly stops ... turns 140 degrees ... and accelerates away from Earth at over 200,000 g's (!) ... as

This image includes captions by Richard C Hoagland. We don't really know what the streak was, even though Hoagland suggested it was a "Particle Beam Weapon."

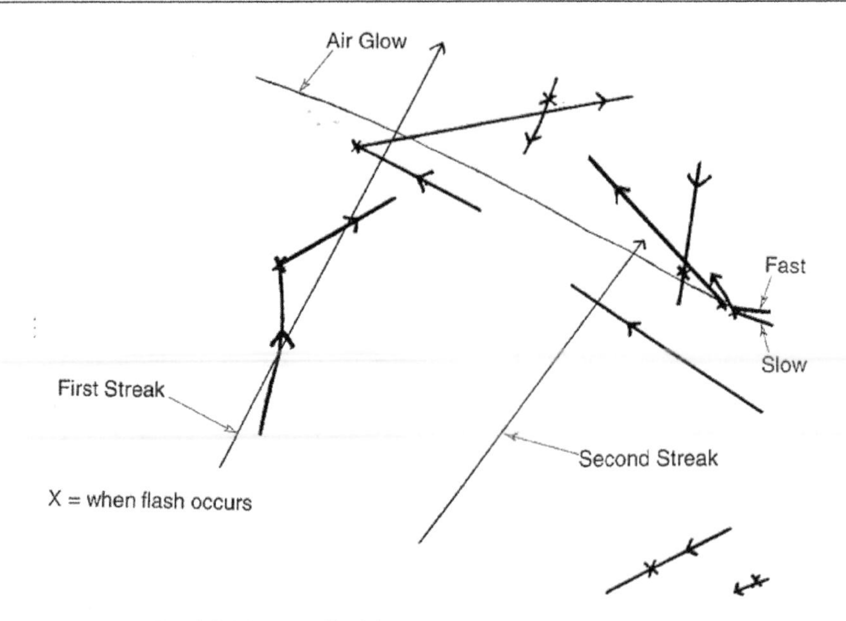

Figure 1: Trajectories traced by placing a transparency over a 13-inch television screen.

Dr Jack Kasher's 1994 tracing of the trajectories of the objects.[513]

In the image above, the faint line running between the 7 O'clock / 1 O'clock positions appears in the video to be something which is fired at the object travelling towards the 10 O'clock position from the 4 O'clock Position.

Donald Ratch, thought the sequence was unusual enough to alert his congresswoman, Ms. Helen Bentley, requesting that she look into the matter. Ratch also contacted Vincent DiPietro, a NASA engineer at Goddard Space Flight Centre in Greenbelt, Maryland (who I mentioned in Secrets in the Solar System[104]), and he found it interesting enough that he contacted his congresswoman, Ms. Beverly Byron, asking her to examine the recording. The congresswomen then sent the tape on to George E. Brown, Jr., chair of the Congressional Committee on Science, Space, and Technology and Martin P. Kress, an assistant administrator for legislative affairs at NASA. It was then looked at by various scientists and letters were sent back to the people that enquired about the tape.

Three years elapsed and then the clip was independently analysed by Dr Jack Kasher who is/was a professor of Physics and Astronomy at the University of Nebraska at Omaha. In April 1994, Kasher published a detailed report about the brief STS-48 incident[514] in a MUFON journal and I will quote portions of this report below. For a similar analysis, you can also view a 20-minute video by Richard C Hoagland, which was originally released in 1993, sometime before Kasher's report.[515] (So, we are talking about events that happened 27 years ago and were analysed extensively 25 years ago…)

In his report, Kasher describes what was recorded:

> *The TV camera located at the back of Space Shuttle Discovery's cargo bay was trained on the Earth's horizon while the astronauts were occupied with other tasks. A glowing object suddenly appeared just below the horizon and slowly moved from right to left and slightly upward in the picture. Several other glowing objects had been visible before this, and had been moving in various directions. Then a flash of light occurred at what seemed to be the lower left of the screen; and the main object, along with the others, changed direction and accelerated away sharply, as if in response to the flash. Shortly thereafter a streak of light moved through the region vacated by the main flash, and then another streak moved through the right of the screen, where two of the other objects had been. Roughly 65 seconds after the main flash, the TV camera rotated down, showing a fuzzy picture of the side of the cargo bay. It then refocused, turned toward the front of the cargo bay, and stopped broadcasting.*

It is worth watching this clip a few times and it becomes even stranger if watched at a higher playback speed.

Dr Jack Kasher included the response to the letters of enquiry to the congress women, mentioned above, in an appendix of his report. In summary, the objects were simply labelled "ice particles" which changed direction because they were affected by the small shuttle "attitude adjuster" rockets being fired as part of a manoeuvre. It appears that the fast-moving streaks were not really discussed in the responses though.

Kasher points out that the ice particles would have to be quite near the shuttle camera for them to have been seen. Other explanations included that the particles were very small, but very close to the lens. The usual explanations of "space junk" and "meteorites" were mindlessly served up, completely disregarding (a) one of the objects' sudden change of direction and (b) the timing of the change of direction being associated with the flash and the "streak." Kasher rules out some explanations:

- Objects could not have been tiny particles close to the camera lens - camera was focused at infinity - the side of the cargo bay is visible.
- Cargo bay was obviously out of focus at first, so any tiny particles near the camera would have been blurred or not visible at all.

Kasher's report includes a more detailed analysis of the video, based on looking at a pixel map that he made, using some specialist equipment at the University he worked at. (This sort of analysis was rather more difficult to perform 25 years ago.) He looks at how quickly an ice particle could have accelerated, based on the action of the shuttle's thruster and the distances and times measured in the sequence of frames in the video. Following further detailed analysis, Kasher concludes:

> *I realize that I have spent a great deal of time proving that the main object in the videotape was not an ice particle (perhaps some would think an excessive amount of time). But I think that it is absolutely necessary that I do so. As I showed above, there are only two possible explanations--that the objects were ice particles near the Shuttle, or that they were spacecraft maneuvering out in*

> *space away from the Shuttle. The ice particle theory must be shown to be completely and thoroughly out of the question, because the ramifications are truly extraordinary if the objects really were spacecraft. ... Since the shuttle was about 355 miles above the Earth when the film was taken, the horizon would be slightly more than 1700 miles away. Obviously, the craft was somewhere between the Shuttle and this distance.*

Kasher calculates a range of accelerations and velocities, depending on how far the main object was from the shuttle:

Distance (Miles)	Acceleration	Final Velocity (mph)
1710 (horizon)	18,000	430,000
1000	10,500	250,000
100	1,050	25,000
10	105	2,500
1	10.5	250

Kasher therefore concludes:

> *Except for the one-mile distance, these numbers are far beyond the capability of any known earthly craft. Even if the ship were as close as 10 miles, the 100 g acceleration would absolutely flatten a human pilot.*

Kasher then speculates that the two streaks are missiles being fired and that, of course, this can either be interpreted as a hostile act, or as some kind of incredible test of military hardware. The likely conclusion that something was being fired would almost certainly imply that whatever was fired was many miles from the Space Shuttle, due to the risk it would have been to the shuttle. Kasher then comments about the SDI/Star Wars programme, as others have speculated the event was a test of that system. Kasher wrote:

> *The Star Wars program instituted by President Reagan in 1983 has not been developed to the point of being operational, and probably would not be so for at least ten years, if it will be at all. So any top secret program that might have been seen on the Shuttle videotape would have to have been under development for some time, perhaps even decades. This implies a secret level of scientific achievement that is much beyond what is normally attributable to the military today, or to any scientific establishment. It also implies that the current Star Wars system has been a tremendous waste of money, since equivalent defense systems have already been secretly developed over the years to the point of being operational.*

I find this, as the last point, quite interesting, in view of what Kasher stated in an interview he gave for "Riddle of the Skies" – a 1999 documentary[516] broadcast on Channel 4 in the UK and Discovery Channel elsewhere. (The documentary shows the STS-48 clip, along with a summary of Kasher's analysis. Of course, the "snow/ice" particle explanation is put forward, but even the person talking about it admits it does not explain the movement of the object!) Just before the analysis is shown, Kasher states:

> *I studied the physics of the effects of nuclear weapons, among other things. I specialized in electromagnetic theory, electromagnetic pulses...* ***did some work on the Star Wars program****, out there. In 1991, I started doing research*

> *with NASA. My clearance was up to the secret restricted data.* ***In fact, one paper that I wrote was so classified that I couldn't read it myself*** *- I had to get special permission to read it.*

So, Kasher claims he had a top secret clearance, though he expresses this in a peculiar way. Was his analysis something of a "limited hangout?" It would be interesting to learn how Kasher came to be asked to do the video analysis. He does not state this in his report, though it appears it was done through MUFON (Mutual UFO Network), which some have said is run by the intelligence agencies anyway – or at least is heavily steered in certain directions by them.

SOHO "Stealth" Object

Another interesting set of images was posted – as a video - by Peter Dunn in May 2012[517]. I include a single still image below, but this is really only for reference, as you can only see the significance of what is being shown by studying and understanding the sequence of images in the video.

First, we must explain what SOHO is – I wrote a whole chapter about other anomalous SOHO images in Secrets in the Solar System[104], so please review that too!

The image below came from the NASA/ESA Solar and Heliospheric Observatory – a probe that was launched in 1995 to study the Sun – and it is still in operation today (May 2018). This mission is not widely known about – as the area of research that SOHO is concerned with is not of that much interest to people in general. Also, none of the pictures returned by SOHO resemble normal optical photographs, so many people would not understand what they are looking at if shown these pictures "cold."

The images of interest were taken between 06 April and 03 May 2012 – almost a month! What is of interest is a 2 x 2 pixel portion of the image below.

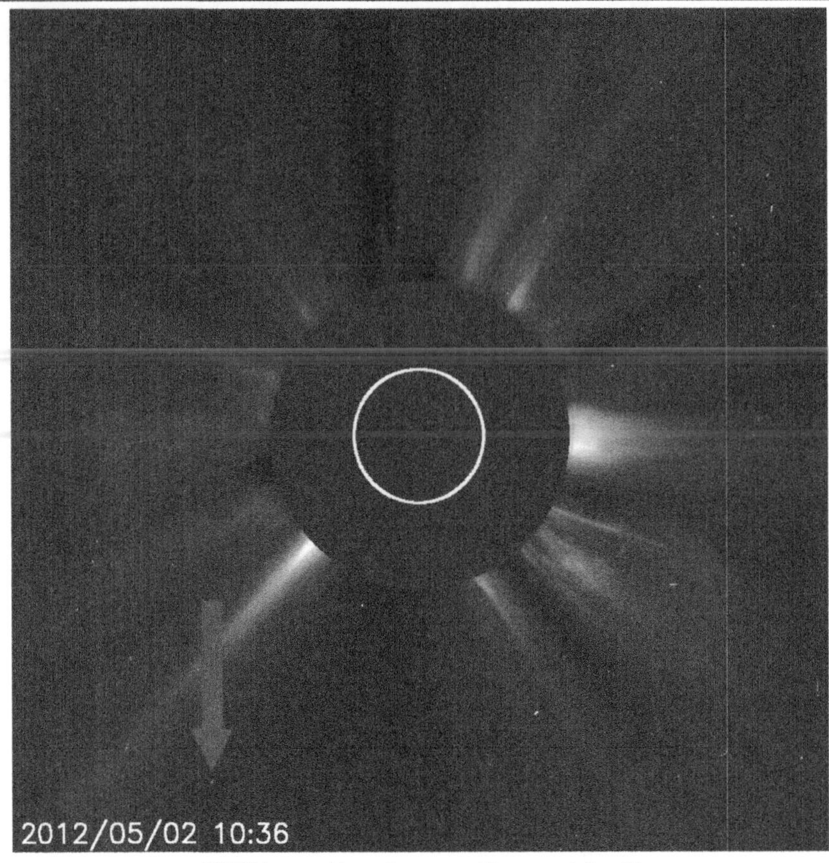

SOHO image with small mystery object – see below.[518]

You can find the original images via www.helioviewer.org. Over the 16-day period, the small portion appears to fade in and out of view – and seems to be influenced by the effect of Coronal Mass Ejections[519] from the sun. I have enlarged the relevant section of the image below and enhanced the contrast.

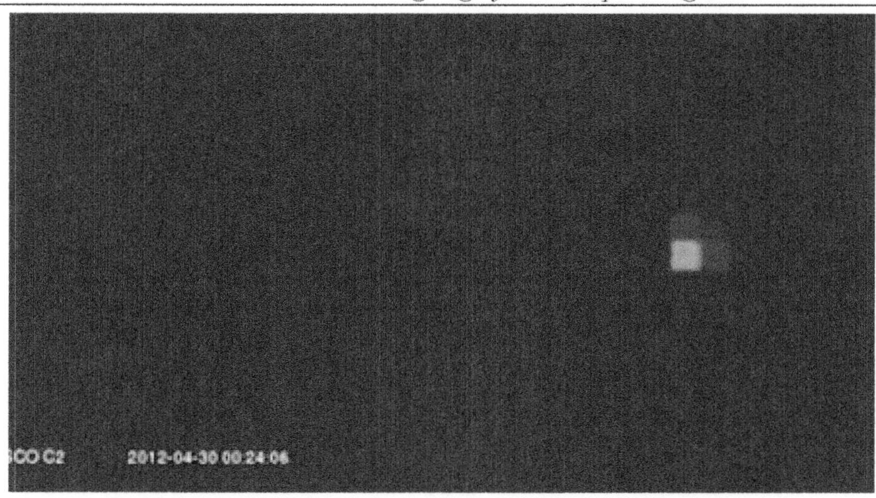

The object moves, very slowly, relative to SOHO over a 2-week period. The images above are from 17 April 2012 and 30 April 2012 (find them on Helioviewer.)

To understand what these images might be showing, we need to know that, as I discussed in chapter 20 of Secrets in the Solar System[104], the SOHO probe occupies a particular point about 1 million miles from the earth, where it has an uninterrupted view of the sun. It sits in the "First Lagrangian Point (L1), where the combined gravity of the Earth and Sun keep SOHO in an orbit locked to the Earth-Sun line."[520] It is *always* pointing at the sun and sits in the same orbital plane as the earth. It is about four times further away than the moon. In other SOHO images, other planets and bright stars are seen, as are comets and the afore-mentioned Coronal Mass Ejections (CME). You can look at over 20-years-worth of images now and establish that what is shown in the sequence above isn't a planet, a star or anything else like that. The object also changes its appearance in relation to the occurrence of a CME. This means the object is probably in between the earth and the Sun – or it could be much further away, in which case it would be extremely large – and therefore should be easily detectable from the Earth in some manner (even by occultation). The likelihood, then, is that it is in a similar orbit to SOHO – perhaps even sitting in the same Lagrange point.

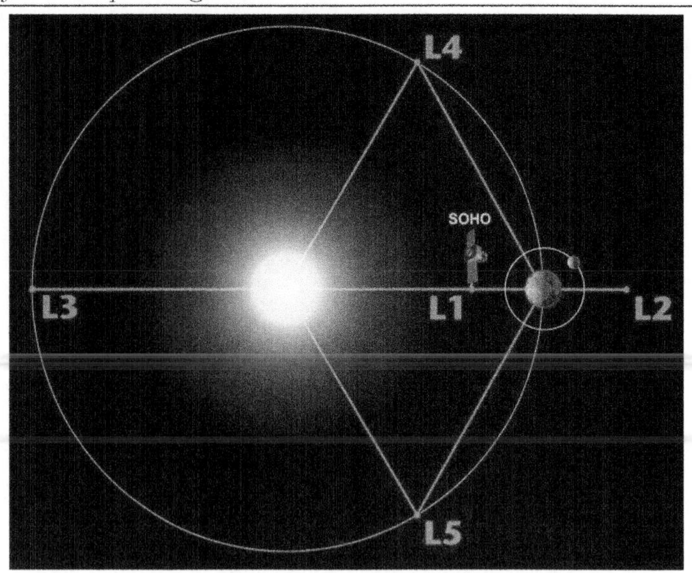

Position of SOHO Graphic adapted from NASA Website.[521]

As the name of the video implied, it is a "Stealth" object – at times it is completely invisible. My guess then, is that what we are looking at is some type of Space Platform. Could this be one that is used as part of a Secret Space Programme?

Part 4

Maintaining Truth's Protective Layers

27. Who Was Edgar Mitchell?

Dr Edgar Mitchell

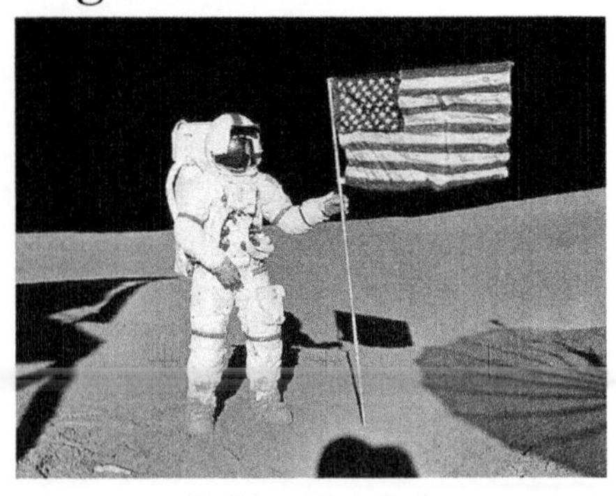
Dr Edgar Mitchell…?

In the past I have, for reasons discussed at length in chapters 14 to 18, described Dr Edgar Mitchell as "The Sixth Man NOT to walk on the Moon." For several years I believed him to be a genuine whistle-blower, in relation to the ET/UFO issue. In this chapter, we will learn that Mitchell was not telling and perhaps could not tell us the truth about the important things he spoke of. I originally wrote this chapter as an article for my website in June 2015[522]. Edgar Mitchell passed over on 04 Feb 2016[523].

Edgar Mitchell on Aliens/ET's

In the fields of ufology and paranormal research, the name of Dr Edgar Mitchell is one that comes up quite frequently. Since at least 2003, he was heard to make incredible statements about events which happened in 1947 near Roswell, New Mexico. For example, in an interview for the documentary "Out Of The Blue,"[524] which was produced by James Fox and shown on the sci-fi Channel, Mitchell "started off gently" and said

> *In the current period, until the last 30 years it was conventional wisdom, both in science and theology, we are alone in the universe – the single repository of life, anywhere in the known universe. Well, no one believes that any more….*

Later, in the same documentary, he went on to say:

> *Well, if the accounts are to be believed, yes, there have been ET visitations, there have been crashed craft, there has been err material and bodies recovered and there is some group of people somewhere that may or may not be associated with government at this point, but certainly were at one time that have this knowledge.*

In Feb 2004, he was quoted as making similar remarks in St Petersburg, Florida. His words were included in a local newspaper[525] (now called the Tampa Bay Times[526]) in an article entitled "Former Astronaut: We've Had Otherworldly Visitors"

> *"A few insiders know the truth . . . and are studying the bodies that have been discovered,"*

Also, in a July 2008 interview with Larry King on CNN, Mitchell appears to validate the stories of witnesses to the Roswell incident[527], claiming that he was told certain things by them before they died and he states that an "alien craft" did indeed crash there. However, he does not give any specific details about who spoke to him, or what they said. Hence, Mitchell's statements were essentially useless in helping us to find out any more about the Roswell incident than has already been uncovered by other researchers.

Another instance of Mitchell making comments (or rather not making comments) about what appear to be intelligently controlled objects in space, was included in David Sereda's excellent documentary called "Evidence: the[528] Case for NASA UFOs[529]". In this documentary, David Sereda completes an extensive analysis of the STS 75 "tether incident," where many strange objects are seen in the field of view of the camera on the space shuttle as it attempts to video the results of an experiment. The experiment involved unravelling a 12-mile long, thin wire to perform an energy related experiment[530]. (It is strongly recommended that you review this footage for yourself.) This wire was the "tether" for a satellite that was released from the shuttle's payload bay. As part of his investigation into this extremely unusual NASA footage, Sereda decided to contact Edgar Mitchell[531] and send him a copy of the NASA video. Sereda states that, in response to his query about the "tether" footage, on November 24, 1999 Edgar Mitchell wrote the following letter to him[532]:

> *David Sereda,*
>
> *Yes, I have received your film and reviewed it and the info package you provided. I see utterly nothing about the tether incident that is particularly interesting. If there is more revealing footage then I will look at it. However, I have looked at many feet of space film, and have yet to see only one that has anything worth looking into regarding UFO appearances. Edgar Mitchell.*

Sereda, unhappy with Mitchell's response wrote to him again and Mitchell responded further, thus:

> *David,*
>
> *At first glance they are just particles--------perhaps outgassed, I do not know. But I certainly would look for a more prosaic explanation than UFOs. If you have a better film, I will look, but I am not very optimistic they are anything exotic. EDM.*

Still troubled by this new response, Sereda wrote to Mitchell again to ask him to study the footage more closely. Mitchell then responded:

> *David,*
>
> *Okay, I will look at the tape, but be hard to convince that anything of significance is occurring. I am not impressed by physicists from M.I.T. or anywhere else in this arena. Blobs of light routinely appear on film if being reflected from some small object in space. Look up in the sky in the early morning and see the reflections from a satellite as the sun comes up--------the reflection is many times larger in apparent size than the actual object. Star light is the same. Distant stars would not be visible at all except for the spreading and diffusion of the light, which makes them visible. EDM*

To understand how ridiculous Mitchell's responses are, one needs to study the "tether" footage for oneself (so please do!) Of course, Mitchell needn't have responded to David Sereda at all – it is as if Mitchell appears to be working hard to show himself as "a good guy." That is, he has shown himself to be approachable and responsive to Sereda's letters and questions. However, there is something amiss…

IONS

Mitchell was also well-known as the founder of the Institute of Noetic Sciences (IONS) – an act he allegedly took as a result of "an Epiphany" which he experienced on the return journey from the moon[533]. IONS includes the following statements on its website[534]:

> *The Institute of Noetic Sciences™, founded in 1973 by Apollo 14 astronaut Edgar Mitchell, is a 501©(3) non-profit research, education, and membership organization whose mission is supporting individual and collective transformation through consciousness research, educational outreach, and engaging a global learning community in the realization of our human potential. "Noetic" comes from the Greek word nous, which means "intuitive mind" or "inner knowing." IONS™ conducts, sponsors, and collaborates on leading-edge research into the potentials and powers of consciousness, exploring phenomena that do not necessarily fit conventional scientific models while maintaining a commitment to scientific rigor.*
>
> *The Institute's primary program areas are consciousness and healing, extended human capacities, and emerging worldviews.*

This all sounds quite attractive and benevolent and rather appealing to people (like me) who have become disillusioned with the straitjacket of mainstream science and feel a lack of fulfilment in following mainstream religions. However, reading the text above I am immediately disconcerted by the inclusion of a "trademark" after the name of the Institute is written. What's that all about?

It is clear from the above that IONS concerns itself with topics relating to consciousness. The study of consciousness is far more important than most people seem to realise – not least because understanding consciousness more fully can have potentially sinister outcomes. One simple example is the use of subliminal messages to implant things in our subconscious mind. And then there is Mitchell's admission of his interest in the alleged psychic Uri Geller… Indeed, Mitchell himself explained this interest in Geller in the uncut version of a 2013 BBC documentary called "The Secret Life of Uri Geller"[535]. (Mitchell's interview did not appear in the uncut/TV version of the documentary.)

> *An American physician by the name of Andrea Puharich called me after I came back from the moon and said "I've found this rather amazing Israeli young man who can do these sorts of things. Would you like to study him? And I was intrigued and I said "Yes, bring him to the States and let's take a look." And then I set out to arrange the testing that we did with him in the Stanford Research Institute where I met Hal Puthoff and Russel Targ. What I did not know at the time was that the work at SRI was also being funded by our CIA. I only discovered that later.*

Later, in the same documentary[536], he goes on to state:

> *As we finished our work with Uri at SRI, I was called by the head of the CIA and asked to come to Washington and brief him on what we had learned - and that head happened to be George HW Bush. He was head of the CIA at that point in time and it turned out that it was all part of the same big study. The Russians were studying it, we were studying it, the CIA were studying it, I was studying it and so it was all a part of the study of so-called study of parapsychology. I didn't know, at that time, all that the CIA was involved in…*

One could naïvely conclude that Mitchell really did not have any knowledge of the CIA's involvement in research into parapsychology, but further (rather troubling) research draws one to the conclusion that Mitchell was, indeed, working for the CIA - along with the afore-mentioned Hal Puthoff and Russel Targ.

Following a study of the sordid catalogue of Apollo anomalies (a few of which we covered in earlier chapters), it becomes obvious that Mitchell could never have walked on the moon as he has claimed[537]. Earlier, we referred to Bart Sibrel's film "Astronauts Gone Wild."[538] This film features some "ambush" interviews with several astronauts, which are uncomfortable to watch, but are, nevertheless, most revealing. For example, in the interview with Mitchell, the following exchange can be heard:

> *Sibrel: would you swear on the Bible that you walked on the moon?*
>
> *Mitchell: You bet your sweet ass.*

Sibrel then reads out the oath, which Mitchell repeats, with his hand placed on a Bible. Mitchell then states his view of the Bible, and he is clearly very uncomfortable with what has just transpired. He then repeats:

> *We did go to the moon. You bet your sweet ass we went to the moon.*

There is an edit in the interview and then Mitchell states that Sibrel should turn off the camera as he is "done" with the interview. Sibrel then (perhaps unwisely) tries to exchange pleasantries, but Mitchell becomes rather agitated and says:

> *I don't say it's been a pleasure… Please get your ass out of my house and you came here under false pretences and I think you're an asshole - and if you continue this and if you press this I will personally take you to court.*

Sibrel responds:

> *"I hope that you do… I invite you to."*

Following a further exchange, in which Sibrel says:

> *Lying about walking on the Moon is a satanic lie.*

Mitchell then says:

> *I don't hit people but you're going to be on the deck unless you get the heck out [Mitchell knees Sibrel in the rear]*

As Sibrel is leaving, we can hear that Mitchell still has on the radio microphone and his camera is still rolling. Mitchell is heard to say to his son (who has joined in the exchange)

> *You want to get a gun Adam and we can shoot at him before he leaves ...*

As Sibrel and his cameraman are getting into their car, the radio microphone is still recording and Mitchell's son is heard to say:

> *Want to call the CIA and have them waxed?*

It seems pretty clear to me that a very dark side of Mitchell's background is revealed in this brief and disturbing clip. Who would say such a thing?

Sadly, it has become apparent that this wasn't the first time that Mitchell had talked about guns and people being shot. In chapter 7, we discussed the work of Bruce De Palma in relation to anomalous effects on gravity seen in rotating objects. In other research, De Palma developed something he called "The N Machine" which was a closed path homopolar free energy generator. De Palma had an encounter with Mitchell in 1980, which he discusses in an interview[539] when he states:

> *One of the pivotal people I encountered early on in my career was Edgar Mitchell the astronaut... He said to me that if I tried anything on my own in California I would get my head blown off - so I was scared to death.*

DePalma goes on to state:

> *The CIA operates through various innocent looking fronts – to find out what people are thinking and what they're inventing. Now, what's more innocent than a benign institute – founded on transcendental principles to help New Age inventors bring free energy into the world?*

For those that have not yet made the connection between Edgar Mitchell and the large deceptions relating to the space program and free energy which are being perpetrated on the general population, we can say that, in the words of Richard D Hall[540], they are "losing the information war."

Fortunately for us, Bruce DePalma has left us something of a legacy of information on his website and even on the home page you can find a little bit more about his dealings with Edgar Mitchell[541].

> *The first threat came from Edgar Mitchell, the astronaut who told me that the Government had said there "never was any doubt that the N machine was the Free Energy machine they were looking for, and if I tried anything on my own in California I would have my head blown off. And the CIA warned me via Mitchell that I should not leave the country because I would be kidnapped. This was back in 1980, when these efforts frightened me out of going to Hanover, Germany for Dr. Nieper's first Gravity Field Energy Conference, which is where P. Tewari brought his N machine from India and got the prize for the most clearly observable Free Energy phenomena to date.*

He also states:

> *In fact, Mitchell's contract with me was a deal he was trying to set up where he was going to have to raise 150 – 300 million dollars and start a company which would be the size of General Motors in three to five years, and from then on, who knows? Out of this I would get a one year employment contract and $30,000 and the possibility of maybe owning one or two percent of the operation; something I turned down because of my lawyers advice. And Mitchell, of course, blew up.*

"What Did It Feel like to Walk on the Moon?"

We have now seen evidence of the strange - and even threatening - behaviour of Edgar Mitchell. On the page about him on the IONS website[533], it describes the "realisation" which he had on his alleged return journey from the moon:

> *He knew that the beautiful blue world to which he was returning is part of a living system, harmonious and whole—and that we all participate, as he expressed it later, "in a universe of consciousness."*

Yet even after this realisation, we have two instances of him talking to people directly about guns and violence – when both of the people that he was talking to appeared to be trying to establish the truth about the matters they were investigating. How can we explain this contradiction?

A clue to the answer may lie in what he said during an interview on the Art Bell show on Wednesday, May 15, 1996 1 AM to 4 AM PDT. This interview included Richard C Hoagland and the transcript of it is posted on Richard Hoagland's website[542]. In the interview, Mitchell discusses how he has often been asked what it felt like to walk on the moon.

> *Oh sure. Basically, when people asked me, "What did it feel like to be on the moon?" Being a super rationalist and a Ph.D. and all of that, I didn't think it was a germane question.* ***I thought if you ask me what did I do on the moon, or what did I think about on the moon, I could have told you. But what I feel, I didn't know.*** *And so I set out to... I started thinking about that question. First of all, it irritated me because I didn't have an answer to it, and eventually* ***I asked myself, "Should I know what I felt like on the moon?" So I went to some good friends of mine, Dr. Jane Houston and her husband Bob Master, and said help me find out what I felt like on the moon,*** *and that began the investigation of inner experiences for me back in 1972 and led to the approaches that I have taken in understanding experience and the psychic experience and all this whole subject matter of consciousness that we've been looking at for 25 years.*

But this doesn't seem to tally with his story about having this "revelatory awakening" on his alleged journey back from the moon. The description given above about how the world is "a harmonious whole" does not seem to fit with this idea that he didn't know what it felt like to walk on the moon. Also, his statement about "needing help" to "know what it felt like" is most peculiar, if you ask me.

Since I originally wrote this chapter, I learned of the details of the reaction of Buzz Aldrin, in 1971, to a similar question. I discussed this in chapter 15.

MK Ultra

Perhaps we have to consider what else was happening at the time the Apollo Programme was current. An interesting observation is made on a Webpage titled "Apollo Truth - The reason behind NASA's fake Moon missions." [543](author unknown, but an email address is listed[544])

> *Now you know why CIA director Richard Helms, had the MKULTRA files destroyed in 1973[545], immediately after Apollo finished. Anyone who was*

> *around during the mid to late 60s would know that references to brainwashing and mind control were becoming quite the norm with TV programmes like "The Prisoner" and films such as "Clockwork Orange" and "The Ipcress Files". I think if most Americans knew what the CIA had been up to since WW2, they would rebel against their country anyway. As Richard Helms once said "We can create false memories now" That is certainly true for those who still believe in the Moon landings.*

The page goes on to say:

> *The most baffling aspect of all this, is how astronauts like Armstrong and Aldrin could have been led astray by such a scenario. ... How could they possibly go along with such a scam, which would undoubtedly be exposed in later years? Many believe they, and the other astronauts, were influenced by mind expanding drugs, which were extremely common about this time, or it could be that they too, were under some form of mind control by the CIA. This may be true as Buzz Aldrin has suffered mental illness from psychological brain damage.*

If we consider the reaction of Mitchell to Bart Sibrel, and his statement about briefing the CIA director George HW Bush regarding investigations into the abilities of Uri Geller, it seems to me that Mitchell was quite probably subjected to some form of mind control to make him believe that he really did walk on the moon. When he is challenged over this issue, we can see that his reaction is one of great discomfort, rather than one of being able to rationally explain how and why Bart Sibrel (and people like me) are wrong to state that he could not possibly have walked on the moon during the Apollo 14 mission (and neither could any of the other publicly named astronauts).

Please tell the Truth About Edgar Mitchell

From the other evidence discussed in this article, I argue that Mitchell became a key asset in intelligence work, which is being done in the alternative knowledge community and within the "new age" movement. In other words, I am in complete agreement with the statements of Bruce DePalma and feel strongly that Bruce DePalma's death was neither natural nor was it an accident.

For those in the UFO/ET community who have now read this chapter, I now regard it as your duty to tell the truth about Edgar Mitchell and his true role as a gatekeeper and someone who did not (or was not able to) tell the truth about the Apollo Programme and what he knew about the UFO/ET issue. I hope you don't react like famous Crop Circle Researcher Colin Andrews did, when I posted a comment on his Facebook profile, following the death of Edgar Mitchell. A far more appropriate reaction, in my view, would have been "Oh, gosh Andrew! I never knew about that. Sad. But thanks for the information!"

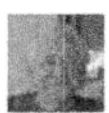

Colin Andrews added 3 new photos.
5 February ·

My friend, Apollo 14 Astronaut, Edgar Mitchell has passed away at 85 years of age. I so enjoyed my time with him, on stage making presentations together and privately over a beer. This man was an extraordinary person, a pilot, scientist, astronaut, educator and many other skills and loves. He has done more for human-kind than most have yet to know. Synthia and I send our deepest sympathies to his family and friends. More on my website, including NASA announcement:

http://www.colinandrews.net/

Donna Dignan Ingoglia RIP Edgar.
Like · Reply · 6 February at 03:24

William Mavers Hero...for TRUTH !!!
Like · Reply · 2 · 6 February at 03:33

Sanya Korlaet RIP Edgar, the brave soul.
Like · Reply · 6 February at 05:51

Pat Dee
Like · Reply · 6 February at 06:48

Andrew Johnson A shame he suppressed important truths. I've collated that information on my website if anyone wants a link. People don't seem to realise he wasn't (by his own admissions and statements) the person they thought he was.
Like · Reply · 1 · 6 February at 08:11

Carl Read The Right Stuff ! A very sad day. RIP Edgar Mitchell
Like · Reply · 6 February at 08:20

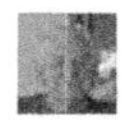

Colin Andrews ▸ **Andrew Johnson**
9 February · Guilford, CT, United States ·

Andrew,
I am surprised, actually rather shocked that you would make such unsubstantiated and repugnant statements about Edgar, a brave and intelligent, sensitive soul who stepped out from the crowd to try and make a difference and improve our understanding of our Universe, our planet and ourselves. Face to face contact and discussion and yes, research is the minimum requirement to enable such conclusive statements you have made here. As a friend of his but not in blind faith, my own assessment of Edgar is that he has been completely honest and transparent. I feel it is my duty to cover his back, when he is no longer able to defend himself. It's a very sad reflection upon our lack of humanity that we should so seemingly attack one another so much more quickly that support each other. There is a time and place for all things and in my view, when our fellow man departs is not the time to pile on more of our own differences. We are losing the respect and caring for each other here on Mother Earth, which makes any future off planet advances worthless, whether you believe him or not.

Like   Comment

 1

Andrew Johnson Colin, not unsubstantiated at all - repugnant yes, but backed by video evidence and reliable testimony. Here it is for you. Note Mitchell's admission in 2013 that he worked for the CIA in the 1970s. www.checktheevidence.com/cms/index.php?option=com_content... I hope you will watch all the video, painful though it is.

This exchange with Colin Andrews in 2016 illustrates how secrets can be kept.

28. SAIC and Limited Hangouts with the "Two Steves"

The expression "limited hangout" means, in the context of topics of this book, "telling some of the truth but not all the truth that could be told." This was essentially the theme of my two 9/11 books and, as I wrote at the beginning of chapter 30, it is important to understand that a technology exists which can be used in a weapon system, for propulsion and for energy production. The technology is real – and we know the basic principles of its operation – it uses a mixture of electrostatic/electric fields, a magnetic field and microwaves, arranged in a way that creates some type of resonance effect in the aether and/or materials. This can *all* be established by studying the evidence pertaining to the destruction of the WTC and the events which took place in the period before and after its destruction on 11 September 2001. This chapter is, therefore, quite similar to chapter 6 of "9/11 Holding the Truth."

SAIC - Science Applications International Corporation

To understand the significance of this company, let me explain that SAIC were defendants in Dr Judy Wood's 2007 9/11-related Qui Tam Case[546]. This company was mentioned by several Disclosure Project Witnesses as significant benefactors of Black Budget Programs. In 2013, SAIC split into 2 companies – and so sometimes now goes under the name **LEIDOS.** From SAIC's website (page 31 of the referenced PDF), we read[547]:

> *Following the September 11, 2001, terrorist attacks, we responded rapidly to assist a number of customers near ground zero in New York City and in Washington, D.C*

SAIC are involved in research or manufacture of Directed Energy Weapons Systems/Components[548] (Listed as LEIDOS). SAIC oversaw ground zero security from 13.11.01[549]. Another specialty of SAIC is Psychological Operations (PsyOps)[550]. The "two Steves" should know this – and they should talk about it.

Dr Steven Greer, 9/11 and SAIC

As I described at the start of this book and have written about elsewhere, it was my discovery of Dr Steven Greer's research that led me to find out the truth about 9/11. It is therefore ironic that I should later find evidence that Steven Greer himself has become part of the 9/11 cover up and the energy cover up. I have written about his apparent role in the energy cover up in an article on my website called "Something in the Aero.[551]" However, I had also observed Greer's reluctance to talk about 9/11 – such as when he wrote, in 2007, in an article entitled "Transformation of Risk" originally posted on his Aero2012 Website[552]:

> *This overview explains how the current risks of environmental global warming, air pollution and public health challenges, energy resource scarcity and competition, global terrorism and current electric grid obsolescence and vulnerability are transformed by these out-of-the-box technologies.*

Years earlier, Greer posted an article soon after 9/11, on his Disclosure Project Website[553]. He wrote:

> *This is not to excuse in any way the evil, monstrous and inhuman acts of Osama bin Laden or others of his ilk. There can be no rationalizing such horrific deeds. But we can understand it: Why him, why us, why now: why.* ***Fanatics like bin Laden are hell-bent on running America out of the Middle East*** *because they view our presence there as a virtual invasion of their land, culture and values. They view us as an imperial power colonizing their region* ***in order to secure cheap oil, and it is resented****. To a lesser extent, they are concerned about our support for Israel, but bin Laden himself has made it very clear in numerous speeches that their main concern is getting the US out of Saudi Arabia, the land containing the most holy sites in the Islamic world.*

However, on 21 November 2015, at the end of a 3 hour + talk[554], Steven Greer faced a question from a seemingly informed member of the audience. Greer's response seemed to clearly underline his support for the cover up of what happened on 9/11. He did not discuss any real evidence - nor did he wish to. Instead, he said very little of note. One might even perceive what he said as a warning - to NOT look at the available evidence…

> *Question: There's compelling evidence that not just the towers, but the entire World Trade Centre Complex was destroyed by perhaps a directed energy type of weapon something that was operating on a technology of molecular dissociation - that seems to be very much related to what you've been talking about. Have you looked at that?*
>
> *Answer: I have looked at it - you know, I have bigger fish to fry... 9/11 was terrible - um we lost 3000 people... it lead to 2 huge wars and trillions of dollars... [Pause] I will tell you this. I'm not going to get into specifics. I mentioned] a gentleman named Richard Foche 3 highest guy... Richard Foche at the Naval Research Labs he was in meetings with the then Vice President of the United States - Cheney - and he said "Absolutely it was known about in advance - and there was involvement at that level." And he told me - and I have witnesses, multiple witnesses to this - that if he were to speak of this... (Now he's passed away...) That he, his wife, his children and his grandchildren would all be killed. That's all I am gonna say on the 9/11 issue. But there is a lot to that story - I've never shared publicly what I just said... but the man has passed away. So it's not hard to ... and I think most people... get into these* ***grand conspiracy theories...*** *Basically all you have to do is stand down. There's constant threat - there is terrorism out there... you don't even have to hoax an event in that case ... You just have to stand down the systems that would've intercepted it. You understand what I am saying? So it's like when you're doing a 'code blue' ... "Boom! Clear!" You say "clear" before you do the electric shock. So you just clear the system and let it come in and hit... Now, what I think happened on 9/11 is more like what I just described and I have very good reason to believe that because someone of that integrity and rank had a front row seat - and knew about it - and I've never spoken of this before right now. But I think that .. it's disturbing, it's enraging, but it's a minor*

> *rounding error compared to what's coming if we don't end this cartel's hegemony on the secrecy related to UFO's, ET's what have you. 9/11 is going to be a very minor blip on the screen. So that's my warning to people - you can take it or not.* ***But I've known what you've just asked about since it happened. My whole security team lit up*** *like you wouldn't believe when that happened...*

If Greer had "known what you've just asked about since it happened" (i.e. the use of an energy weapon), why was he pushing the Bin Laden fable in his 2001 article, referenced above? I think Steven Greer knows a lot more than he said here. Even going back to 2005, in an interview for Hustler magazine[555], he says:

> *Q: Who else is making these phony UFOs?*
>
> *A: The companies involved are SAIC [Science Applications International Corporation], TRW, Northrop, Raytheon and EG&G. We have enormous intelligence on this. I know the buildings where this stuff is going on. This needs to come out so people know the truth.*

Later in the same interview he states:

> *Q: Who's in this covert group?*
>
> *There's a committee of 200 to 300 people who are on the policy board for this issue. Admiral Bobby Ray Inman, who went from head of the National Security Agency to the board of* ***SAIC*** *- which is one of the crown jewels of this covert entity - is a member. So is Admiral Harry Trane. George Bush Sr., Cheney and Rumsfeld are involved, as is the Liechtenstein banking family.* ***The Mormon corporate empire has an enormous interest in this subject****; they have much more power than the White house or the Pentagon over this issue. And there are secret cells within the Vatican.*

I will just briefly mention here that Steven E Jones is a Mormon.[556] Please read my other 9/11 books to find out more about Jones.

Greer says some interesting things on P157 (or thereabouts) of his book Forbidden Truth, Hidden Knowledge[557]:

> *At a subtle level of electromagnetism, you can transmute elements, and also transfer something from one place to another - and infect or harm someone electronically. This is a very lethal application of a science that could be used to heal people. Unfortunately, right now, the worst elements of humanity currently possess these technologies.*
>
> *When people worry about these technologies being disclosed, I say: "Forget about it. The worst elements already have them!*

From "Testimony of Denise McKenzie, former SAIC employee, March 2001", Page 308 – 309, "DISCLOSURE PROJECT BRIEFING DOCUMENT"[558]

> *Essentially SAIC is one of the crown jewels of the super-secret, black project world and is connected to UFO technologies and covert funding. Former NSA head Adm. Bobby Inman is heavily involved with SAIC, it should be noted. Here again, we see the revolving door between military and corporate projects described by Dr Rosin.*

Greer, Goldwater, SAIC and Bobby Ray Inman

In the UFO literature, the name of Senator Barry Goldwater[559] (1909-1998) crops up from time to time. He was a prominent politician – he lost the 1964 campaign for the presidential candidacy to Lyndon B. Johnson, for example. He became interested in the UFO/ET cover up and "started asking questions." In interview (date unknown, but probably early 1990s), Goldwater stated he had asked Curtis LeMay (mentioned in chapter 23) about the UFO/ET cover up:[560]

> *I think the government does know - I can't back that up but - I think that at Wright-Patterson Field [AFB], if you could get into certain places, you find out what the Air Force and the government knows about UFOs. Reportedly a spaceship landed and was all hushed up... quieted and nobody [has] heard about much of it. I call Curtis LeMay and I said "General, I know we have a room at Wright-Patterson where you put all this secret stuff. Can I go in there?" and I've never heard him get mad - but he got madder than hell at me! Cussed me out and said "Don't ever ask me that question again!"*

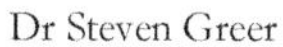

Dr Steven Greer | Barry Goldwater | Admiral Bobby Ray Inman | Gen. Curtis LeMay

On 13 April 2012, Dr Steven Greer was interviewed on PUJA radio[561]. At that time, Grant Cameron had posted documents relating to Barry Goldwater's affairs that had been released after an FOIA[562]. The documents confirmed the Greer had spoken to Goldwater. Greer was discussing the role of Admiral Bobby Ray Inman in UFO Secrecy. In the interview, he mentions SAIC when talking about his meeting with Senator Barry Goldwater.

> *These meetings I had with Senator Goldwater have now been confirmed by the release from the Goldwater library and in the last month - back in March ... Now here's the "piece de resistance" ... After this discussion, he started asking me... "Who have you been briefing on this?" I said well you know "I briefed President Clinton's team and his CIA director James Woolsey and I've met with the head investigator for the Senate Appropriations Committee... And he says "Well they were in the same boat you were in senator. This kind of secret Collegium of folks running this [haven't] him told him a thing...*
>
> *Even when the president made inquiries, they wouldn't tell him anything and when the CIA director made inquiries, they wouldn't tell him anything - because it is so secret. So he looked at me, he says "Well, who is involved in this secrecy?"*
>
> *So I started naming some people that I had learned from very reliable sources were members of this so called majestic committee - it used to be called back*

> *in the Truman and Eisenhower era.* ***One of them was Admiral Bobby Ray Inman. Admiral Inman had been head of the National Security Agency, but before then had been deputy director of the CIA. But at that time had gone on to the board of directors of Science Applications International Corporation SAIC*** *- which is a huge government contractor that has dealt with some of this stuff.*
>
> *And... when I landed on Admiral Bobby Ray Inman, [Barry Goldwater] says "Oh I know Bobby Ray I've known him for years!" and I said "Well, Senator Goldwater, would you pick up the phone and give him a call and see if we can meet to see if we can get some cooperation on making a change and disclosing this information bringing this science and technology out so we can get out of Mideast oil and we can make a civilization here that is that works and that is free...*
>
> *Senator Goldwater was very enthusiastic about doing this. So he did it, and let me tell you, after that, Senator Goldwater contacted me and he says "I can't make any more phone calls like that." And then his daughter Joanne ... and I spoke on the phone and she just says "I don't know what Bobby Ray Inman said to Daddy (meaning her father Senator Goldwater), but he cannot make another phone call like that - it put the fear of God into him."*

Greer and the Apollo Lunar Module

In many interviews, and on his websites[563], Greer has stated:

> *My uncle [Macon Epps][564] was the senior project engineer for Grumman (now Northrop Grumman) that built the Lunar Module, that landed on the moon in July of 1969.*

As Greer should have known, following deep exploration of his subject area, this was not possible. However, he still protects the lie of Apollo to this day.

Greer and the CIA

In 2017, a YouTube video was posted of a radio/podcast interview with Dr Greer, in which he said the following[565]:

> *"In the 90s, I had 3 people on my team assassinated including a... very close friend.*
>
> *[What has shifted now...]*
>
> *There were people in the agency who really wanted us to succeed – and I was gonna shut down this whole project when this happened... and they said "NO! You have to keep going... we need* ***someone on the outside this has to be known...*** *our planet that's going off the rails here...*
>
> *I'm an emergency doctor working in an ER, taking care of shootings and all that stuff and I said "You know what? This is just getting to be too ridiculous.." and they said, "****Do you want executive protection?****" and I said "Whatdya mean?" and they said... "****Do you wanna be have a [assisted?] protection - you and your team?****"*
>
> *This was after the* ***CIA Director Bill Colby - who was going to help us bring out one of these free energy devices and bring it out to the world –*** *was found floating down the Potomac river – assassinated. And I said "Well, I really don't like to play that game... it's a very dangerous game – it's called wet*

> *works...... But if they're targeting my people, I said 'I believe in self-defence – make it happen.' So, I won't say who this person is... still a very senior person – science director at the CIA – offered this protection and some other folks and I said 'yeah let's do it.' So, since then, we've had no problems.*

Is this an admission that Steven Greer is working for the CIA? In the "Sirius" Film he made, released in April 2013, he talked about having a "dead man's trigger" document[566] – to be released if he was ever assassinated. We see, then, some serious contradictions here. It seems Dr Greer is controlled by the CIA and will not tell the truth about 9/11 and SAIC's clear connection to the event. It is also likely that SAIC know a great deal about the technologies of which we are speaking – that is, those that were used to destroy the WTC and that are used as part of a Secret Space Programme.

Steve Bassett and "Layer Maintenance"

Steve Bassett has, in the past, spoken about a possible secret space programme – at least in 2014.

Steve Bassett in April 2017

From his website[567]:

> *Stephen Bassett is the executive director of Paradigm Research Group founded in 1996 to end a government imposed embargo on the truth behind the so called "UFO" phenomenon. Stephen has spoken to audiences around the world about the implications of formal "Disclosure" by world governments of an extra-terrestrial presence engaging the human race. He has given over 1000 radio and television interviews, and PRG's advocacy work has been extensively covered by national and international media. In 2013 PRG produced a "Citizen Hearing on Disclosure" at the National Press Club in Washington, DC. On November 5, 2014 PRG launched a Congressional Hearing/Political Initiative seeking the first hearings on Capitol Hill since 1968 regarding the extra-terrestrial presence issue and working to see that issue included in the ongoing presidential campaign.*

Though some people are uncomfortable connecting the UFO/ET/alien issue to 9/11, I speak about this publicly and expect researchers and knowledgeable people like Steve Bassett to point out connections when they are obvious. Clearly, he is interested in the weaponisation of black technology. This came up in a round table discussion on Sat 28 June 2014, (at 21:40 in)[568] at the 2014 "Secret Space Conference" held in San Metino California[569]. He said

> *One of the very possible reasons why the secrecy has been so profoundly maintained is that the power of the energy systems and technologies being worked on as well as the antigravitic technology – **if it gets in the wrong hands could be a serious problem**. I'd like to mention something you may have forgotten about. Back in 1991/92 there were articles which I read that the military kind of had a problem in that they had been developing a huge range of sophisticated weapons and were very frustrated that there wasn't a war they could use them in....*

He then mentions Saddam Hussein and Kuwait and he talks about the Gulf War and how it was shown on TV in graphic detail. Then at about 24:30 into the discussion, he says:

> *Maybe the secrecy is because these weapons, if they get in the wrong hands, could be dangerous. I would like to make the suggestion that they are already in the wrong hands…*

Of course, if Steve Bassett had referenced or described Dr Judy Wood's book at this point, he would have been able to illustrate that he was, indeed, correct! *He would have been making an observation, not a suggestion.* Would that not have increased his credibility – referring to a comprehensive, publicly available investigation by a scientist, part of which had been submitted in a court case[170]? This investigation (contained in "Where Did The Towers Go?") even includes reference to certain antigravitic effects discovered by John Hutchison and inspected by Col John Alexander in 1983[252]. I mentioned Col John Alexander, because Steve Bassett should know about him as a UFO commentator – Alexander has strong military ties and knowledge[570].

Here is what Steve Bassett said when asked if he believed the official account of 9/11 at the UFO truth Southern Conference on 29 Apr 2017[571]:

> *Q: Steve, do you believe in any way the official version of [the events of] 911?*
>
> *A: Well the answer is… do I believe that the investigation of 911 – which they didn't want to do until they were forced to by the women victims …. Which was an utter insult to the American People – unbelievable – the investigation is not full, is not adequate and there are major discrepancies. We do not know the full story. We have a 100 theories, all of which cannot be right. So essentially, one of the most important events in American History. And again, just like so many other things – left unresolved - because "you don't need disclosure" – you're just people. As far as what's going on and my principal focus here….* ***I know Richard Gage very well...*** *The number one problem for the government is Building 7…*

Why would Bassett highlight that he knows Richard Gage? To confirm that they are both members of the same "cover up crew?"

As with Dr Steven Greer, Bassett should be honest about what is known about 9/11 and how it connects with the UFO/ET issues. As with Greer, Edgar Mitchell and many others, he seems to be deliberately (or based on instructions) adopting a "limited hangout" position. Not only that, but in relation to 9/11, at least, he promotes disinformation.

Similarly, Bassett has promoted the CIA-run fake disclosure initiative called "To the Stars Academy"[572], which I wrote about in further detail on my website in October 2017.[573] He has also been reported in the mainstream press as saying that aliens prevented the onset of World War III.[574]

Ed Fouché and SAIC

In conversations with Edgar Fouché, he mentioned to me that he had worked for SAIC in the past (I think he mentioned that he was a director of Engineering). He repeated this statement in a posting he made on the

"AlienScientist" forum[575] on 14 Sep 2011 - over 2 years before I came into contact with him.

> *I have had hundreds of employees under me in my life as a Director of Engineering and Senior Program Manager in the USAF and DoD. I've worked for companies like SAIC, PCA, CEA, and SSAI after my retirement from the USAF.*

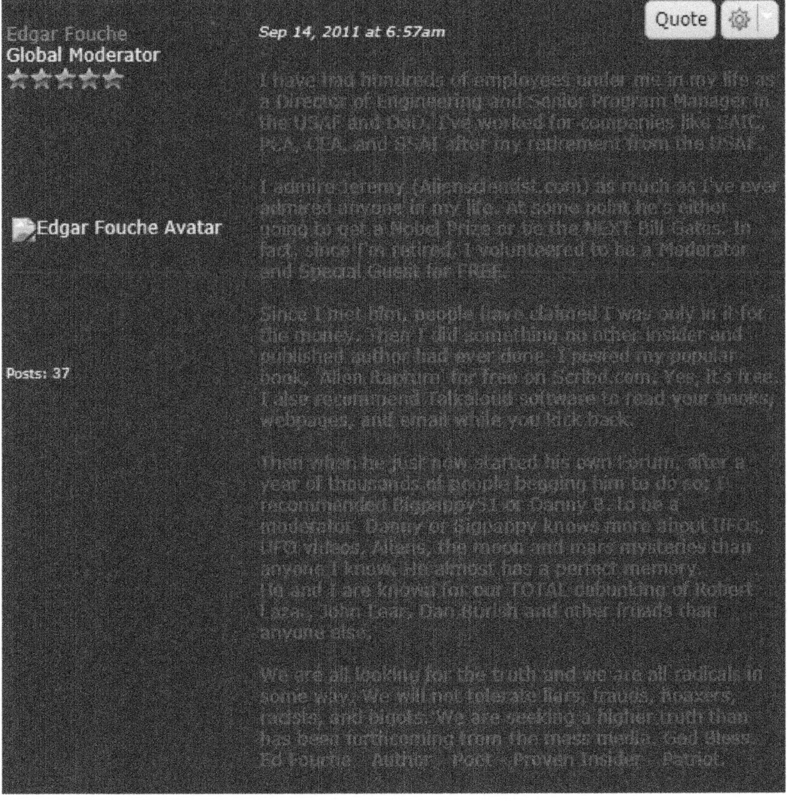
Edgar Fouche
Global Moderator
Edgar Fouche Avatar
Posts: 37
Sep 14, 2011 at 6:57am
Quote

I have had hundreds of employees under me in my life as a Director of Engineering and Senior Program Manager in the USAF and DoD. I've worked for companies like SAIC, PCA, CEA, and SSAI after my retirement from the USAF.

I admire Jeremy (Alienscientist.com) as much as I've ever admired anyone in my life. At some point he's either going to get a Nobel Prize or be the NEXT Bill Gates. In fact, since I'm retired, I volunteered to be a Moderator and Special Guest for FREE.

Since I met him, people have claimed I was only in it for the money. Then I did something no other insider and published author had ever done. I posted my popular book, Alien Rapture, for free on Scribd.com. Yes, it's free. I also recommend Talkaloud software to read your books, webpages, and email while you kick back.

Then when he just now started his own forum, after a year of thousands of people begging him to do so; I recommended Bigpappy51 or Danny B. to be a moderator. Danny or Bigpappy knows more about UFOs, UFO videos, Aliens, the moon and mars mysteries than anyone I know. He almost has a perfect memory.
He and I are known for our TOTAL debunking of Robert Lazar, John Lear, Dan Burish and other frauds than anyone else.

We are all looking for the truth and we are all radicals in some way. We will not tolerate liars, frauds, hoaxers, racists, and bigots. We are seeking a higher truth than has been forthcoming from the mass media. God Bless. Ed Fouche - Author - Poet - Proven Insider - Patriot.

Ed Fouché's posting mentioning SAIC

Interestingly, on "The Outpost Forum," someone accused Ed of leaking classified information in relation to his work in SAIC[576]. Here is his response.

> *Re: "Edgar Fouché Admits to Leaking Classified Information on Abuses at SAIC to Reporter in San Antonio"*
>
> *They are grasping at straws. My General, Doran, my boss, knew all about this. The Fraud, Waste, Abuse, and Corruption was rampant at Kelly AFB in San Antonio Tx where I had my Liaison office for many years.*
>
> *It wasn't SAIC, it was SA-ALC. San Antonio Air Logistics Centre, a USAF Depot. Get it correct. You are grasping at straws.*
>
> *I am PROUD of my going after these people. NO Classified or Confidential documents were ever released. Even if I had, the statute of limitations would have run out well over 20 years ago. Ha, nice try David Hilton and Jeremy Rys. You guys are doing a lot of things that smack of a smear, disinformation, and trolling. You'll love it when I start my new TV and YouTube videos.*

> *I testified to the Congressional Base Closure committee. This is one of the many reasons Kelly AFB was shut down.*
>
> *I saw contractors exchange drugs, money, other forms of bribes, women, etc. in order to get contracts from the government!*

Again, the essence of what Ed said checks out – the base was closed in the 1990s.[577] In another video (foul language warning!) Ed reports what happened in this regard.[578]

Due to my earlier investigations into SAIC and their connection to the events of 9/11, I did not discuss the company much with Ed. However, we can observe that, Steven Greer and his witnesses have said that SAIC as a company are a benefactor of the black budget. Ed stated he had worked for SAIC. He also stated he had worked on black programmes. Once again, what Ed Fouché stated is consistent.

29. Relevant Disinformation

Fake Flying Triangle and Other UFO Videos

There are probably more fake UFO/Alien videos on the web than there are genuine ones. Proving that a video is fake can sometimes be just as difficult as proving it is genuine. However, in judging a video, consider that to be genuine, these factors need to be known:

- Who took the video/photo?
- When was it taken – date and time needed?
- Where was it taken (location of camera and direction pointed)?
- What equipment was used?
- What sounds were heard?
- Does the witness account match what is seen/heard in the video?
- Is there more than one video/photo? How well do the photos match?

The less information that is available, the more difficult it becomes to validate a video or photo. If a witness is not forthcoming with information, then the video or photo is more likely to be fake. So, for any given video or sighting, find out as much information before judging it!

As an example, the video in which this FT is seen is often cited as being filmed in Paris.

Still from fake FT video.

There is no information about "who, what, when or where." It becomes clear from watching the video that the backdrop is a still image, with a CGI model being animated in front of it. So, be very careful about judging such videos.

Flying Triangle Disinformation

In May 2014, an image claiming to be that of a landed, black flying triangle was circulated. At first glance, the image is attention-grabbing, but it is not difficult to see problems with it. It was posted in a "report" by Ted Twietmeyer.

rense.com

Black Triangle On The Ground Being Serviced?

By Ted Twietmeyer
5-13-14

Below is a photo of a black triangle on the ground which was sent to me. Bottom photo credit reads "TheObjectReport.com." It appears to be on a runway at a unknown location:

Fig. 1 Original photo, untouched and unedited by author.

Black Triangle photo posted on Rense.com[579]

Here is a biography of Ted Twietmeyer[580]:

> *Ted Twietmeyer, author of "What NASA Isn't Telling You ...About Mars" ... is a former defense contractor who has spent over 25 years working with numerous government agencies including many of the NASA installations.*

It seems he has been writing about and discussing the same sorts of subjects as I have, but, of course, I am just a lowly part-time OU tutor and Ted has run "numerous (unspecified) government agencies."

When I first saw this image, I knew it was too good to be true. Fortunately, Google reverse image search was "clever" enough to find the original image – in about 30 seconds.

Dryden Flight Research Center EC90-065-16 Photographed 1990
Dryden Shuttle recovery convoy. NASA photo

Original image of shuttle Atlantis, which was used to create a fake image of an FT.[581]

So, was someone creating the image as a bit of "harmless fun"? Or was it deliberately created as disinformation, due to the disclosures being made by

Edgar Fouché (which were renewed by him around the same time this image appeared).

TR-3B In Space! Seen From Shuttle!

It almost seems possible that if the TR-3B was flying in space from the 1970s onwards, perhaps it could be seen from the shuttle. After all, look at what we discussed in chapter 26. Let's have a look at an "amazing" pair of images from STS-61:

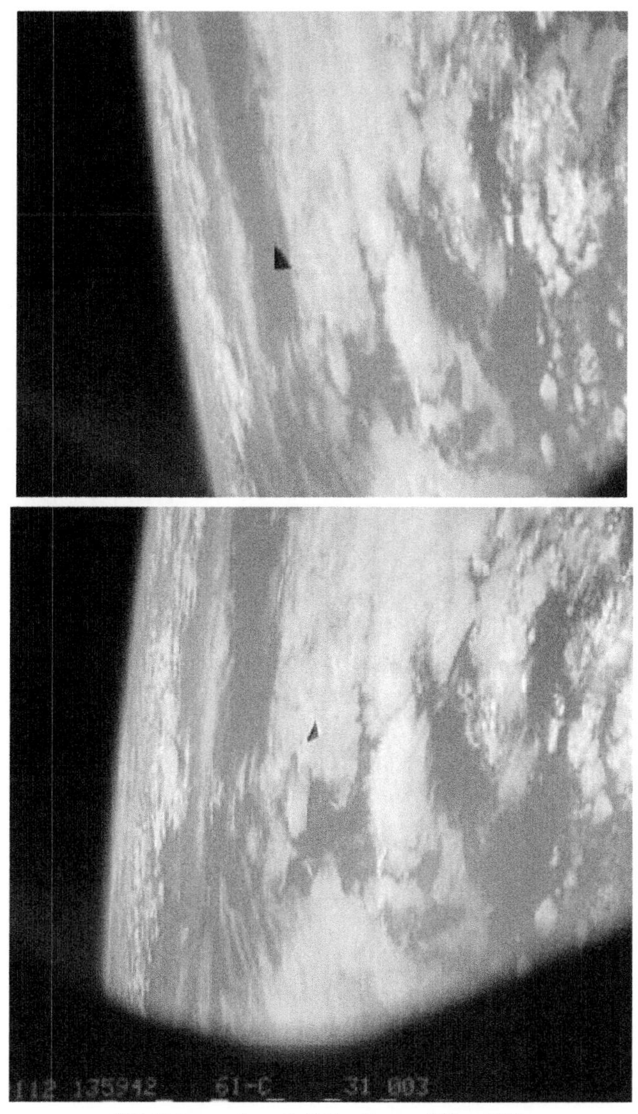

STS-61 Images from Roll 31 – Frames 2^{582} and 3^{583}.

These images do look spectacular, but for those wishing to know the truth, they can download the 100 megabyte high-resolution versions of these images and zoom in…

Contrast enhanced, rotated and zoomed portion of STS-61 Image - Roll 31, Frame 2[582]

This is clearly a piece of debris – either a detached shuttle tile, or some other piece of orbiting debris, within a few hundred feet of the photographer.

"Solar Warden" - Codename for Secret Space Program?

In 2014, a series of History Channel Documentaries was broadcast in the USA called "Hangar 1: The UFO Files." Episodes of this series covered some of the material mentioned in this book. I only saw one episode, however – which I think, from the description, was Episode 6[584]. This episode discussed Ronald Reagan's UFO sightings. In a 10-minute segment, it also discussed sightings of the flying triangle in Belgium and elsewhere. It included interviews with Michael Schratt, Richard Dolan and others. However, it did not mention Edgar Fouché – nor was he interviewed. The segment repeated questions and statements several times, while excluding almost all of the technical details Edgar Fouché disclosed. The programme asked (at least twice) if there was a link between the Flying Triangle/TR-3B and "Solar Warden". It is unclear where the "Solar Warden" information came from:

Dr Richard Boylan writes[585] (and he is not always known to be a reliable source!)

> *In actuality, since the late 1980s, the U.S. has joined with a handful of other nations to create a secret international Space Fleet, Its program, code-named "Solar Warden", has now grown to 59 small disc-shaped "scout ships" and 9 cigar-shaped motherships [each longer than two football fields joined end to end]. The Solar Warden Space Fleet is operated by Space and Naval Warfare Systems Command's Space Operations Division (SNWSC-SO), headquartered on the Virginia seaboard.*
>
> *SNWSC-SO has approximately 300 personnel on Earth and in space – it operates as one part of the UN's Central Security Service.*

> *Because the Space Fleet has the job of being Space Policeman within our solar system, its program has been named Solar Warden.*
>
> *The "Solar Warden" Space Fleet vehicles were constructed by international aerospace Black Projects contractors, with contributions of parts and systems by the U.S., Canada, United Kingdom, Italy, Austria, Russia, and Australia. Solar Warden Space Fleet's mandate is to keep space peaceful and free from misuse by any Earth country trying to conduct war-like or illegal activities in space.*

UK Figure Darren Perks (also unreliable) also claimed to have received information about it in 2011.[586] (Perks made up stories about an incident at Cosford[587] and he also sent hoax emails to British researcher Phil Hoyle)

It is my opinion, therefore, that the History Channel documentary was an attempt to either promote disinformation or it was just a "limited hangout" presentation – replete, of course, with fancy CGI graphics and a rhythmic percussive sound backdrop and many "flash edits."

UFOs/ETs and the Apollo Moon Hoax

I have lost count of the number of times that UFO researchers, authors and speakers have spoken of the Apollo missions as if they really happened as shown. It is less excusable that such researchers continue to do this than those in the "mainstream" do, because UFO researchers should, really, have "broken free" of mainstream propaganda. I discussed a good example of this sort of situation at the end of chapter 27.

Another example of accidental promotion of disinformation is by respected author Timothy Good, in his important 1988 book "Above Top Secret". On pages 383 – 384, in a segment entitled: "Did Apollo 11 Encounter UFOs on the Moon?"[588] Goode writes:

> *According to hitherto unconfirmed reports, both Neil Armstrong and Edwin "Buzz" Aldrin saw UFOs shortly after their historic landing on the Moon in Apollo 11 on 21 July 1969. I remember hearing one of the astronauts refer to a "light" in or on a carter during the television transmission, followed by a request from mission control for further information. Nothing more was heard.*
>
> *According to a former NASA employee Otto Binder, unnamed radio hams with their own VHF receiving facilities that bypassed NASA's broadcasting outlets picked up the following exchange:*
>
> *NASA: What's there? Mission Control calling Apollo 11...*
>
> *Apollo: These "Babies" are huge, Sir! Enormous! OH MY GOD! You wouldn't believe it! I'm telling you there are other spacecraft out there, lined up on the far side of the crater edge! They're on the Moon watching us!*
>
> *In 1979, Maurice Chatelain, former chief of NASA Communications Systems confirmed that Armstrong had indeed reported seeing two UFOs on the rim of a crater. "The encounter was common knowledge in NASA," he revealed, "but nobody has talked about it until now."*

There are numerous other examples of this sort of thing. For example, in the previously referenced (in chapter 26) Dr Jack Kasher interview in the Channel 4

"Riddle of the Skies" documentary, he shows and discusses a photo of a UFO allegedly taken near the moon during one of the Apollo missions. Of course, we now know this could not have happened.

Mark and Jo Ann Richards and the Earth Defence Headquarters

For a number of years, Jo Ann Richards had claimed that her husband, who had been incarcerated for over 30 years, had knowledge of a secret space programme. A page about the case, discusses an interview by Alfred Webre[589] (whom I wrote about in my book "9/11 Finding the Truth"[590]). This page states:

> *Navy Captain Mark Richards was an officer involved in the Dulce Battle as well as very active in the Secret Space Program and U.S. Space Command for many years prior to his being arrested and convicted of a murder he did not commit.*

I have not spent time fully investigating the case, because it seems to be completely anecdotal. It is possible that Mark Richards does not even exist. The page discussing the interview reads:

> *This is an interview with Captain Mark Richards conducted at Vacaville Prison on November 2, 2013. This is the first time in the over 30 years of his incarceration, that any journalist has interviewed him. I was not allowed to document this face-to-face interview with camera or any recording devices or to take notes during the interview. Everything you are about to see has been recalled from memory immediately after my meeting with him. Just prior to the interview I spoke briefly with his wife Jo Ann who also was present during the interview.*

Try reading this a couple of times. It appears there is no evidence that the interview even took place, let alone any evidence that whatever was claimed to have been said was true! If you were sat next to his wife for the whole interview, why would you only speak to her "briefly?" The information is 100% second hand, with no other corroborating testimony, images, documents or recordings of any kind. Webre is not a reliable source of information, in my experience.

Jo Ann Richards runs a Website with the title "Earth Defense Headquarters" and she has a registered 501(c)(3) non-profit organization, to accept tax deductible donations.[591] The website has little or no information about the Mark Richards case. The home page gives a Mission Statement thus:

> *It is the goal of EDH to present alternative truths that are not found in mainstream textbooks or media. EDH wants to help affect positive changes in the community and beyond.*

However, the Richards' story is "a good yarn" - and people do like "a good yarn," as we can discover on a page, mentioning one of Jo Ann Richards' interviews:[592]

> *What suddenly grabbed my attention was mention of Angkor Wat in Cambodia, as my other half just returned from there and Vietnam from a photo shoot in a recently discovered massive cave.*

The story of giant spider beings the size of a Volkswagen Beetle who wiped out the population at Angkor Wat in a couple of days could very well be another hidden aspect of Earth history. The story is that the extraterrestrial spiders used mind control on the inhabitants, convincing them there was no water despite the river nearby, and once weakened through dehydration, the creatures consumed them. Richards said the monstrous spiders were the reason for the Vietnam war. Did they believe Agent Orange was necessary to defoliate the jungle to eradicate them?

Project Serpo

Project Serpo[593] was "revealed" in 2006 when anonymous whistleblowers claimed they had knowledge of an "exchange" programme where a group from earth were sent to the planet Serpo to live there for a period of time. I listened to the story, but concluded it was an exercise in releasing disinformation through people like Richard Doty, Bill Ryan and Victor Martinez. Doty had previously admitted deliberately feeding disinformation to the UFO community and my experiences in presenting verified evidence regarding the ET/UFO issue to Martinez and Bill Ryan was not a positive one. Ryan got involved with Kerry Cassidy and developed Project Camelot[594] and then later he "split" and ran Project Avalon[595]. Ryan promoted the Rebekah Roth disinformation.[596]

It appears the Serpo story was designed to muddle up and discourage interest in a number of documents that have been released over the years – such as some of the MJ-12 documents and, for example, the 1954 Special Operations Manual[597].

Project Pegasus

Andrew Basiago[598] claims he was part of a project which involved teleportation technology, which allowed him to travel to Mars. You can read about his claims via the Coast to Coast Website[599]. His story seems to have changed over the years, and he behaves like he believes his own story. On his website, he has little or no supporting evidence for his case and he writes:

A "peak experience" is how Andy described his lecture at the Ramtha School of Enlightenment (RSE) in Yelm, Washington on February 12, 2011. Andy received standing ovations when he entered the Great Hall at RSE to the resounding chords of Vangelis' "Chariots of Fire," when he called for public teleportation by 2020, and when he concluded his five-hour talk. His address was attended live by 1,250 people, viewed online by 250 others, translated into nine languages, and streamed live to 25 nations.

The website doesn't seem to contain any links to verifiable evidence – it merely comes across as more story-telling. The point is that aspects of the technology he (and others) discuss in his stories may well be real, but they are then discredited by being referred to in "fantastic tales" which are then often indistinguishable from science fiction. A good example of this are his claims relating to Barack Obama (Barry Sotero), which are written up in a posting on Dr Michael Salla's website:[600]

> *Basiago further claimed that in 1980, he and Barack Obama actually were roommates while being briefed about upcoming jump room teleportation to Mars: "So, Barack Obama was not only aware of his Presidency when we were being trained in 1980, when he was 19 years old – and just turned 19 in August of '80 – but when we were rooming together briefly, a couple of days – at the College of the Siskiyous. He was reading briefing documents that they were giving him to groom him for the Presidency."*

Perhaps the truthful part of this story is that Obama (Sotero) was indeed groomed for being the president, and he was made aware of this (for example, he was the only presidential *candidate* that I am aware of, who went on a *European* tour during his presidential campaign.[601]

Captain Kaye / Randy Cramer

Dr Michael Salla reported about this case on April 15, 2014[602], which I again, hold little confidence in.

> *A new whistle-blower has come forward to claim that he spent almost three years serving in a secret Space Fleet run by a multinational organization called the Earth Defense Force. In exclusive testimony released today on ExoNews TV, the whistle-blower, who uses the pseudonym, Captain Kaye, said that the Earth Defense Force recruits personnel from the military services of different nations including China, Russia and the U.S. He claims that he was recruited from a covert branch of the U.S. Marine Corps called "Special Section," after having served previously for 17 years on Mars protecting five human colonies from indigenous Martian life forms.*
>
> *Captain Kaye's testimony describes three different types of space fighters and three bombers that he was trained to fly. Training took place on secret moon base called Lunar Operations Command, Saturn's moon Titan, and in deep space. The space fighters and bombers had different propulsion systems ranging from fission and fusion nuclear energy, an electro-gravitic system, temporal space drive and more traditional thruster systems. He also reveals the existence of at least five large cigar shaped space carriers that are capable of interstellar flight that house the smaller fighters and bombers. He claims he served on one called "EDF SS Nautilus" which was up to ¾ of a mile long.*

Again, feel free to study the details for yourselves, but it sounds too much like science fiction to me.

Corey Goode "Spent 20 Years in Multiple Secret Space Programmes"

Goode came onto the scene in 2014 and did a number of interviews. His case has been followed by Dr Michael Salla[603] who has written a number of detailed postings. Salla admits that Goode has no evidence to support his claims.[604] I will now just list the titles of some of the relevant articles on Salla's Website:

> *Secret Space Programs, Sphere Being Alliance & Corey Goode Testimony, Written By Admin On June 18, 2015*
>
> *Military Abduction & Extraterrestrial Contact Treaty – Corey Goode Briefing Pt 2, Written By Dr Michael Salla On June 16, 2016.*

Corey Goode on Nordic Extra-terrestrials working with Religious Leaders for Disclosure, Written By Dr Michael Salla On June 27, 2017.

The Coming Solar Flash & the Galactic Federation – Q&A with Corey Goode, Written By Dr Michael Salla & Corey Goode On April 4, 2018.

Again, Goode might even believe what he is saying, and parts of his story might be true and be based on elements of a secret programme. This all gets mixed in with fantasy and story-telling, and we have no way of establishing what is true and what is false – because we have no evidence to judge.

John Lenard Walson – Far Above Space and Time

In about 2007, some videos started appearing which showed strange, shimmering objects in the night sky. The videographer used the name "Gridkeeper" but later it became known they were owned by someone called John Lenard Walson … (You can find the various videos on YouTube now). As I recall, one of the videos were sent to me by a friend called Nick Jedrezjewski and I said I wasn't sure what to make of the videos. I was surprised when, the next day, I received a telephone call from Walson assuring me that the videos were genuine – and that he would be revealing more information about them. I didn't immediately think they were hoaxed and it did look like they were objects that had been filmed through a long focus lens or telescope. Apparently, Walson had developed some technique for setting the shutter speed of the video camera, which allowed these objects to be recorded, when normally, they would be invisible.

I studied a few more of the videos and then began to consider that Walson really had developed a technique for revealing hardware that was in orbit and was normally invisible (stealthy). I even mentioned this in a few of the talks I gave in later years.

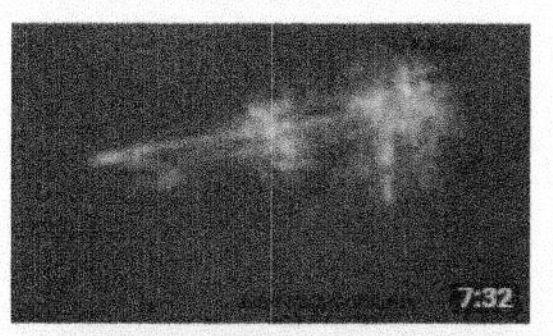

Far Above Space and Time - Early Episodes
GRIDKEEPER • 13K views • 10 years ago
Clips from the first few episodes from blip.tv http://www.youtube.com/gridkeeper http://www.youtube.com/johnlenardwalson Big ...

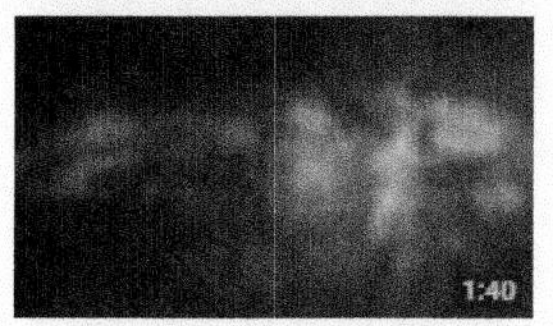

Far Above Space & Time -Ready?
GRIDKEEPER • 8.8K views • 10 years ago
http://www.youtube.com/gridkeeper http://www.youtube.com/gridkeepermusic.

John Lenard Walson - Far Above Space and Time
dl0j0lb
Looking up at the night sky with John and Gridkeeper (Stars and Venus) • 8:01
Space Craft UFO - JLW - Part 1 - The Early Years • 8:05
VIEW FULL PLAYLIST (37 VIDEOS)

YouTube Channel Listing for "Gridkeeper"[605]

However, further research revealed that Walson had been using another name, and this fact came up in connection with a probable intelligence asset by the name of James Casbolt. Casbolt appeared in Truro court in December 2013 and was eventually sent to prison for 12 years because he blackmailed his wife, who was the daughter of a wealthy businessman, Hank Meijer.[606] Casbolt claimed to work for MI6 and despite a previous offence, was able to enter America and join the US Army. Casbolt also used a pseudonym. In an article on "Truthseeker.co.uk" (seemingly yet another controlled site), Michael Prince/James Casbolt writes:[607]

> *"I met **a former military man called Simon Anderson** at the end of 2007 in Chichester. Many people will have heard of him from the internet as he was responsible for taking the photographs of huge spacecraft some of which were over **two miles long and positioned in the outer atmosphere as we speak**. He did this using advanced photography and telescopic equipment and these photographs were featured on the Jeff Rense website which is one of the biggest sites on the planet dedicated to this kind of material. Santiago Garza also featured some of Simon's footage at one his conferences.*

The posting goes on to note how Walson and Casbolt knew each other. Some time after I discovered this posting, I also discovered that Walson/Anderson had secretly recorded a conversation with Astronomer/Lecturer Dr John Mason and then edited it and posted it without permission:[608]

> *From: Dr John Mason*
> *To: webmaster@rense.com[609]*
> *Sent: Sunday, January 13, 2008 4:18 PM*
> *Subject: Re: Claims made by John Lenard Walson*
>
> *To Jeff Rense:*
> *Dear Jeff,*
> *It has been brought to my attention that John Lenard Walson is using my name to support his claims about images of "space machines in Earth orbit." I am afraid that Mr Walson is seriously misrepresenting my views on his work. In addition, the video clip of us chatting together at the Planetarium was made without my knowledge or permission and has been edited considerably.*
>
> *It is true that I spoke to John Walson several times and examined his images and videos. I was very sceptical about his observational methods, but in the interest of being helpful and constructive, I attempted to give him guidance as to what he should do. I warned him about the dangers of interpreting incredibly over-magnified images. I don't know what his images show and, in my opinion, he hasn't yet provided anywhere near enough background information for anyone to be able to work it out.*
>
> *I told John Walson that he must explain carefully to others how he was obtaining his images. I said he must describe his observing techniques and give precise pointing directions, dates and times of the images obtained to enable others to confirm his observations and check that the results were reproducible. I wanted to provide gentle encouragement in an attempt to get him to adopt more scientific practices.*
>
> *In one instance, I tried to explain to him that at the time of year he told me one of the pictures was taken, an orbiting spacecraft would not be sunlit at 3.30 am*

> *unless its altitude was very high, but he misunderstood my meaning. I then attempted to explain what I meant but he may have edited that bit out.*
>
> *There was one image that broadly resembled ISS but differed in certain aspects, but it was impossible to check precisely because I couldn't ascertain any detailed information about the circumstances under which the image was obtained. Indeed this applied to all the images he showed me, and I told him so. Again, that exchange was either not recorded or has been edited out.*
>
> *Since I did not know Mr Walson was recording our conversation, he does not have my permission to show any video of me on any website, and the vast majority of my general comments to him, made in the interest of being helpful, seem to have been ignored or have been edited out! I must therefore respectfully ask that you remove the video clip from your website.*
>
> *I look forward to hearing from you.*
>
> *Kind regards,*
>
> *John Mason (Dr)*
> *Principal Lecturer*
> *South Downs Planetarium*
> *Chichester*
> *UK*

So, regardless of what Walson's (or Anderson's) images actually show, would an honest person behave in the way that John Mason illustrates in his candid letter? What a shame that Walson abused the trust of a pleasant, open-minded astronomer and scientist. Perhaps that was the goal of the meeting?

Conclusion

In this chapter, we have seen just a few brief example facets of the edifice of disinformation that has been constructed in relation to a secret space programme. We have had a glimpse of how some figures promoting this disinformation have a questionable background, or simply promote or tell their stories without providing enough substance to back them up. Fake imagery regularly appears and sometimes, the appearance of such imagery seems to coincide with the release of "good" information.

30. Perception Management of Edgar Fouché and The TR3B

In chapter 5 of my book "9/11 Holding the Truth,"[610] I wrote about a character called Jeremy Rys, who uses a pseudonym "Alien Scientist" on his website[611] and YouTube channel[612]. In that chapter, I wrote about his connection to Edgar Fouché (see chapter 25 of this book) and his discussion of topics such as antigravity technology, 9/11 and exotic technology. There is something of an overlap between what I wrote in my previous book and what is written in this chapter. This reason is that, to understand what happened on 9/11, you have to understand that an undisclosed technology was used to destroy most of the WTC complex. This technology operates on principles only properly understood by people in the black world. The technology that the weapon system is based on could be used for propulsion – and it can also be used to create limitless free energy. *The technology used also affects gravity.* For more information on these issues, in relation to 9/11, please read Dr Judy Wood's book "Where Did the Towers Go?"[76] Also, watch the videos on my website[613], read my free 9/11 eBooks and study other related resources.

I would now like to spend some time discussing "Alien Scientist," as he is a key figure in the workings of the cover up and "muddle-up" of important evidence and disclosures related to 9/11 and antigravity technology.

Alienated Scientist and the "Case for Antigravity"

I wrote about this Jeremy Rys/Alien Scientist character in an article that I posted in October 2011[614]. The main issues which that article documents were Jeremy Rys' repeated lies about Dr Judy Wood's research[2]. Rys stated that Dr Judy Wood had claimed "space lasers" had destroyed the WTC. In one particular video, he used a defaced image of Dr Judy Wood's book[615] "Where Did the Towers Go?"[76] Readers can also note the image at the top of a page on Rys' website entitled "The "Directed Energy Weapons" Hypothesis vs. Real Science[616] and see if the image has been drawn up by a "real scientist." In my 2011 posting, I included screenshots of where I politely pointed out errors in his videos. Also of note is Alien Scientist's Facebook Page[617], where it states:

> *A non-profit educational foundation teaching science, logic, and critical thinking skills and how to apply them to complex topics with an overall goal of bettering humanity by promoting the creation of a peaceful, sustainable, space faring society.*

It is unclear what the components of his "educational foundation" actually are, however – beyond a YouTube channel and a website (i.e. perhaps I have inadvertently created my own "educational foundation" by creating a website and YouTube channel. That is to say, the description is meaningless in this context).

The essence of this book dates back to 2004, when I compiled some of the information I had come across into a PowerPoint presentation. I then narrated this PowerPoint, using a desktop microphone, to create a video – called "The Case for Antigravity"[5]. Oddly enough, one poster that re-uploaded the video was Jeremy Rys "Alien Scientist"[618]. Interestingly, Rys included my original description/summary and then added the following text:

> *This was one of the videos I stumbled upon in my initial research into antigravity which lead me into researching Ed Fouché and first revealed to me the existence of Metamaterials and Quasicrystals (which I had no idea about at the time)* ***I credit this video for inspiring me to take the research a step further and create the AlienScientist YouTube Channel. I am now paying my respects to the author of this video by re-uploading it here to YouTube*** *so that more people can learn about the work that inspired me.*

It was somewhat ironic to later discover that this "AlienScientist" who had been "inspired" by my 2005 video, would not "pay me respect" when I pointed out errors and mistakes in one of his other videos. It became clear that his motivations were significantly different to my own. "Alien Scientist" was not interested in "the truth, the whole truth and nothing but the truth," as I am. He later demonstrated that his main interest in the truth was so he could help cover parts of it up.

Alien Scientist – Connections

As I wrote in "9/11 Holding the Truth," in researching Jeremy Rys, two significant connections came to light. Firstly Rys had a connection to Richard Heene, the father of "balloon boy." The Balloon Boy Story[619] received blanket coverage in the US media for a short period in late 2009. Richard Heene claimed that, without him knowing, his son was carried off inside a helium filled balloon that he had made in his back garden – a home video showed them launching the balloon[620] then panicking after realising their son was "in the balloon" as it floated upwards. The "media drama" revolved around the story of the child being endangered by this situation – as no one knew where it would land. However, Richard Heene was later charged with wasting police time, as he lied about his son being in the balloon. He pleaded guilty to the charge and went to prison in early 2010[621].

The peculiar thing is, a photo on the American Antigravity Website[622] (archived here[623]) shows Jeremy Rys filming Richard Heene - with what seems to be a good quality video camera. So what, exactly, is the connection between these two people (apart from them both being proven liars)?

The other interesting connection is Jeremy Rys' father - Richard A Rys. A page on his website[624] documents:

> *SUMMARY OF MAJOR PROJECTS*
>
> *June 2010 – present - Working with* ***Invensys*** *to provide* ***control systems for 8 PWR style Nuclear reactors*** *to be located in Fuqing, Fangjiashan, and Hainan China. The projects total about 7300 MW of power. The main role has been to test the control system software and hardware. The control platform is*

> *a combination of IA systems, Triconex, and ATOS for the graphical user interface.*

Further inspection of this web page reveals that Richard A Rys works in the energy industry, with a company called Invensys – in a project engineer's role. A quick look at the Invensys Website[625] reveals

> *Invensys works with: **23 of the top 25 petroleum companies,** 48 of the top 50 chemical companies, **18 of the top 20 pharmaceutical companies, 35 of the top 50 nuclear power plants**, all of the top 10 mining companies, 7 of the top 10 appliance manufacturers.*
>
> *Invensys enables:, **20% of the world's electricity generation**, 18% of the world's crude oil refining, 37% of the world's nuclear energy generation, **62% of the world's liquefied natural gas production**, 23% of the world's chemical production.*

Having established certain facts about Jeremy "Alien Scientist" Rys, let us return to his association with Edgar Fouché, which I wrote about in the afore-mentioned "Alien Scientist and His Alien Science" October 2011 article[614].

Alien Scientist supports Edgar Fouché!

On the same day I posted my article – 11 Oct 2011, Jeremy Rys posted a new YouTube video[626] to promote his new internet discussion forum, linked to his website. In this video, starting at 0:32, Jeremy Rys states:

> *Former Area 51 Employee Edgar Fouché is a special guest on my forum, so you have the rare opportunity to directly talk with a former employee of the top secret military base at Groom Dry lake and ask him any questions you'd like.*

If Jeremy Rys is using this to promote his new forum, he must think that Ed's information is important and truthful, must he not?

The Alien Scientist Forum and Its Demise - Edgar Fouché Writes to Andrew Johnson

The Alien Scientist Forum[627] ran for almost 2 years - it was taken offline for a period of time, starting in early November 2013. Around 20th November 2013, Edgar Fouché contacted me via my YouTube channel[628] and we arranged to talk on Skype. He acknowledged our earlier correspondence back in 2010 regarding Jeremy Rys and we agreed to do some interviews. I can only presume that Ed Fouché decided to contact me because I had mentioned him in earlier videos that I had made and I had also some knowledge of Jeremy Rys and how he was "operating". Perhaps his decision to contact me was connected to the demise of the Alien Scientist forum, which Ed had apparently invested a considerable amount of time posting information on (all of these posts are no longer accessible).

In my conversations with Ed Fouché, he told me that Rys had been to prison twice. This seems to be indicated in this report[629] from the area where Rys lives[630], which mentions probation,

At 10:23 p.m., Jeremy Rys, 1 Cherry St., was put in custody on an arrest warrant. Police went to his home at 9:15 p.m., and his family said he was not home. Police received a call from Rys who said he forgot his phone charger when he went to work, and will call his probation officer in the morning. Dispatch called probation who asked for Rys to be picked up by police.

Edgar Fouché – Disclosing Secret Technologies (Video Series)

As I discussed in chapter 25, in early January 2014, Ed and I recorded about six hours of interviews. By the middle of February 2014, I had edited together these videos and posted them on my website[478] and two YouTube Channels[631]. I later added images of some of the documents he sent electronic copies of in relation to his military career and training. I did not make any effort to promote these interviews – I just posted a link to them to my email list and on my Facebook page[632].

In Feb. 2014 I discussed, with Richard D Hall[540], the possible connection between Ed's disclosures about the TR3B and the flying triangle cases that Richard had documented in his film "Almost Identified Flying Objects"[500]. He later decided he would like to film an interview / discussion with me about Ed's disclosures and this was recorded in May 2014, posted on Richard's Website on 14th Jun 2014[633] and it was broadcast on Sky Satellite Channel 192 (UK and Europe) on **21st June 2014**. (Note the date!)

David Hilton writes to Andrew Johnson

David Hilton has a YouTube channel[634], which was created on 28 June 2012. He describes himself as an Anthropologist, Historian and Researcher. His avatar and description are of interest, so I included it below and we will discuss this later.

He has several interesting videos, among the 20 or so that he has uploaded. He has three videos relating to Charles Hall's account of meeting Tall White Aliens in Area 51 while he was working as a Weather Man there in the 1960s[635] (my friend Richard D Hall interviewed Charles Hall in 2011[636]).

Out of the blue, on **22 June 2014** *(the day after Richard D Hall's programme was broadcast on UK TV)* David Hilton wrote to me, sending me a link to a video he had posted called "**Edgar Fouché - Faked Documents**".

Recipient: info@checktheevidence.com

Message text: Mr. Johnson,

I suggest you view this video. https://www.youtube.com/watch?v=QppycPjHzXk

Regards,

David Hilton

David Hilton's name was already familiar to me, as it had come up during the interviews I did with Ed Fouché.

As you can see, David Hilton made no reference in his message to the videos I had made with Ed and neither did he make any reference to the interview with Richard D Hall. Presumably, David Hilton had become aware of Richard D Hall's programme and decided to write to me, *the day after* Richard D Hall's programme had aired – he seemed to have posted the video around the same time Richard D Hall's programme aired. However, it is worth noting that I had posted 6 hours of interviews with Edgar Fouché on my YouTube channels and website 3 – 4 months earlier. (i.e. why didn't David Hilton write to me then, and send me his video?)

When I watched the video he had made, I noticed he had used a clip of the interview between Richard D Hall and myself and then Hilton went on to make claims that Ed had faked one of his documents. I was uncomfortable that David Hilton had used this clip, without permission, in what was to me, essentially, a "hit piece."

I therefore asked him to re-edit the clip and post a new version excluding the clip of myself and Richard Hall. Initially, he was reluctant to do this and stated:

> *I respectfully disagree. The Richard Hall clip is indicative of the community at large, and of media in general. The clip, and almost all other sources place great emphasis on Edgar Fouché's documents as establishing his credentials. In fact the clip states "he worked at Area 51 which is proven with lots of documents that he has released." If fact only one document has anything to do with Area 51, as per Edgar Fouché's claim, and that is the AF Form 77a document. Most of his documents are mundane documents which prove nothing except that he was in the US Air Force. I'm sorry, but I must regretfully decline your request to remove the clip. Removing the current video would destroy all the current links the video has gotten, take away an establishing point about media, and its only about a 45 second clip so it is not overly long. I really hope that this point does not sour communication as I have enjoyed our exchange.*

In conversation with Richard he said that normally, he would not have minded David Hilton using the clip, but Richard added that as he was talking to me about Ed Fouché's disclosures and the interviews I'd done with Ed, he did not grant permission for Hilton to use the clip. Richard shared the email exchange he had had with Hilton about use of the clip, in which David Hilton said:

> *Actually Mr. Johnson is only seen for a couple of seconds. The clip features you and your words. However, fine I'll change the video to unlisted, so as to preserve it for those that have posted it on forums. In the mean time I'll re-edit and re-post it without your footage.*

I wrote back to Hilton saying that making the video unlisted "so that he could post in on forums" was not really in the spirit of what I had asked. He then made the original video (including a clip of myself and Richard D Hall) private and he re-posted a new version (without the clip of Richard D Hall and myself) on 26 June 2014[637]. His description for this video read

Edgar Fouché's Air Force Form 77a is a fake document. He also says he wants to hide his tampering on camera.

So, what of the claims that Hilton had made in his video? I decided to ask him some questions and he responded in an email dated 22 June 2014, a portion of which I have pasted below, along with his answers.

> ***From:*** *David Hilton*
>
> ***Sent:*** *22 June 2014 18:06*
>
> ***To:*** *ad.johnson@ntlworld.com*
>
> ***Cc:*** *Richard Hall;* ***Jeremy Rys; Dan B; MW***
>
> ***Subject:*** *RE: Edgar Fouché "Fake Document" Video*
>
> *.....*
>
> *AJ: Did you get Edgar Fouché's permission to post this?*
>
> *DH: No. His permission was not requested or required.* ***Jeremy Rys found this video in his file server, and provided it to me. It is a conversation between Edgar Fouché and Jeremy Rys. Jeremy Rys recorded it in 2011, gave me a copy of it, and gave me permission to use it. I enlarged the view of Edgar Fouché to concentrate on him. When you see slight flips or flashes in the Fouché footage, it is because of adjusting the video to maintain concentration on Edgar Fouché.***
>
> *....*
>
> *AJ: Do you have any other main areas (go into as much detail as you like) where you think Ed has "made stuff up"?*
>
> *DH: After knowing Ed on a personal level since February of 2012, it is my opinion that he made up/lied about everything. I've caught him in so many lies and contradictions that I can't keep up with them all. Its not really something that can be completely hashed out in an email. I may be over-reaching here, but I'd like to invite both you and Mr. Hall to a skype conversation on the matter with myself, Dan B and Jeremy Rys, if they are willing. Another friend of ours MW can give you many examples of Fouché lying. I know that both you and Jeremy have very differing opinions between the two of you, however I hope that we can put that aside for now to address this Fouché issue.*

I was particularly interested in the parts I have emboldened in his response and note that his response was cc'd to Jeremy Rys (and Jeremy Rys never responded to any of our exchange). Clearly David Hilton had made some bold accusations regarding Ed's document and I now point the reader at Edgar Fouché's public response to the accusations in the video[638], which I have pasted a portion of here:

> *This document was created by Typewriter or Electronic Typewriter, no admin had ANY type of computers in 1979! Didn't see Admin PCs until the LATE 1980s! Also Jeremy Rys altered the contrast and definition as he told us. I have the original draft, the corrected second version, and the final version written by my boss and typed by his secretary.*
>
> *Instead we concentrate on the quality of the original. This can be a problem for the following reasons:*

> *1. the original is old, and has suffered physical degradation.*
>
> *2. the original was produced on a manual typewriter, so the individual characters can show variations in pressure and position.*
>
> *3. the original is a carbon copy produced on a typewriter.*
>
> *4. the original is a low quality photocopy, and shows variations in toner density and character spread, as well as general copier "grunge" caused by a dirty glass or background.*

I noticed that David Hilton had another video on his channel entitled Edgar Fouché Confesses - Liar & NSA Agent - posted 8th June 2014[639] (only about 3 weeks after Richard D Hall had posted our interview about Ed Fouché and the TR-3B on his website). Those who watch the video may be fooled into thinking Ed Fouché was indeed confessing – but the video is clearly not one that was an "organised" or "ordered" discussion of matters. David Hilton has given this description for the video:

> *Recorded 2-11-2013. This video was originally going to be about an F-15 crash I was researching, but the conversation never turned to it. Edgar Fouché claims he did not know he was being recorded. Edgar Fouché was fully aware he was being recorded. It was a planned recording that he wanted to do. He had a huge red button on Oovoo that advised him that he was being recorded, and told him to press it to end the call if he did not want to be recorded.* ***AlienScientist made a video on his YouTube Channel using parts of this****. I'm posting the unedited version here for all to see since many have requested it, and since Edgar Fouché himself decided to slander and libel me.*

Again, Edgar Fouché has told his side of the story[640] – which is basically that he did not know he was being recorded, he was "not in good shape" and he was being sarcastic at various points in the late-night conversation and at other times he was being deadly serious. As David Hilton states above, this video was recorded in February 2013 – so why did it take 9 months or more for it to become "important", if what Hilton claims in the description is true? Again, I have emboldened the reference to Alien Scientist. Now let us compare this to an earlier comment David Hilton made about the "forged document" video, where Edgar Fouché appears towards the end. David Hilton says:

> *Jeremy Rys recorded it in 2011, gave me a copy of it, and gave me permission to use it.*

So, both David Hilton and Jeremy Rys recorded Ed and posted videos of him without his permission. If you watch both videos carefully, neither of the videos support the claims that Rys and David Hilton have made against Ed.

I also asked David Hilton about the FOIA he had done for information on Ed Fouché – and he explained he had written to the Air Force but had not yet received a response.

On 27 June 2014, I wrote back to David Hilton with some further questions about his videos.

> *Ed admits on camera he doesn't want people to think he's forged/tampered with it - he repeats this in the email to Jeremy dated 7 Feb 2011. Why would he admit that he'd changed it in writing if he was trying to hide something?*

Where you compare your Uncle's document with the colour curves, it's not exactly a fair comparison - as you use a colour scan of your Uncle's document and a grey-scale of Ed's document, so I think this makes the comparison less useful really.

When Ed is reading the document he actually reads the words "cryptological", even though he reads some of the other words slightly differently, so I didn't really see this as evidence of tampering with the document either. (He said he was "skimming through" anyway).

As a general "concept" Ed has already said that his work on black programmes turned him into a liar - because he had to lie to his wife and tell her that he was somewhere doing something when he was somewhere else doing something else. So, by your logic, this makes absolutely everything he says untrustworthy - so why even bother listening to him in the 1st place?

As I alluded to in an earlier message, I find that other parts of Ed's account corroborate with completely separate areas of research (and I know these have been discussed on the AS forum and other forums - but sadly those discussions degenerate into "slanging matches" all too easily and quickly and any objective of establishing the truth about these things is quickly submerged). I won't repeat those thoughts here, because they are mostly covered in the video interviews I did with him that I posted a couple of months or more ago.

Has Ed been used as a "carrier" to promulgate disinformation about the TR-3B and related technologies? I am not totally sure. But what I can say with certainty is that 100s or 1000s of people have seen craft near and far which seem to closely resemble what Ed described.

I received no response from David Hilton to these questions and comments.

At this point I would ask - why would someone want to post these videos of Ed, without permission? What was the objective? Does it truly contain reliable evidence that Ed has deliberately given false information in his disclosures about the TR3B and related topics? What exactly has David Hilton proved with his videos and illegal recordings? If you consider the "fake document" video and Ed's response to it, and Hilton's lack of response to my questions, can we make any conclusions about his true reasons for editing and posting these videos?

Observations about David Hilton's Channel and Other Videos

A further video posted by David Hilton on the Outpost forum[641] is entitled "Edgar Fouché's Lie". This video, in which David Hilton appears briefly is, however, set as "unlisted" on David Hilton's YouTube Channel. The video is about an alleged Skype conversation with Edgar Fouché, but I could not see the screen clearly enough in Hilton's video to verify the claims he was making.

David Hilton has posted no other videos about Ed Fouché or the TR-3B – not one! He has, however, posted 3 videos featuring Jeremy Rys - one was about the termination of Rys' YouTube channel[642]. In the description of this video, he wrote

> *AlienScientist talks about how his channel is being* ***attacked by Judy Wood's people.***

So, who are *Judy Wood's people*? Can we have some names? Not likely!

Hilton fails to point out that Jeremy Rys was lying about **Dr** Judy Wood's research and he, like Hilton used material without permission, in order to denigrate the character of the person he was making the video about (i.e. this is not fair use). Another video by Hilton is entitled AlienScientist Channel Terminated[643], and another one was AlienScientist Is Back[644]. For me, at least, this makes the nature of what David Hilton is supporting very clear.

Avatar

Sometimes (not always), on internet forums and on YouTube channels, the use of a particular avatar can reveal something about the poster. In my own case, I often use a picture of an apparent artificial dome on Mars[645] as photographed by Mars Global Surveyor and Mars Reconnaissance Orbiter. I use this avatar in the hope that someone will ask me "what the hell is it?" I.e. I use it because I am very curious about this image and want to encourage others to "check it out."

I would now like to further consider the Avatar image on David Hilton's YouTube channel[634]. Perhaps David Hilton is interested in Science Fiction – but the image is one associated with a particular Science Fiction story – H.G. Wells' "The Invisible Man[646]".

I note this because David Hilton wrote in an email to me that he "does not have a Web Camera" and he only seems to appear in one of his videos. So is this how David Hilton would like to think of himself as "The Invisible Man"? I can only guess…

Ed Fouché on David Hilton

In July 2014, Ed Fouché sent an email to a few people, including me, containing further information and thoughts about Hilton, some of which was originally contained in Skype conversations Ed had been advised of. Ed pointed out the similarity of Hilton's forum handle to something used in a USAF programme.

> *He goes by names like Tham, Thamulas and Themulas which is very close to the Classified US Military USAF Program called Team Themis! When he was on the Alienscientist Forum as Admin, he had six IP addresses! One of them was Washington D.C.*

This program is discussed in an article that Ed referenced:[647]

> *"The hack also revealed evidence* ***that Team Themis was developing a "persona management" system — a program, developed at the specific request of the United States Air Force****, that allowed one user to control multiple online identities ("sock puppets") for commenting in social media spaces, thus giving the appearance of grass roots support. The contract was eventually awarded to another private intelligence firm."*

> *And remember, this is the same military that says political activists are terrorists and need to be targeted. At the highest levels, combating 'terrorism' simply means going after law-abiding citizens and journalists — especially so-called 'leakers'. With propaganda scripts that run 24/7 and are intended to discredit people like Edward Snowden, top level intelligence agencies are teaming up with the military to combat whistleblowers through such phony means.*
>
> *An excerpt from a particularly concerning summary of a recent German report on how political activists are targeted reads:*
>
>> *"The targets of these attacks are scientists... It does not stop at skirmishes in the scientific community. Hackers regularly target various web pages. Evaluations of IP log files show that not only Monsanto visits the pages regularly, but also various organizations of the U.S. government, including the military. These include the Navy Network Information Centre, the Federal Aviation Administration and the United States Army Intelligence Centre, an institution of the US Army, which trains soldiers with information gathering."*

Ed also had some more information about David Hilton and some strange things he said, in June 2012, in a Skype Conversation to another friend of Ed's called Dan Benkert (whose handle was Bigpappy51). Was Hilton playing a game? Was he inflating his own ego? Or was he telling the truth? Does my observation about the avatar that Hilton uses and Ed's observations about him – and a consideration of Hilton's own words, written in a 2012 Skype conversation - prove that Ed was being handled, following his re-emergence onto the "disclosure" scene?

> *[12:10:16 AM] David Hilton: actually i work for an agency that monitors things you shouldn't do. go ahead and tell people. they won't believe you. why do you think i should up, and won't my face.*
> *[....]*
> *[12:12:47 AM] David Hilton: you've been monitored for a while due to your associations, and I'm not the only one on the forum.*
> *[....]*
> *[12:14:51 AM] David Hilton: others are being monitored too at alienscientist. and i got a shocker for you. ed wont won't believe you if you tell him any of this.*
> *[....]*
> *[12:16:11 AM] David Hilton: all info at alienscientist is archived. ty filing some FOIA requests and you'll see.*
> *[12:17:49 AM] David Hilton: we don't need to look for you people. you all show up at these forums. you are all so easy to manipulate. ed is very easy to manipulate because he is so trusting of people like me and you and he shouldn't be.*
> *[....]*
> *[12:20:51 AM] David Hilton: Since no one is going to believe you, when this F-15 deal is completed we plan to bring Ed in for breaking his secrecy oath.*
> *[....]*
> *[12:21:55 AM] David Hilton: we've let Ed get away with his bullshit for many years because he doesn't know if his friends are telling him the truth, but we do.*
> *[12:22:22 AM] David Hilton: Ed is way to gullible and trusting, and that will be his downfall.*

[....]
[12:22:58 AM] David Hilton: You seriously have to ask about what people will believe? anything you say about this conversation will not be believed.
[....]
[12:32:12 AM] David Hilton: Ed was far deeper than you ever suspected, but his memory has been "processed." Be sure to send all of this to Ed. He won't believe any of it, and he'll think you're crazy.
[....]
[12:32:44 AM] David Hilton: I now control Ed not you. Funny how things work out isn't it. The exposer becomes to tool of control.
[....]
[12:40:04 AM] David Hilton: Ya, but my power is far greater than Jeremy's.
[....]
[1:22:21 AM] David Hilton: Tell you a little secret. I know Ed's entire history. He doesn't even remember it.

Car Accident

Sadly, on 28 June 2014, about 1 week after the Richard D Hall programme aired, Edgar Fouché was involved in a car accident – about which he later sent me the following message, which was typed by a friend of his:

> *I'm here at his home, writing this to only two people using one of his comps. Ed asked me to write you this. I wrote it down, here it is. Friday early afternoon Ed was returning home from the grocery store. A dark US made pickup truck came a top a hill in front of him halfway in his lane. Ed made evasive manoeuvres and hit a tree. Breaking two ribs, and fracturing 2. Ed head knocked out the front window. He had/has a mild concussion. No tickets were issued. The driver of the pickup was not identified, nor did he stop. His head is a bloody mess still.*
>
> *You don't know Ed like I do. Some of us think someone tried to murder or shut him up.*
>
> *I'm the one picked him up at hospital. His car was total totalled.*

Summary and Conclusion

Previously, in "9/11 Holding the Truth," I had documented how Jeremy Rys had lied about Dr Judy Wood's research (and he continues to do this to this day). Here I have discussed how David Hilton tried to mischaracterise what Ed Fouché has said (i.e. he has not fairly described the circumstances under which Ed made these statements). Further, David Hilton has made dubious claims about **one** document that Ed Fouché has produced, and Ed Fouché has explained why these claims are unfounded. Let me now summarise some observations and data in this article.

- Jeremy Rys has had a past conviction for which he has been on probation. He has had an association with convicted fraudster Richard Heene.
- Both David Hilton and Jeremy Rys have admitted to recording conversations with Ed Fouché and posting these recordings on the internet without permission.
- David Hilton has made 3 videos about Jeremy Rys and his YouTube channel (in relation to its termination due to copyright violations).
- David Hilton has made zero videos about the TR-3B or any of the technology that Edgar Fouché has talked about.
- David Hilton made comments, in 2012, about Ed's activities and his memories etc.

Hardly anywhere in this lengthy article have I discussed *the actual evidence and testimony* that Edgar Fouché has repeatedly and consistently made since 1998 – and the fact that parts of that testimony – such as that regarding the MJ12 documents[648] researched, completely independently, by Stanton Friedman et al, add significant weight to the validity of Ed's testimony. Additionally, the information Ed brought forward regarding the "Alien Autopsy"[649] seems to fit with what we know about the Santilli film.

Jeremy Rys has attacked Edgar Fouché[650] and Dr Judy Wood – both of whom have brought forward credible information relating, in some way, to secret energy technology and secret antigravity technology.

Due to the length and complexity of this chapter, and others that I have written of a similar structure, few people will read it. The doubt and uncertainty introduced and promoted by the likes of Jeremy Rys, David Hilton and many, many others - is effective in preventing the development of an accurate understanding of these stupendously important technologies. With all the back-biting, character smears and internet forum noise, few if any people will be able to work out what the real truth is. Mission accomplished.

31. Perception Management in Fact

In this chapter, we will look at two examples of mainstream documentary programmes which have presented information about the subjects of antigravity technology and the secret space programme. I will try to illustrate how they both exhibit what can be called "refutation bias" – or, more simply, debunking tactics. This chapter is similar to chapter 24 of "9/11 Holding the Truth." It documents the same sort of features and techniques of an ongoing cover up and "muddle-up" of important and world-changing information.

Channel 5, UK: Faking the Moon Landings

In October 2015, the UK's Channel 5 broadcast another programme in the series "Conspiracy," called "Faking the Moon Landings."[651] Like this book, the documentary concerned itself with the issues of the Apollo Hoax and the Secret Space Programme. In March or April 2015, I got in touch with a researcher who was involved in this episode of the series. The upshot of this was that I agreed to meet two of the people who were putting this programme together, to offer them information. I did not agree to be interviewed, as I had already "seen what happens" with these sorts of programmes but, I thought it was worth at least meeting them and offering them information – particularly in relation to Ed Fouché's disclosures, for example.

On 30 Jun 2015, I therefore travelled to London at my own expense to meet a researcher and a producer (whose names you can probably find out if you really want to).

When I met them, I started by relating the experiences of Scott Felton, a man who had done some detailed research into the Berwyn Mountain UFO Landing case from 1974. This case is covered in depth by Richard D Hall in his "Berwyn UFO Cover Up – Exposed" documentary, in which Scott is interviewed regarding his experience with the Channel 5 researchers.[652]

I needed to relate Scott's experiences (described below) to the Channel 5 guys, to set the scene for why I did not really want to participate in the documentary.

In relation to the Berwyn Case, Scott Felton met with people from a company called "Firefly Productions" and they spent the best part of two days filming with him for a 2008 documentary called "Britain's Closest UFO Encounters."[653] Scott took them onto the mountain and demonstrated to them where the events actually took place. He then demonstrated that the standard explanation for the Berwyn events – that is, that lights seen on the mountain were from poachers with lamps – could not possibly be true. The researchers agreed with Scott. However, the day before the broadcast of the documentary, Scott received a phone call from the producer Ian Levison to tell him that Scott (and the explanations he gave) would not be featured in the documentary. Instead, they featured several "debunkers" and according to the witnesses that were interviewed, they were then made to look like bad observers or idiots – or both (and therefore, they were not happy).

Scott also explained that Channel 5's commissioning editor interfered with the documentary from a higher level, and had Scott's material completely removed.

I told the producer of the "new" documentary all this, and explained to him this was why I did not want to participate. After this, the producer said he couldn't guarantee that he would use what anyone said in the programme – or whether they would be featured in it at all. I explained to him that I was "not bothered about being in it." I wanted the information I presented to be put in the documentary, in the correct context.

Around this point in the conversation, the producer said:

> *... The way I look at these things, to be honest, I don't actually have an opinion as to whether these stories are true or not - actually the way I approach them is... "I am here to present the arguments on both sides." Basically some people say, "Okay so you know there* ***is a programme called Solar Warden*** *and we've found something interesting." Some people say "Gary McKinnon made that story up" or "doesn't know what he's talking about," or "there's never been such a thing" and "no one knows why he said that." Those are the two sides of that coin. What we aim to do in this story is say just that, "This is what one side says and this is what the other side says - and it's up to you, the audience, to decide what you believe."*

What was initially interesting here was that he mentioned Solar Warden early in the conversation which, as I said in chapter 28, is probably disinformation – it is not even clear where this term came from. What was less of a surprise, though, was the producer's description of "telling both sides of the story." This, of course, in the strictest sense is what one should aim to do – and they claim that the way they make these documentaries is "balanced." However, as we will see, this isn't really true. I said in response that often, the process of establishing balance often involves telling lies, or at least misrepresenting the truth – and distorting or omitting evidence.

Next, in relation to secrecy and the notion of a secret space programme, we discussed military supremacy and national security issues in broad terms (i.e. perceived "good" reasons for secrecy). We agreed that of course there have been secret satellite launches and so forth. I pointed out to him that it wasn't this sort of secrecy or secret space programme that I was concerned with. Echoing what Edgar Fouché said (and repeating what I have said in the past), I said to the prodcuer:

> *"I've come to know that really, what is used within parts of the secret space program, if it was not secret, would completely change everything on the planet - the way that we live. Essentially I'm talking about things like energy and propulsion systems."*

I then spent the next 2 hours going through parts of the evidence I have included in my presentations – and in this book – with the researcher and gave him a DVD with many files and video clips etc that I hoped he would use for reference. He still wanted me to appear in the documentary and gently suggested it "would be great" if I would do so. I declined.

A few months later, the documentary was broadcast. I have, below, included the quotes which I expected would be included, in some form.

> *Narrator: Professor of Science and Media, Chris Riley, has been studying the moon landings for a decade he believes the conspiracy theorists are living in fantasy.*
>
> *Riley: Any **sane person** that studies the science behind the conspiracy theories can conclude for themselves that there is absolutely no doubt that 12 human beings walked on the moon and that's in part backed up by the fact that 400,000 Americans worked on that programme, for the best part of 10 years. That's four million human years of effort to achieve that goal and with that kind of ambition and aim everything is possible really.*

Yes, folks… this is "balance" – suggesting that people who question the Apollo evidence are "not sane." The flawed logic of Riley's statement was covered in chapter 18. That is, almost 50 years further on in time, a "re-run" of Apollo should be far easier – but nothing even close has been done.

Riley continued:

> *The flapping flag point is really a gross misunderstanding of how things behave in a vacuum. Now as soon as you've got something hanging in a vacuum and you're twisting the pole that the flag is hanging off - you're going to set the flag swinging and when you set something swinging in a vacuum, when you let go of it, it will carry on swinging because there's no air to slow it down. It looks somewhat otherworldly - and that's because is otherworldly.*

This is just one point of the evidence (not even mentioned in this book by the way, as it's something that is equivocal – and difficult to test without a vacuum chamber.)

The Channel 5 documentary ended with two statements from academics. The first one was from Professor Chris Riley:

> *Every year, NASA has to fight for their budget and one of their great ambitions is this search for evidence of intelligent technological life. If they knew of any kind of evidence for this, they will be singing it from the rooftops, because their future budgets would depend on the success of those discoveries. The fact that anyone **can imagine** that such things are known by any agency or government on earth [and] are not disclosed is **frankly bonkers.***

Another example of measured, balanced reasoning. Of course, this book is not about *imagination*, it is about evidence. I describe those, like Riley, who refuse to study or acknowledge evidence as being in denial (which, in other circumstances, is often addressed through some type of psychotherapy). Or, I would describe them, similarly, as being in a state of cognitive dissonance - I would not describe them as "bonkers."

Instead of a summary of the evidence on each side of the issue, the programme closes with these remarks:

> *Narrator: Still whatever sceptics say, people **continue to believe** there is a secret world of cover-ups, dating back to the very dawn of the Space Age.*

Professor Gerard De Groot:[654] *These* ***conspiracy theories*** *will always exist. They have existed throughout history. But I think the other reason why perhaps a conspiracy theory takes root is because space is otherworldly. It's the perfect scenario for carrying out something secret, because you limit the number of witnesses. You know, it's not just a case of going to a place in the Arizona desert. You can carry it off because it's out there.*

Narrator: In the world of ***conspiracy theories*** *space remains the final frontier.*

So, the programme ended talking about conspiracy theories and beliefs and people who "imagine" things being "bonkers." This is what passes for serious "balanced" commentary in the mainstream media in the 21st century. Switch it off, folks!

Following the broadcast of the programme, which contained the very sentiments and statements I was hoping it would not, I obviously realised that my decision not to participate was a good one. With this in mind, I wanted to get the reaction of the researcher that had suggested I appear in the documentary. Did he think the documentary was "balanced"? Were the final comments fair?

From: Andrew Johnson <ad.johnson@ntlworld.com>
Sent: 31 October 2015 11:56
To: Name <name@channel5.com>
Subject: RE: "Secret Space Programme" - Follow Up after our Meeting

<Name>,

How did you feel about the programme that went out last night? (I watched it this morning).

Were there any things that you wanted to be shown that weren't?

Andrew

The following response came back.

-----Original Message-----
From: Name <name@channel5.com>
Sent: 31 October 2015 11:56
To: Andrew Johnson <ad.johnson@ntlworld.com>
Subject: Automatic reply: "Secret Space Programme" - Follow Up after our Meeting

Thank you for your e-mail.

I have finished working on 'Conspiracy' so please contact the production manager, Alison Higgins, for any enquiries. E-mail: alison.higgins@channel5.com[655]

Kind Regards,

Name

BBC Horizon – "Project Greenglow - The Quest for Gravity Control"

In completing the research for this book, I came across a BBC Horizon program which was broadcast on 23 March 2016.[656]

The programme was centred around the research of Dr Ron Evans. This, of course, was the subject of part of chapter 11 of this book.

This photo has been lit or re-coloured to give it a green cast. Did the BBC want Dr Ron Evans to look like an alien? What educational purpose does this image serve?[657]

The Horizon programme contained very little in the way of useful scientific information and was essentially a glossy, superficial package which offered no definite answers, in the way this book does, for example. For science fiction fans, they would have immediately recognised that the voice of the narrator was that of Peter Capaldi – who, at the time this documentary aired, was playing the role of Doctor Who[658] – as the timelord's twelfth incarnation. Two minutes 10 seconds into the documentary, the tone was set, with the narrator, Peter Capaldi, saying[659]:

> *If the dream of gravity control ever came true, it would revolutionise the world and could send us to the stars.*

Comparing this with the Channel 5 documentary that I discussed in the previous section, a different approach was needed - because the work of scientists was being discussed, so they couldn't apply the "conspiracy theory" memes quite as liberally. However, this did not stop them from putting out false information. For example, about 6:40 into the programme, Capaldi narrates thus:

> *Years earlier, Ron had watched a gravity experiment bring down one of Britain's best-known scientists... professor of engineering at Imperial College London, Eric Laithwaite.*

This is untrue. Of course, Laithwaite was heavily criticised, but he continued to do scientific work and research up to his death. From the Imperial College Website[660]:

> *Eric Laithwaite was Professor of Heavy Electrical Engineering at Imperial between 1964 and 1986, Emeritus since 1986 and Fellow of Imperial College since 1990.*

You would also know that the "bring down" statement was a lie if you'd read chapter 7. Not long after this false statement came another one:

> *Still hoping to make gravity control a subject of serious research, Laithwaite acknowledged his mistake. Yet his reputation was irreparably damaged.*

Laithwaite never "acknowledged his mistake" about the unusual properties of gyroscopes. He made it clear that spinning objects have a "different property" – and this was also discussed in chapter 7. Why did the documentary have to include lies? It could have, instead, quoted from The Times obituary for Laithwaite, from 1997. Laithwaite said:

> *"I thought my fellow scientists would be genuinely interested, so I wasn't prepared for the utter hostility of their reaction".*

The centrepiece of the documentary seemed to be an attempt to debunk Podkletnov's "force beam" experiment. This is done by Dr Martin Tajmar, Professor of space systems at Dresden University. They show Tajmar's experimental set up, of which few details are given, being "charged up." Below, I have included a picture of two of the "electrodes" used in the experiment.

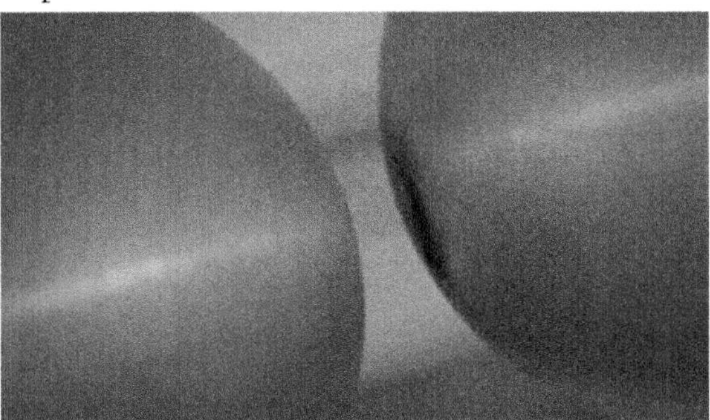

BBC Documentary – "Reproduction" of Podkletnov's "Force Beam" experiment.

In my estimate, these are metallic spheres about 4 inches in diameter. They are moved together slowly – until they are less than ½cm apart. Then, there is an electrical spark. It does not appear that the experiment is being performed in a vacuum. This means that the potential difference between the two electrodes is less than 15000 volts. Judging by the final shot, the discharge occurs when the spheres are less than 2mm apart. This would mean the voltage across the spheres is about 6000 volts. Referencing the 2012 paper that Podkletnov published with G. Modanese[221], we read:

> *The light attenuation was found to last between 34 and 48 ns and to increase with voltage, up to a maximum of 7% at **2000 kV**.*

This means Podkletnov was using up to 2 *million* volts – over 300 times the voltage used in the experiment shown in the BBC's "debunk." So, to disprove an experiment, all you need to do is pretend to reproduce it, ignore the details of the experiment, find that it didn't work and then pronounce that your reproduction failed.

Following this demonstration, there is a brief appearance by Professor of Physics, John Ellis of King's College London. He states:

Professor of Physics, John Ellis of King's College London.

> *"So this guy had the idea that by, you know,* ***messing around*** *with superconductors, he could change the strength of the gravitational field. Crap."*

It seems clear to me that Ellis has not read Podkletnov's papers, is not familiar with the special techniques Podkletnov used and he has nothing of value to say in critiquing any aspect of the research. Instead, he is simply rude and insulting. Is this the best that the BBC and King's College, London, could offer?

Also on 23 March 2016, an article about the programme was posted in the BBC online magazine entitled "Project Greenglow and the battle with gravity[661]". This article is peppered with such phrases as "Wouldn't it be good..." and "If only..." and "It **would** send us to the stars..."

So, again, it's just a dream, right? The documentary ends in a most similar way, thus. Ron Evans is walking through a fairground looking up at the rides as they whirl around...

> *Peter Capaldi: For Ron Evans, gravity control is just something we haven't learned to do... yet.*
>
> *Ron Evans: I'm sure we will one day. It's just a matter of time.*

Of course, BBC! You know best! There's no evidence that working antigravity propulsion systems have already been developed – and I hereby retract the first 25 chapters of this book, right?

Conclusion

The mainstream media is controlled or influenced by "the secret keepers" to the point where truthful information about the technologies involved in a secret space programme will not be fairly discussed or revealed. Evidence may be shown, but it will then be ridiculed or "debunked" by "experts" who only have one option open to them – to "toe the line" and maintain "truth's protective layers."

32. Perception Management in Fiction

This chapter is mainly concerned with some aspects of mainstream media perception management of the Apollo Hoax in fictional material. We will briefly look at several different aspects of the way knowledge of the hoax has been portrayed. Further examples of fictional works that revolve around the Moon Hoax are "Operation Avalanche[662]" and/or "Moon Walkers,"[663] both of which have been written about by other researchers.[664]

Stanley Kubrick with his "Eyes Wide Shut" Because of "The Shining"?

The Shining[665] (released in 1980 - some years after Apollo was "popular") is a disturbing (and somewhat perplexing) psychological thriller about a man who goes mad when he looks after a hotel in a fairly remote location in the mountains, during a long winter shut-down. It has nothing to do with the Apollo missions! It is based on a story, of the same name, by Stephen King.

On his website, Jay Weidner provides a detailed discussion of the film[666] and notes a key scene. As Danny (Jack's son in the story) stands up, the answer is revealed in an instant. Danny is wearing a sweater with a crudely sewn rocket pictured on the front. On the rocket clearly seen on Danny's sweater are the words: "APOLLO 11 USA."

Weidner writes:

> *The audience watching the film literally sees the launch of Apollo 11, right before their eyes, as Danny rises from the floor. It isn't the real launch of Apollo 11, it is, of course, the symbolic launching of Apollo 11. In other words - it isn't real.*

There are several other scenes in the film which seem to make encoded references to the Apollo programme. A number of researchers have pointed out similar things and noted how, for example, one of the themes of "The Shining" is telepathic communication.

In 2015, Carl James presented a lecture about Stanley Kubrick and his films.[667] James has written a two-part body of work called "Science Fiction and the Hidden Global Agenda."[668] He analyses the films of Kubrick and others to try and understand what is "below the surface." In relation to a secret space programme, it's possible that Kubrick had some knowledge, perhaps as a result of research into US Military hardware for the 1964 film "Dr Strangelove."[669] It is then interesting to note, at this point, that in the film Dr Strangelove, it has been suggested that General Curtis LeMay, who we mentioned in chapter 28, was the basis of one or two of the characters.[670]

In Stanley Kubrick's brilliant and frightening 1964 film, Dr. Strangelove, LeMay is seen in two characters: first the cigar-smoking General Jack D. Ripper, who sends his wing of nuclear armed B-52s against the Soviet Union all on his own because he has become completely paranoid; and then as the George C. Scott character, General Buck Turgidson, the head of the Air Force in the Pentagon who, instead of seeing Ripper's act as a disaster, sees it as an opportunity.

What if Kubrick found out about the sorts of information that caused LeMay's reaction to senator Barry Goldwater?

As many others have written, "Eyes Wide Shut" is another dark Kubrick film[671], which features several scenes about Satanic Rituals in groups within "High Society," was released (at Kubrick's insistence) on 16th July 1999. This was exactly 30 years to the day after Apollo 11 was launched. Kubrick died on March 7, 1999 – 4 days after the final print of "Eyes Wide Shut" was delivered. As alluded to above, and by Jay Weidner and others[672], Kubrick's films seem to be filled with much hidden meaning – or even "disclosures."

One possible indication that this is true happened in December 2015, when an alleged "tell all" interview with Kubrick was released.[673] This was quite soon after the Channel 5 "Faking the Moon Landings" documentary (that I discussed in chapter 31) had aired and it was also soon after Carl James had appeared in an interview with Richard D Hall.[674] This "Kubrick tells all" interview was very quickly shown to be a hoax, employing an actor. The whole episode was rather strange and stories even appeared in the UK mainstream press.[675] It created a very similar feeling, in my view, to the "Dark Side of the Moon" documentary (covered later). That is to say, an obvious hoax was created and played out to protect ideas/information about a bigger hoax. In his 2016 tour, Richard D Hall[676] noted that Bart Sibrel had claimed that Stanley Kubrick had written an affidavit, which was to be published a certain number of years after Kubrick's death. Hall posed the question that if the affidavit was about Kubrick's knowledge of the hoax, the mainstream media might need to "second guess" its release and therefore put out "prophylactic disinformation."

MOON LANDINGS 'FAKE': What Stanley Kubrick's family say about 'hoax admission' video

RELATIVES of Stanley Kubrick have spoken out after a shocking video supposedly showing the movie legend make a deathbed confession that he helped NASA fake the moon landings emerged.

By JON AUSTIN
PUBLISHED: 22:29, Tue, Dec 15, 2015 | UPDATED: 22:29, Tue, Dec 15, 2015

SHARE TWEET 227

The alleged Stanley Kubrick in the video and the iconic moon landing image

"Dark Side of the Moon" – A 2002 Mocumentary.

We will now consider a very strange mocumentary called "Dark Side of the Moon,"[677] that was made in 2002. Initially, is seems to present the evidence that Stanley Kubrick was involved in making the hoax footage. In 1968, "2001: A Space Odyssey" – a very expensive movie – was released, showing incredible (for the time) effects and sets on the moon. As mentioned earlier, several researchers, including Carl James and Jay Weidner have discussed the "coincidental" timing of the production of the Kubrick film and the Apollo missions, and for example.

The "Dark Side" documentary states that Kubrick needed a very special type of camera for the film "Barry Linden" (a historical drama) and only NASA had one with the right optics. In exchange for use of the camera, it is implied Kubrick worked for NASA on the "fake moon shot."

The film features intimate footage with General Haig, Donald Rumsfeld and Henry Kissinger – seen laughing and joking. At the end, many of the "whistleblowers" in the story are "unmasked" as actors, or real people reading from a script…

At around 35:22, the mocumentary describes how a former KGB agent, named "Muffley," disclosed there were two "undeveloped" Apollo photos on a film taken in a studio. These are then described with the following narration:

> *On the Apollo 11 mission, Muffley pointed out, everything went perfectly according to plan except for one thing – there were no pictures. The film was unusable and none of the photos came out. But everything depended on these pictures, eagerly awaited all around the world. Muffley showed us 2 photos from the end of the reel of film which had never been developed. They had been slumbering away in NASA's archives. On the first, the shadows make no sense, spreading in all directions. On the second, a photograph of Stanley Kubrick taken during the filming of 2001 – can be seen left abandoned on the studio's fake lunar surface…*

The following image is then shown.

Still from "Dark side of the Moon." Observant readers should be familiar with this image:

Apollo 16 Photo AS16-117-18841[342] – Kubrick's face was "photoshopped " onto this image. Needless to say, I find this choice of image interesting.

What reason would there be for *anyone* to go to the trouble of faking this image and including it in a "spoof documentary"? Does it indicate that the image, is indeed, one which reveals the Apollo Hoax? Is it so compelling that it had to be used as a point of focus in a "mind mangling" disinformation exercise? Someone went to an awful lot of trouble and expense to create this mocumentary. Can we consider that it was "a hoax to cover up a hoax"?

Capricorn One and Bill Kaysing

The plot of this 1978 film involves astronauts being put in a rocket to go to mars, but at the last minute, they are removed and told there is a fault which would have killed them and they are then coerced into participating in a fake-up.

Set for the Mars Landing from the 1978 film "Capricorn One"

This film, which was made with the full support of NASA, is often discussed on the web in relation to the Apollo Hoax. However, in November 2015 (again, around the same time that the fake Kubrick interview was released), I was sent information from, "The fastest pen of the West – biography of Bill Kaysing"[678] by Albino Galuppini[679]. Albino wrote:

> *In 1977, Ruth and Bill bought a caravan in Las Vegas for the sole purpose of being a mobile office for developing a screenplay inspired by the book "We Never Went To The Moon." For this purpose, the Eden Press, which had published the first edition of the text, hired screenwriter Ken Rotcop, entrusting him with the task. They all worked hard, assisted by Bill Butters, who felt certain of being able to be funded in the project by Sir Lew Grade, a famous Ukrainian born, London based film producer.*
>
> *There was no doubt, however, that their initiative was closely watched by unknown entities. Randy Reid, co-author of the first edition and printer of the book, said that their correspondence was monitored. Such ectoplasmic forces did not wait long to materialize, implementing clever countermoves.*
>
> *Lew Grade did not give his support to financing the film. In that precise period, producer Paul Lazarus, with director Peter Hyams, produced a movie whose plot traced, in a stunning way their screenplay. With the only expedient that the simulation of space travel took place on Mars instead of on the Moon.*

> *Hyams film was titled "Capricorn One" and counts among its protagonists famous actors including OJ Simpson, Elliott Gould, James Brolin and Brenda Vaccaro. The movie premiered in 1978, achieving some success, and then disappeared from circulation.*
>
> *However it had obtained its purpose: to "beat to the punch" their idea and film project.*
>
> *The writer (Kaysing) then filed a lawsuit against Universal Films and director Peter Hyams, unsuccessfully, at a court in Beverly Hills for copyright infringement. According to the thesis proposed by his lawyer, the script was largely borrowed from the book on the lunar plot and identified 16 elements in the film that violated copyright.*

Albino continued:

> *A private investigator, hired by the legal office in support of the suit made a shocking discovery. In Washington, someone in connection with the Capricorn One script had falsified the date of filing for copyright in the office in charge – to predate Kaysing's claim of copyright. [He said:]*
>
> *"Bill, I have a wife and two boys, I do not want to oppose people so powerful as to be able to alter the filing date at a patent office."*
>
> *Bill realized in that moment that for many common people, including his attorney, their physical integrity and that of their families are the priority. But he did not fall into despair: "That which endures, is victorious".*

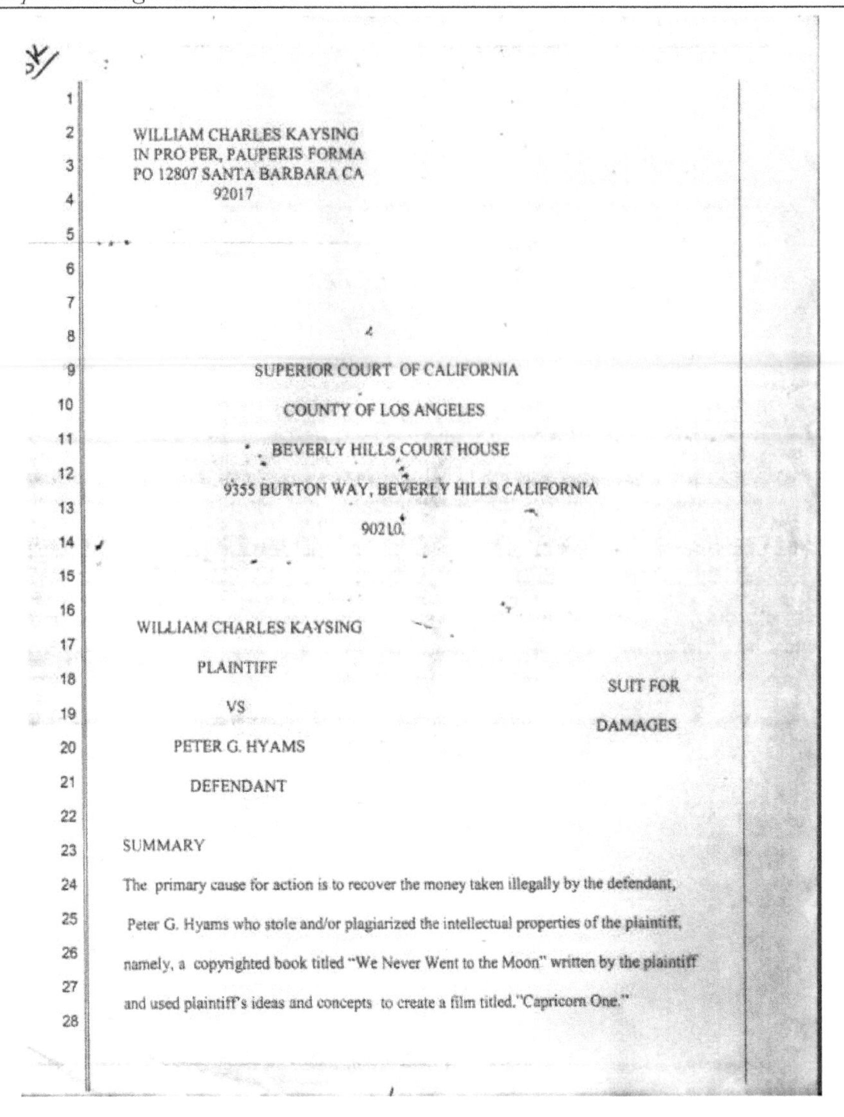

WILLIAM CHARLES KAYSING
IN PRO PER, PAUPERIS FORMA
PO 12807 SANTA BARBARA CA
92017

SUPERIOR COURT OF CALIFORNIA
COUNTY OF LOS ANGELES
BEVERLY HILLS COURT HOUSE
9355 BURTON WAY, BEVERLY HILLS CALIFORNIA
90210.

WILLIAM CHARLES KAYSING
PLAINTIFF
VS
PETER G. HYAMS
DEFENDANT

SUIT FOR DAMAGES

SUMMARY

The primary cause for action is to recover the money taken illegally by the defendant, Peter G. Hyams who stole and/or plagiarized the intellectual properties of the plaintiff, namely, a copyrighted book titled "We Never Went to the Moon" written by the plaintiff and used plaintiff's ideas and concepts to create a film titled."Capricorn One."

Miscellaneous Apollo Hoax references

Many people noticed a very peculiar reference to the hoaxing of the Apollo missions in the 2014 Christopher Nolan film "Interstellar."[680] In one scene, which has almost nothing to do with the main plot of the film, one of the main characters, Cooper, visits his daughter's school, to discuss her progress there with a teacher or school head. The following dialogue is included.

> *School Official: It's an old federal textbook we've replaced them with the corrected versions - corrected explaining how the Apollo missions were faked to bankrupt the Soviet Union*
>
> *Cooper: You don't believe we went to the moon?*

> *School Official: I believe it was a brilliant piece of propaganda - that the Soviets bankrupted themselves pouring resources into rockets and other useless machines. And if we don't want a repeat of the excess and wastefulness of the 20th century then we need to teach our kids about this planet.*

Jay Dyer has written this up at length on his website.[681] Though mention of the hoax seems very significant, the stated reason (in the plot) does not seem to be the most likely one. Also, the official's response about "teaching kids about the planet" echoes some of the "Agenda 21" propaganda, which I wrote about in "Climate Change and Global Warming… Exposed: Hidden Evidence, Disguised Plans."[682]

Red Bull Commercial

A 2006 Red Bull is amusing, and perhaps is poking fun at those questioning the veracity of the Apollo record.[683] However, it is significant that a large multi-national company would use this as a theme for one of their advertisements. In the commercial – a cartoon - the LEM is seen landing and an astronaut descends to the surface, uttering Armstrong's "giant leap" words, but then, he starts to float up – and he has a pair of "angel wings" protruding from his back pack. The dialogue is as follows:

> *Astronaut: Houston we have a problem.*
>
> *Houston: Oh gee, you drank a Red Bull – Red Bull gives you wings!*
>
> *Astronaut: What shall we do?*
>
> *Houston: Well, nothing. Come back to Earth and we'll shoot the whole thing in a studio.*
>
> *Astronaut: OK!*

R.E.M – "If You Believe They Put A Man on the Moon…"

This song is about[684] comedian Andy Kaufman[685], who used to play the innocently complex immigrant Latka Gravas on the hit TV Show "Taxi." Kaufman often went on "Saturday Night Live" to do Elvis Presley imitations. He also, at one point, became a comedic wrestler and directed a short film about wrester Fred Blassie called "My Breakfast with Blassie." Kaufman's life story was told in the film "Man on the Moon" starring Jim Carrey (1999)[686].

The REM song also mentions Elvis – who, some people claim, is still alive…! It appears REM mixed together the two ideas of the Apollo Hoax and Kaufman faking his own death – he died at 35 of a rare form of lung cancer, but had previously talked about faking his own death.

Science Fiction Worlds, UFOs, ET's and Star Trek

In this book, I have tried to consider many types of evidence which strongly suggest that a secret space programme exists. I have also illustrated some of the techniques used to cover up the existence of such a programme. We have already considered the role of mainstream media in discouraging people from understanding certain facts about what has been discovered. However, I think it is also important to consider other aspects of the output of mainstream media - just as people like Carl James[687] and Dave McGowan[688] have done. With some research, it is relatively easy to establish unexpected links between the Military Industrial Complex and the entertainment industry – primarily in Hollywood. This area is also touched on, in several places, by Peter Levenda in his "Sinister Forces"[689] series of books.

At one level Science Fiction can inspire and stimulate the mind and the intellect. It can encourage "bigger" and "different" thinking – it certainly had a big effect on me when I was growing up! Well-written science fiction literature and well-made films and TV shows, with a good script and a good story can "transport" the reader or the viewer into a completely "different world." It offers new perspectives on current and age-old problems and issues. Yet, in most cases, Science Fiction fans, authors and people involved in producing shows, books and so on, seem to have little or no knowledge or no interest in the topics and evidence covered in books like this.

Below, I present an example of this, although let me also state that for many people involved in these programmes, though they may enjoy what they do, they might admit, at the end of the day, "it's just a job." Having said this, it is not uncommon for Star Trek Cast members to point out the loyalty and enthusiasm of fans is often much stronger for science fiction shows (and particularly the various incarnations of Star Trek) than it is for other genres. Star Trek (and shows like it) have a big impact on the fans' psyches – their consciousness, even.

Cast Members at Star Trek Next Generation Cast Re-Union is Las Vegas Nevada, 04 Aug 2017. From left to right – Levar Burton, Brent Spiner, Gates McFadden, Patrick Stewart, Michael Dorn, Marina Sirtis, Denise Crosby, John De Lancie.

On 04 Aug 2017, in Las Vegas Nevada (an appropriate place considering what we have covered in this book), a large convention took place – a cast reunion of those who starred in "Star Trek – the Next Generation." This was to celebrate the show's 30th anniversary. Many of the presentations were recorded by fans and posted on YouTube. In one such presentation, involving a panel of eight members of the original cast, a question was asked at 40:53 into the recording.[690] The question refers to "Gene." Gene Roddenberry was the creator of Star Trek.

> *Question from the audience: But if I may just for a brief moment ask, in your opinion, we are, I think as a planet possibly on a greater a knife edge [than] we have [been] for a very long time. Excuse me for a moment for being a little bit serious... Do you have an idea how we can get from here to Gene's vision of the 24th century in some form? I don't mean to rest too much responsibility at your feet - I'm sorry it's quite a big question, but I just wonder if you have any ideas and* ***I know Levar asked me to hold on to this question until now and maybe you regret that Levar****... Do you guys have any ideas how we can use the love in this room for what Gene created and actually get to what he envisaged?*

Levar Burton (who played Geordi La Forge) answers first:

> *The only thing that comes to mind - and I'm glad you said it – is if we can take the love and the passion for the values and ideals that Gene has gifted us with, and really put them into practice in our everyday lives and then pass those values on to our children. I do believe we will get there.*
>
> *We have no choice – it's either... and I believe you - and agree with you, when you say that we're sort of at a tipping point. We are either going one way or another and I've always been optimistic, in the sense that, in the face of contrary evidence,* ***I still want to bet on the human beings****.*

Gates McFadden (Dr Crusher) responds with a discussion of what might be called "personal responsibility:"

> *A lot of times I think it's maybe important to go "what can I do" and rather than be concerned about what everyone else can do, and I think as I get older I realize that when I feel vulnerable, that's sort of the birthplace of love and empathy and when I fear and I get defended and I become guarded is when I'm less open to actually perceiving another person and trying to actually accept who they are. I think that's to me has been a really interesting journey and I would love to continue down that road of being open and being vulnerable – in a good way – in a strong way… I mean having courage to not be afraid and to love…*

No other cast members answer the question. There is no mention of anything related to any of the topics mentioned in this book.

It is interesting to note, as Carl James has in his "Star Trek Agenda"[691] research, that Burton has a connection to the Institute of Noetic Sciences[692] (but he didn't mention this in his answer to the question above.)

> *Burton is the host and executive producer of a documentary titled The Science of Peace, which was in production as of 2007. It investigates the science and technology aimed at enabling world peace, sometimes called peace science. The film explores some of the concepts of shared noetic consciousness, having been sponsored in part by the Institute of Noetic Sciences.[18]*

As we mentioned in chapter 27, the institute of Noetic Sciences was founded by the CIA Operative and supposed Apollo Astronaut Edgar Mitchell. Why didn't Burton talk about technology and other aspects? He had a golden opportunity to educate the audience about what he knew…

Another of the Star Trek (Voyager) cast members, Robert Picardo – who played the holographic "Emergency Medical Programme" is a member of the Planetary Society[693]. On their website, it says of him, "Robert Picardo joins the Board of Directors, excited to help influence the world of space science and exploration."

Before moving on, I want to mention a peculiar episode, from the "Star Trek - Voyager" TV series, called "The Voyager Conspiracy,"[694] which aired in November 1999. In this episode, the Borg/Human character "Seven of Nine" has been fitted with a new cortical processing subunit and this allows her to access the ship's databases directly. Over the course of the episode (which is well-written), Seven becomes convinced that the senior crew members – Janeway and Chakotay - are involved in one or more conspiracies to use the ship for "nefarious" purposes. The plot line is interesting, but there is a "subtext" which suggests that Seven, having had access to a lot of new information becomes "mentally ill" – her paranoia becomes overwhelming. She then removes the "subunit" which gave her this new ability and all is well again. Is the character's situation in this episode meant to have been synonymous with that of the many curious and intelligent people during the late 1990s, who had, for the first time, begun to gain free access to vast quantities of information on the World Wide Web? What subconscious messages were embedded in this episode?

Social Engineering?

So, is Star Trek aimed at being part of a social engineering project which "herds people" mainly into a realm of fiction? Is the realm of fiction and fantasy the only place these folk can "deal with" the ideas of advanced energy, propulsion and weapons technology existing? I would also like to pause and consider the role of Star Trek actors such as Leonard Nimoy[695], James Doohan[696] and Jonathan Frakes[697], who have also narrated or hosted documentaries which cover "alternative knowledge" topics. We can then note, as we did in chapter 31 when discussing Peter Capaldi, how the association of a "science fiction character" with the narration of a serious documentary can colour the perception of what are, after all, extremely serious matters.

Meeting Star Trek Cast Members in September 2005

As I alluded to earlier, TV shows like Star Trek can have a very strong influence on the way matters discussed in this book are perceived by "the masses." As a small example of evidence of this influence, I will now relate an experience I had in September 2005.

I've been (almost) a lifelong fan of Star Trek and I have even attended several UK conventions over the years – briefly meeting James Doohan (Scotty) in 1992, at one of these events.

In 2004, I came into contact with Lloyd Pye, due to his research into the very unusual "Starchild" Skull.[698] We became good friends and I even organised two or three speaking events for him, over the years. The Starchild Skull is very much worthy of your study because all of the evidence suggests it is not from a human being, nor did it belong to a being with enough DNA similarities to any earth species of primate or mammal.[699]

In August or September 2005, Lloyd emailed me to say that he had been in touch with someone that was coming to the UK and she was interested in talking with me about the Starchild Skull, and related matters.

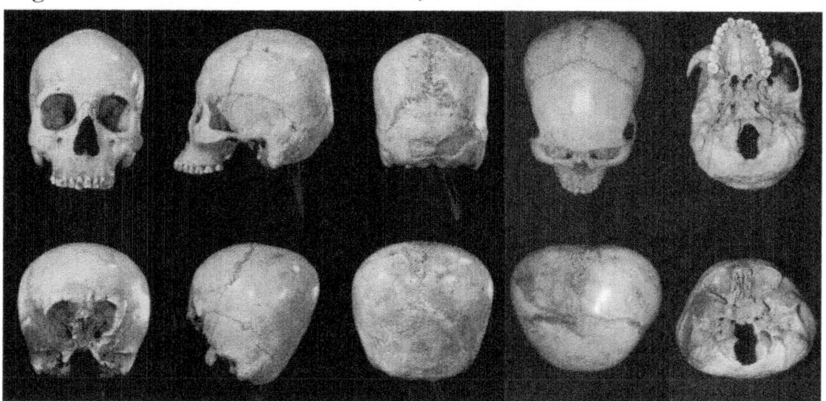

Human Skull (Top) and Starchild (Bottom) Comparison.

This was a lady called Celeste Yarnall[700] – who had portrayed a character in one of the 1960s Star Trek episodes. She was to attend a Star Trek convention/event in Leicester, UK on the weekend of 10th September 2005. There were some synchronicities in the timing of this meeting, in that on the very day she would be in Leicester, I already had an appointment to see students at De Montfort University there. It also turned out that the location of the Star Trek event was the Holiday Inn, which was a short walk from where I was working on that day.

I met Celeste at the Holiday Inn in the early evening of Friday, 10th September 2005 (the only day I had free on that weekend) and she introduced me to two of the other well-known cast members. One cast member was from "Star Trek: Next Generation" and the other was well known in "Star Trek: Deep Space Nine." I had brought a small Sony Vaio Laptop with me, on which I had PowerPoint files containing images and information about the Starchild Skull and some of the anomalous images from Mars. (Of course, I later included many of these Mars images in "Secrets in the Solar System."[701])

Celeste encouraged me to show the images and explain them to the other cast members. However, after about 10 or 15 minutes, it became clear they weren't really that interested in what I had to show them, and they wandered off and talked to other people.

Celeste also spoke to the organisers and explained to them why I was here and suggested that they, too, should have a look at some of the images. Again, I could tell they weren't really that interested in this sort of evidence. Celeste further suggested that the organisers might organise an event based on the sorts of topics Celeste and I had discussed, and that perhaps, I could give a presentation for them. The organisers did not comment on this idea and so, Celeste and I continued our conversation privately – for about 3 hours. Celeste was very kind and accommodating and gave me a signed photograph.

This episode illustrated to me that "Trekkers" were by and large, much more interested in the "fantasy world" of Star Trek than they were in evidence regarding life on Mars, physical evidence of an alien being having lived on the earth, or evidence of a secret space programme. Hence, in my view, the perception of these people has been successfully managed – and their hopes, dreams and desires have been oriented around a realm of fantasy rather than a realm of fact and evidence.

33. Conclusions

History with a Security Classification

Much of what I have covered in this volume could be described in this manner – i.e. "history with a security classification." Like many others, I consider that the bulk of the events and actions I have written about here were initiated following the Roswell event which happened in July 1947. Like Edgar Fouché, I think that a secret group was formed to manage the hardware, information and intelligence related to that event and presidents Truman and Eisenhower knew a great deal more than they ever admitted publicly.

By the time of the launch of Sputnik in October 1957, the USA's security apparatus had become much more extensive and robust and covert development of various technologies was well underway. I argue that examples of these technologies include the MFD technology that Edgar Fouché described in chapter 25 and possibly even the engine inspected in 1971, described by David Adair in chapter 23 (although I find many of the details of Adair's story less compelling).

Similarly, there seem to be elements of truth in the Colonel Corso and even the Bob Lazar story, which indicate secret projects have been successful in developing advance propulsion technologies.

Recurring statements and accounts regarding Area 51 and Nellis AFB in Nevada – and the presence of extremely efficient security around its borders – strongly indicate that "big secrets" are kept there. Similarly, repeated mentions of Wright Patterson AFB indicate that personnel there do know much more about aspects of a secret space programme than they will likely ever reveal.

The influence of the "secret keepers" reaches far beyond Nevada, Ohio or any areas where similar clandestine research and development has taken place. They are able to strongly influence the world's largest media outlets and even have the ability to "micro-manage" certain aspects of them, to prevent evidence being shown in an appropriate context and in an accurate and honest way.

The Secret Space Programme is still Secret!

Some years ago now, when I was first becoming familiar with a little of the history of electrogravitics, someone has suggested to me the following equation may give an answer to what has been produced by enormous budgets and top secret, protected black programmes.

(Tesla+Thomas Townsend Brown) × 60 years × ? billion dollars = **SPACESHIP!**

With the permanent grounding of the US Space Shuttle in 2011 and no earnest effort to explore the solar system (or further afield) with manned spacecraft, I can only conclude that the "Powers that Be" have already achieved most or all the objectives of any Secret Space Programme and there is no requirement for them to initiate a new "public" space programme that can be compared to the old one in any meaningful way. Perhaps their "media and social engineering" machinery is now so effective, they can divert peoples' attention in any way they need to.

I am well aware that the morsels of information I have written about here do not prove that a Secret Space Programme exists though I think one of the clearest glimpses of a part of such a programme may have been the 1991 STS-48 shuttle mission video footage, discussed in chapter 26. Perhaps you will agree with me that the preponderance of all the evidence strongly suggests the programme has been in existence for several decades.

It is worth remembering the phrase, "Those that know don't talk and those that talk don't know." However, Edgar Fouché was something of an exception to this "rule". Everything that he has said that I have been able to check out has "checked out." Even Ed Fouché did not claim knowledge of a secret space programme, only of secret technologies.

To truly "get a handle" on all this is extremely difficult for the majority of people – for many reasons, but the main reason is that they have been exposed to a pervasive and comprehensive propaganda programme – which has been in operation since the end of WWII and encompasses all mainstream media and academia. As a result, few people are aware of or care about deception on a massive scale - involving runaway black budgets and decades of deep, dark secrets. The deception regarding events of 9/11 and their cover up is of a similar scale to the Apollo deception, but it is troubling to consider that one event was meant to be an act of Muslim terrorism, while the other was meant to have been humans' greatest technological achievement.

The Military Industrial Complex – including such entities as SAIC, Lockheed, Northrop and the TLA's (Three Letter Agencies) control all the most important information and "call the shots" (no need to pardon the pun). I cannot see this situation changing – particularly when the "secret keepers" also seem to control or strongly influence most of the prominent speakers and researchers in the "Alternative Knowledge" community and any "cracks in the secrecy edifice" have a "paste of disinformation" liberally applied to them, often by people who have committed misdemeanours of one kind or another.

A Definition of "Freedom"

Some years ago, I had a conversation about the level of freedom in our everyday lives. After all, if we have sufficient funds, we can, mostly, get in a car or a train etc and travel to anywhere on the planet in relative comfort and in a reasonable length of time. Or, I can consider my own situation – having had the free time and facilities to enable me to compile this book. But I'd like my own freedom to be even wider. I'd like to be able to have my own flying saucer, in my back garden. I'd like to be able to get in it and fly to Saturn to observe the ring system and its moons. I'd like to have the option of using free energy to power my appliances, and not be co-opted into using a system which damages the environment around me. Unfortunately, someone has "set up a system" whereby these options are unavailable to us. Indeed, most people don't even consider them as being an option at all. Such is the power of global propaganda and social engineering.

In Closing

I hope you have found the information in this volume worthy of the investment of your time. I thank you for your attention and hope that you will share what you have learned with other knowledgeable friends, colleagues, associates and family members, whenever you feel comfortable in doing so.

References

In order to make access to links easier, I recommend you download an electronic version of this volume and access the links therein, or find a copy of the links on http://www.checktheevidence.com/ .

---

[1] http://www.checktheevidence.co.uk/cms/index.php?option=com_content&task=view&id=314&Itemid=55

[2] http://www.drjudywood.com/

[3] http://www.disclosureproject.org/

[4] http://www.amazon.co.uk/Hunt-Zero-Point-Nick-Cook/dp/0099414988

[5] https://www.youtube.com/watch?v=6PIdgFnTbKw

[6] http://www.travis-walton.com/book.html

[7] http://disclosureproject.org/access/briefdoc.htm

[8] http://avalonlibrary.net/Collection_of_193_EBooks/Frank Scully - Behind the Flying Saucers (1950).pdf

[9] https://theintercept.com/document/2014/02/24/art-deception-training-new-generation-online-covert-operations/

[10] http://www.checktheevidence.com/cms/index.php?option=com_content&task=view&id=419&Itemid=62

[11] https://www.britannica.com/science/Michelson-Morley-experiment

[12] https://www.britannica.com/science/relativity/General-relativity

[13] https://www.britannica.com/topic/gravitational-lens

[14] https://www.britannica.com/science/Thomson-atomic-model

[15] https://www.britannica.com/science/Rutherford-atomic-model

[16] https://www.britannica.com/science/Bohr-atomic-model

[17] https://www.britannica.com/science/quantum-mechanics-physics

[18] https://www.britannica.com/science/boson

[19] https://www.britannica.com/science/quark

[20] https://www.britannica.com/science/string-theory

[21] https://www.space.com/828-leaking-gravity-explain-cosmic-puzzle.html

[22] http://www.dummies.com/education/science/physics/string-theory-and-loop-quantum-gravity/

[23] http://www.oregonlive.com/history/2015/05/past_tense_oregon_ufo_photos_t.html

[24] http://jnaudin.free.fr/lifters/main.htm

[25] http://www.rexresearch.com/gravitor/gravitor.htm

[26] https://www.linkedin.com/in/nickcook1/

[27] http://etheric.com/paul-laviolette-bio/

[28] https://gizadeathstar.com/about/

[29] https://gizadeathstar.com/purchase/

[30] https://www.amazon.co.uk/Hunt-Zero-Point-Nick-Cook/dp/0099414988

[31] https://www.youtube.com/watch?v=hKWSEYGlnsQ

[32] https://www.amazon.co.uk/Thomas-Valone/e/B00J3W7GBC/ref=ntt_dp_epwbk_1

[33] http://www.marcseifer.com/nikola-tesla.html

[34] http://earthsky.org/space/what-is-the-tunguska-explosion

[35] http://ericpdollard.com/

[36] https://www.bibliotecapleyades.net/ciencia/ciencia_energy83.htm

[37] https://youtu.be/3uXL4_Yas2k?t=20m25s

[38] https://www.wired.com/1998/03/antigravity/

[39] http://www.rexresearch.com/piggott/piggott.htm

[40] http://www.thomastownsendbrown.com/

[41] http://file.scirp.org/Html/11-2800956_55806.htm

[42]
https://worldwide.espacenet.com/publicationDetails/biblio?DB=EPODOC&II=0&ND=3&adjacent=true&locale=en_EP&FT=D&date=19281115&CC=GB&NR=300311A&KC=A

[43] https://www.britannica.com/science/Millikan-oil-drop-experiment

[44] http://www.thomastownsendbrown.com/tpx/index.htm

[45] http://www.naturalphilosophy.org/php/index.php?tab0=Scientists&tab1=Scientists&tab2=Display&id=912

[46] http://www.thomastownsendbrown.com/stress/wizard1.htm

[47] http://starburstfound.org/aerospace/projectwinterhaven.pdf

[48] http://www.thomastownsendbrown.com/misc/timeline.htm

[49] http://www.thomastownsendbrown.com/misc/index.htm

[50] http://www.thomastownsendbrown.com/tpx/barker/l_brown_11-55.pdf

[51] http://signallake.com/innovation/AntigravityFeb56.pdf

[52] http://www.bibliotecapleyades.net/ciencia/secret_projects/project080.htm

[53] http://users.erols.com/iri/TTBROWN2.htm

[54] https://www.scribd.com/document/232602872/THE-G-ENGINES-ARE-COMING-by-Michael-Gladych

[55] https://www.scribd.com/document/232601410/ANTI-GRAVITY-POWER-OF-THE-FUTURE-by-G-Harry-Stine-Chief-Navy-Range-Operations-White-Sands-Proving-Grounds

[56] https://www.scribd.com/document/225890800/ANTI-GRAVITY-The-Science-of-Electro-Gravitics-by-I-E-van-As

[57] http://www.thomastownsendbrown.com/misc/letters/mark2.htm

[58] http://www.thomastownsendbrown.com/misc/letters/will1.htm

[59] http://www.aztecnm.com/aztec/ufo/crashsite.html

[60] https://www.jp-petit.org/nouv_f/B2/B2_7.htm

[61] http://web.archive.org/web/20020424225226/http:/www.is.northropgrumman.com/videos/b2_tx.wmv

[62] https://arxiv.org/ftp/physics/papers/0211/0211001.pdf

[63] http://jnaudin.free.fr/html/elghatv1.htm

[64] http://jnaudin.free.fr/

[65] http://jnaudin.free.fr/html/liftbldr.htm

[66] http://jnaudin.free.fr/lifters/hexalifter/index.htm

[67] http://jnaudin.free.fr/html/lifteriw.htm

[68] http://www.americanantigravity.com/

[69] http://www.americanantigravity.com/files/audio/John-Hutchison-Apr2004.mp3

[70] http://www.checktheevidence.com/cms/index.php?option=com_content&task=view&id=170&Itemid=60

[71] http://www.presidentialufo.com/old_site/top_secret_2.htm

[72] http://www.checktheevidence.com/audio/WB Smith - Vancouver Lecture 1958.mp3

[73] https://www.youtube.com/watch?v=v7TGTlkrtik

[74] http://presidentialufo.com/old_site/new_magnet_documents.htm

[75] http://www.presidentialufo.com/wilbert-smith-articles/127-binding-forces

[76] http://www.wheredidthetowersgo.com/

[77] http://www.presidentialufo.com/wilbert-smith-articles/340-hardware

[78] http://www.presidentialufo.com/wilbert-smith-articles/429-did-the-canadians-analyze-the-roswell-metal

[79] http://www.rexresearch.com/smith/smith2.htm

[80] http://www.treurniet.ca/Smith/RotorPics.htm

[81] http://www.treurniet.ca/Smith/Rotor1.jpg

[82] http://www.treurniet.ca/Smith/Rotor2.jpg

[83] http://www.treurniet.ca/Smith/Scale1.jpg

[84] http://www.treurniet.ca/Smith/Scale2.jpg

[85] https://www.sunrisepage.com/ufo/files/government/Canada/Suggestions on Gravity Control Through Field Manipulation, by Wilbert Smith.pdf

[86] http://www.checktheevidence.com/cms/index.php?option=com_content&task=view&id=442&Itemid=59

[87] http://www.presidentialufo.com/wilbert-smith-articles/134-gravitational-speculations

[88] http://zonalandeducation.com/mstm/physics/mechanics/framesOfReference/nonInertialFrame.html

[89] http://www.presidentialufo.com/wilbert-smith-articles/131-gravity-day-1960

[90] http://www.presidentialufo.com/old_site/crashed_saucer.htm

[91] http://www.rexresearch.com/smith/newsci~1.htm

[92] http://articles.baltimoresun.com/1996-12-16/news/1996351017_1_mcdonnell-douglas-boeing-lockheed-martin

[93] http://www.checktheevidence.com/DouglasDocs/

[94] http://www.ufoevidence.org/Researchers/Detail29.htm

[95] http://www.checktheevidence.com/DouglasDocs/Douglas - 1968-06-27 - Proposed Vehicle R&D Program (Project BITBR).pdf

[96] http://www.checktheevidence.com/DouglasDocs/Douglas - 1969-12-02 - Electron Gas Analogy Experiment.pdf

[97] http://www.checktheevidence.com/DouglasDocs/Douglas - 1968-08-22 - Gravity Amplification (GA) Propulsion System.pdf

[98] http://www.checktheevidence.com/DouglasDocs/Douglas - 1969-02-18 - A New Communication Mode.pdf

[99] http://www.checktheevidence.com/DouglasDocs/Douglas - 1969-06-03 - Field Data Acquisition Requirements.pdf

[100] http://files.ncas.org/condon/text/contents.htm

[101] http://files.ncas.org/condon/text/appndx-x.htm

[102] http://www.ufoevidence.org/topics/CondonReport.htm

[103] http://www.checktheevidence.com/DouglasDocs/IUR Article Bob Wood and Condon.pdf

[104] http://tinyurl.com/sitsbook

[105] http://basicresearchpress.com/the-mechanical-theory-of-everything/

[106] https://exemplore.com/advanced-ancients/Ancient-Mysteries-Puma-Punku-in-Tiahuanaco

[107] http://coralcastle.com/

[108] http://www.checktheevidence.com/audio2/Energy and AntiGravity/Joe Bullard - Edward Leedskalnin - George Noory - Apr 03 2007.mp3

[109] https://www.amazon.com/Waiting-Agnes-Joe-Bullard/dp/0967313309

[110] http://www.labyrinthina.com/wp-content/uploads/2015/08/Coral-Castle-Book-Carrol-A.-Lake.pdf

[111] https://www.youtube.com/watch?v=nOoCuDnmtyM

[112] http://rense.com/general39/coral.htm

[113] https://www.amazon.com/Enigma-Coral-Castle-Ray-Stoner/dp/B00073A76W

[114] http://www.hyiq.org/Downloads/Edward-Leedskalnin-Magnetic-Current.pdf

[115] http://www.miamidade.gov/water/biscayne-aquifer.asp

[116] http://www.bgs.ac.uk/research/groundwater/shaleGas/aquifersAndShales/maps/aquifers/Chalk.html

[117] http://www.labyrinthina.com/coral-castle.html

[118] https://science.nasa.gov/science-news/science-at-nasa/2004/19apr_gravitomagnetism/

[119] http://curious.astro.cornell.edu/about-us/150-people-in-astronomy/space-exploration-and-astronauts/general-questions/927-can-artificial-gravity-be-created-in-space-intermediate

[120] https://www.diffen.com/difference/Centrifugal_Force_vs_Centripetal_Force

[121] http://rexresearch.com/depalma/depalma.htm

[122] http://www.famousinventors.org/michael-faraday

[123] https://www.youtube.com/watch?v=971tWhaY2R8

[124] https://www.brucedepalma.com/

[125] http://worldnpa.org/abstracts/abstracts_281.pdf

[126] http://hyperphysics.phy-astr.gsu.edu/hbase/shm.html

[127] https://depalma.pairsite.com/Absurdity/Absurdity05/SecretOfForceMachine.html

[128] http://www.timezone.com/2002/09/16/direct-drive-the-remarkable-bulova-accutron-caliber-214/

[129] https://www.brucedepalma.com/articles/NatureOfElectricalInduction.html

[130] http://www.independent.co.uk/news/obituaries/obituary-professor-eric-laithwaite-1288502.html

[131] http://www.youtube.com/watch?v=oAC-2qUFois

[132] https://www.youtube.com/watch?v=B8s6sCzffrA

[133] http://www.youtube.com/watch?v=KnNUTOxHoto

[134] http://www.checktheevidence.com/WBSmith/2 - Smith on Anti-gravity.mp3

[135] https://www.youtube.com/watch?v=W-TOePv0Rig

[136] https://www.youtube.com/watch?v=SJ52MlUpZsw

[137] https://www.youtube.com/watch?v=kRCq3wLMfIM

[138] https://www.youtube.com/watch?v=XPUuF_dECVI

[139] https://web.archive.org/web/20080121024807/http:/www.eyepod.org/Nazi-Disc-Photos.html

[140] https://www.youtube.com/watch?v=R-9tEbooAIU

[141] https://www.outdooractive.com/de/museum/bad-ischl/pks-villa-rothstein/13586466/

[142] https://www.alivewater.com/viktor-schauberger

[143] https://merlib.org/node/4293

[144] http://rexresearch.com/schaub/schaub.htm

[145] http://www.shamanicengineering.org/wp-content/uploads/2014/08/Thomas-Thompson-and-Riley-Crabb-Implosion-Viktor-Schauberger-and-the-Path-of-Natural-Energy-1985-134p.pdf

[146] http://jnaudin.free.fr/html/repulsin.htm

[147] http://jnaudin.free.fr/html/repulsb.htm

[148]
https://www.google.co.uk/url?sa=i&rct=j&q=&esrc=s&source=images&cd=&cad=rja&uact=8&ved=2ahUKEwj5j4CD4fXZAhWhAMAKHQ3OAVsQjRx6BAgAEAU&url=https%3A%2F%2Fwww.pinterest.com%2Fpin%2F549579960751674718%2F&psig=AOvVaw0Do3liimFFD7rGnk25W75X&ust=1521457644211686

[149] http://www.bibliotecapleyades.net/sociopolitica/reichblacksun/chapter06.htm

[150] http://igorwitkowski.com/

[151] https://www.amazon.co.uk/Truth-About-Wunderwaffe-Igor-Witkowski/dp/1618613383/ref=sr_1_1?s=books&ie=UTF8&qid=1521135075&sr=1-1&keywords=igor+witkowski&dpID=519zEFOjKrL&preST=_SY291_BO1,204,203,200_QL40_&dpSrc=srch

[152] https://goo.gl/maps/TfrLTbRKxrx

[153] https://www.flickr.com/photos/timventura/sets/72157627718524883

[154] http://ufohessdalen.blogspot.co.uk/2012/04/roswell-mj-12-anti-gravity-and-nazi.html

[155] https://www.youtube.com/watch?v=Nyiu2ZB_p-c

[156] http://www.bielek-debunked.com/Henge/The Henge.html

[157] https://youtu.be/B2h6RPcpxvM?t=1h31m38s

[158] https://www.agoravox.fr/tribune-libre/article/la-liberation-58-la-fameuse-cloche-96222

[159] https://www.britannica.com/biography/Walther-Gerlach

[160] http://www.wired.co.uk/article/nuclear-island

[161] https://www.thefamouspeople.com/profiles/walter-gerlach-7228.php

[162]
https://books.google.co.uk/books?id=ASbEHcRgGosC&lpg=PA322&dq=peron%20%22atom%20bomb%22%20richter&pg=PA322

[163] http://www.americanantigravity.com/news

[164] http://www.americanantigravity.com/files/articles/The-New-Nazi-Bell.pdf

[165] http://www.americanantigravity.com/files/articles/Einstein-Antigravity.pdf

[166] https://www.flickr.com/photos/timventura/6222317018/in/album-72157627718524883/

[167] https://sara.com/about

[168] http://www.saic.com/

[169] https://www.ara.com/

[170] http://www.drjudywood.com/articles/NIST/Qui_Tam_Wood.shtml

[171] http://www.ufoevidence.org/Researchers/Detail92.htm

[172] https://www.youtube.com/watch?v=kN-lnjZK4u8

[173] http://www.irva.org/conferences/speakers/buchanan.html

[174] https://youtu.be/n5BdplSagUE

[175] https://educate-yourself.org/cn/houndsofheavenoct91.shtml

[176] http://www.americanantigravity.com/files/audio/Nick-Cook-Oct2004.mp3

[177] https://www.amazon.com/Nick-Cook/e/B001HCYOUM/ref=dp_byline_cont_book_1

[178] http://www.checktheevidence.com/audio2/Energy and AntiGravity/Nick Cook Coast to Coast 14_3_2004.mp3

[179] https://www.youtube.com/watch?v=B2h6RPcpxvM

[180] https://web.archive.org/web/20070621150614/http:/www.highfrontiers.com/

[181] http://www.checktheevidence.com/audio2/Energy and AntiGravity/Nick Cook - Coast to Coast - May 15 2005.mp3

[182] https://www.bbc.co.uk/programmes/b00tw6y1

[183] http://www.checktheevidence.com/audio/Punt PI - Nazi Saucers - Radio 4 - 25 Sep 2010.mp3

[184] https://www.psychedelicadventure.net/2010/02/danny-dyer-i-believe-in-ufos-bbc.html

[185] https://www.youtube.com/channel/UCfJHu-yydx1FebruIi0ri5Q

[186] https://www.youtube.com/watch?v=4b4Hvlu7MGU

[187] https://www.youtube.com/watch?v=mxoaAHlC37I

[188] https://www.youtube.com/watch?v=T5lCAwHL0cc

[189] https://www.youtube.com/watch?v=0tJfqMYHaQw

[190] https://www.youtube.com/watch?v=Ah6G1sweLKo

[191] https://www.youtube.com/watch?v=KjApDnCvh2c

[192] http://www.nicap.org/articles/hillzeta.htm

[193] https://www.youtube.com/watch?v=9-qhn6Q1Ps8

[194] https://unitednuclear.com/index.php?main_page=page&id=25

[195] https://www.youtube.com/watch?v=7xDcaUKNl5w

[196] http://www.stantonfriedman.com/index.php%3Fptp%3Darticles%26fdt%3D2011.01.07

[197] https://www.youtube.com/watch?v=377bOibqbzc

[198] http://www.americanantigravity.com/files/audio/Boyd-Bushman-Oct2004.mp3

[199] https://www.linkedin.com/in/michael-schratt-8494a347/

[200] http://www.checktheevidence.com/pdf/TheLazarReport.pdf

[201] https://projectcamelotportal.com/?s=lazar

[202] http://archive.aviationweek.com/issue/20060306

[203] https://www.wired.com/2006/06/chemistry/

[204] https://blogs.scientificamerican.com/the-curious-wavefunction/the-many-tragedies-of-edward-teller/

[205] https://web.archive.org/web/20170407185838/http:/www.stantonfriedman.com:80/index.php?ptp=articles&fdt=2011.01.07

[206] https://web.archive.org/web/20170407190102/http:/www.stantonfriedman.com/index.php?ptp=stans_bio

[207] https://www.youtube.com/watch?v=VA3HV_gfq80

[208] https://www.youtube.com/watch?v=OCtE50rp7nU

[209] https://ufotrail.blogspot.co.uk/2017/05/boyd-bushman-fbi-and-counterespionage.html

[210] https://ufotrail.blogspot.co.uk/2018/01/nsa-glomars-foia-request-on-boyd-bushman.html

[211] http://www.checktheevidence.com/Disclosure/PDF Documents/gravitsapa.pdf

[212] http://news.bbc.co.uk/1/hi/not_in_website/syndication/monitoring/media_reports/2159629.stm

[213] https://www.sciencedirect.com/science/article/abs/pii/092145349290055H

[214] https://www.chemistryworld.com/podcasts/ybco-yttrium-barium-copper-oxide/6148.article

[215] https://www.britannica.com/science/yttrium

[216] http://bp0.blogger.com/_X2tcLFEJnwU/RavF9AEHkZI/AAAAAAAAAAM/ZQ7DaXG1P04/s1600-h/podkletnov.jpg

[217] http://huttoncommentaries.com/article.php?a_id=13

[218] https://www.revolvy.com/main/index.php?s=Eugene%20Podkletnov

[219] https://link.springer.com/article/10.1023%2FA%3A1024413718251

[220] http://www.intalek.com/AV/Eugene-Podkletnov.wma

[221] https://www.researchgate.net/publication/281440634_Study_of_Light_Interaction_with_Gravity_Impulses_and_Measurements_of_the_Speed_of_Gravity_Impulses

[222] http://americanantigravity.com/files/audio/

[223] https://www.youtube.com/watch?v=xrzfBiFCBVk

[224] https://ac.els-cdn.com/S1875389211005803/1-s2.0-S1875389211005803-main.pdf?_tid=b259205a-f936-4f9c-859d-5d037f8acd7f&acdnat=1521647114_c0456866f6ee3ec2d757ad6c95ad1967

[225] https://web.archive.org/web/20020802222642/http:/www.janes.com:80/aerospace/civil/news/jdw/jdw020729_1_n.shtml

[226] http://news.bbc.co.uk/1/hi/sci/tech/2157975.stmhttp:/news.bbc.co.uk/1/hi/sci/tech/2157975.stm

[227] https://www.space.com/businesstechnology/technology/gravity_research_020731.html

[228] https://www.youtube.com/watch?v=J-byfQG3xIw

[229] https://web.archive.org/web/20060207222219/http:/www.americanantigravity.com:80/documents/Boeing-HFGW-Presentation-03-104.pdf

[230] https://www.researchgate.net/publication/316147560_Trip_Report_and_meeting_minutes_from_the_1st_International_HFGW_Workshop

[231] https://www.amazon.co.uk/exec/obidos/ASIN/1784620238/491

[232] http://popsciencebooks.blogspot.co.uk/2015/05/greenglow-and-search-for-gravity.html

[233] https://web.archive.org/web/20000229200929/http:/www.greenglow.co.uk:80/plans.html

[234] https://web.archive.org/web/19990203035339/http:/www.greenglow.co.uk:80/whatis_gg.html

[235] https://web.archive.org/web/19990203003916/http://www.greenglow.co.uk:80/index.html

[236] https://web.archive.org/web/19990204051849/http:/www.inetarena.com:80/~noetic/pls/gravity.html

[237] http://www.electrogravity.com/

[238] https://web.archive.org/web/19990203022754/http:/www.greenglow.co.uk:80/lecture1.html

[239] https://web.archive.org/web/19981205180502/http:/www.lerc.nasa.gov:80/www/bpp/

[240] https://web.archive.org/web/19990508225805/http:/www.lerc.nasa.gov:80/WWW/bpp/WhitePaper.htm

[241]
https://web.archive.org/web/20001109173400/http:/www.space.com/businesstechnology/technology/anti_grav_000928.html

[242] https://web.archive.org/web/20050207212304if_/http:/www.grc.nasa.gov:80/WWW/bpp/TM-2004-213082.htm

[243] https://web.archive.org/web/20110112232117/http:/www.americanantigravity.com:80/documents/HFGW-2003/106-Ning-Li-Prepub.pdf

[244]
https://web.archive.org/web/20011205174447/http:/www.popularmechanics.com:80/science/research/1999/10/taming_gravity/

[245] https://www.scribd.com/document/262240027/Torr-and-Li-1993-Gravitoelectric-electric-Coupling

[246] http://einstein.stanford.edu/highlights/status1.html

[247] https://bigfatfurrytexan.wordpress.com/2010/12/31/possible-new-leads-in-the-missing-dr-ning-li-case/

[248] https://web.archive.org/web/20070101190136/http:/www.americanantigravity.com:80/documents/HFGW-2003/bio-book.pdf

[249] https://www.scientificamerican.com/article/gravitational-waves-discovered-from-colliding-black-holes1/

[250] https://www.prlog.org/10048184-scientists-see-wtc-hutchison-effect-parallel.html

[251] http://drjudywood.com/articles/JJ/JJ8.html

[252] http://www.thehutchisoneffect.com/

[253] https://www.youtube.com/watch?v=TjDhIKgp2KQ

[254] https://www.youtube.com/watch?v=yABGpiYONmo

[255] https://www.youtube.com/watch?v=RnfJn_DkBFg

[256] https://www.youtube.com/watch?v=46vENCTpJG8

[257] https://www.youtube.com/watch?v=to3brXnbBz4

[258] https://www.youtube.com/watch?v=-STQPccIdy4

[259] http://www.checktheevidence.com/pdf/Hutch Letters.pdf

[260] http://www.checktheevidence.com/audio/John Hutchison & SAIC - Art Bell - Coast to Coast - 04 Jun 2005.mp3

[261] https://www.youtube.com/watch?v=q99jJKX-46I

[262] http://www.keelynet.com/shoulders/pdfs.html

[263] https://www.youtube.com/watch?v=nZlCdwJ48SI

[264] http://searleffect.com/free/biodetail.html

[265] http://www.linux-host.org/energy/ssearl.html

[266] https://www.youtube.com/watch?v=yrT8yniVrxc

[267] https://segmagnetics.com/

[268] https://segmagnetics.com/en/about-us/

[269] http://www.searlsolution.com/investing4.html

[270] http://files.afu.se/Downloads/Magazines/United Kingdom/OVNI (Omar Fowler)/

[271] http://csblogg.ufo.se/csblogg3/?m=20120921

[272] https://youtu.be/JFOiQ1wH9Sk?t=26m8s

[273] http://www.coasttocoastam.com/show/2010/04/13

[274] http://www.drjudywood.com/articles/JJ

[275] http://www.aulis.com/films.htm

[276] https://www.amazon.co.uk/Dark-Moon-Whistle-Blowers-Mary-Bennett/dp/0932813909

[277] https://www.youtube.com/watch?v=xciCJfbTvE4

[278] https://www.youtube.com/watch?v=Qr6Vcvl0OeU

[279] https://krishna.org/wp-content/uploads/2013/04/NASA_mooned_america.pdf?7f4541

[280] https://www.scribd.com/doc/156629102/We-Never-Went-to-the-Moon-By-Bill-Kaysing

[281] http://www.moonfaker.com/videos.php

[282] https://www.youtube.com/watch?v=5KsH2M4m4zM

[283] https://www.youtube.com/watch?v=nE34pFtFEDs

[284] https://www.amazon.com/Dark-Mission-Secret-History-NASA/dp/1932595481

[285] https://www.youtube.com/watch?v=SqAT470a6xE&list=PLFL8NBSusWbw6-DSgPo1ifhs4HcsLn1yq

[286] https://archive.org/details/conquestofmoon00vonb

[287] https://www.youtube.com/watch?v=1ZImSTxbglI

[288] http://nasa.wikia.com/wiki/Gus_Grissom

[289] http://history.nasa.gov/Apollo204/zorn/grissom.htm

[290] https://www.amazon.com/Starfall-Betty-Grissom/dp/0690004737

[291] http://nasawatch.com/archives/2005/06/shuttle-telecon.html

[292] http://stateofthenation2012.com/?p=88241

[293] https://vimeo.com/5671999

[294] https://history.nasa.gov/Apollo204/barron.html

[295] http://www.clavius.org/baron-test.html

[296] https://science.ksc.nasa.gov/history/apollo/apollo-13/docs/apollo-13-press-kit.txt

[297] https://www.britannica.com/place/Fra-Mauro

[298] https://celestia.space/

[299] http://news.bbc.co.uk/1/hi/8226075.stm

[300] https://www.caltech.edu/news/remembering-gerald-wasserburg-51175

[301] https://history.msfc.nasa.gov/vonbraun/photo/13.html

[302] http://ccfbllc.com/Using_Lunar_Retroreflectors.html?ckattempt=1

[303] http://www.americanmoon.org/NationalGeographic/index.htm

[304] https://youtu.be/SSZX3mW7Du4

[305] http://thomas.loc.gov/cgi-bin/query/z?r103:E20JY4-119

[306] https://www.youtube.com/watch?v=Znyx2gTh3HU

[307] https://www.biography.com/people/neil-armstrong-9188943

[308] https://www.youtube.com/watch?v=O3h1V3nzn0A

[309] http://www.cbc.ca/news/technology/neil-armstrong-grants-rare-interview-to-accountants-organization-1.1289392

[310] https://www.theguardian.com/science/2012/aug/25/neil-armstrong-last-interview

[311] https://www.youtube.com/watch?v=wfTapJhp_Qw

[312] http://www.hyattanalysis.com/

[313] https://www.youtube.com/watch?v=PtdcdxvNI1o

[314] https://www.amazon.com/Return-Earth-Edwin-E-Aldrin/dp/0394488326

[315] http://www.conspiracyarchive.com/2013/12/03/buzz-aldrin-lies-about-his-masonic-activities-on-the-moon/

[316] https://www.youtube.com/watch?v=r_IKEJaieGc

[317] https://www.youtube.com/watch?v=65zZpRgrRTg

[318] https://youtu.be/SSZX3mW7Du4?t=1h4m40s

[319] https://history.nasa.gov/ap11ann/FirstLunarLanding/ch-7.html

[320] http://www.thekeyboard.org.uk/Did we land on the Moon.htm

[321] http://www.bibliotecapleyades.net/vida_alien/esp_vida_alien_05c.htm

[322] http://www.ibtimes.co.uk/nasa-astronaut-don-pettit-next-logical-step-go-back-moon-then-mars-beyond-1582401

[323] https://www.youtube.com/watch?v=z2CAkf124KY&feature=youtu.be&t=1m53s

[324] http://mikebara.blogspot.co.uk/2008/03/why-moon-hoax-conspiracy-is-crock-of.html

[325] https://www.smithsonianmag.com/smart-news/how-nasa-censored-dirty-mouthed-astronauts-180953470/

[326] https://vintagespace.wordpress.com/2012/01/06/vintage-space-fun-fact-how-to-not-swear-on-the-moon/

[327] http://news.bbc.co.uk/1/hi/sci/tech/2424927.stm

[328] https://www.hq.nasa.gov/alsj/a11/AS11-40-5878.jpg

[329] https://www.lpi.usra.edu/resources/apollo/frame/?AS11-40-5868

[330] https://web.archive.org/web/20120408210954/http:/www3.telus.net/summa/moonshot/fillit.htm

[331] https://www.cs.indiana.edu/~kapadia/moonstuff/rebuttal.html

[332] http://www.aulis.com/exposing_apollo2.htm

[333] http://www.lpi.usra.edu/resources/apollo/catalog/70mm/magazine/?134

[334] http://www.aulis.com/stereoparallax.htm

[335] https://www.hq.nasa.gov/alsj/a15/AS15-86-11601.jpg

[336] https://www.hq.nasa.gov/alsj/a15/AS15-86-11602.jpg

[337] http://www.aulis.com/stereoparallax/appolo_15_S1.gif

[338] https://spaceflight.nasa.gov/gallery/images/apollo/apollo16/lores/as16-107-17442.jpg

[339] https://www.hq.nasa.gov/office/pao/History/alsj/a16/AS16-107-17435HR.jpg

[340] http://www.asi.org/adb/m/03/05/average-temperatures.html

[341] http://coolcosmos.ipac.caltech.edu/ask/168-What-is-the-temperature-on-the-Moon-

[342] https://www.hq.nasa.gov/alsj/a16/AS16-117-18841.jpg

[343] https://youtu.be/41TaYvxKGi0?t=24m49s

[344] https://www.hq.nasa.gov/alsj/a16/AS16-114-18444.jpg

[345] https://www.hq.nasa.gov/alsj/a16/AS16-116-18577.jpg

[346] https://history.nasa.gov/alsj/a16/a16.sta9.html

[347] https://youtu.be/5KsH2M4m4zM?t=21m31s

[348] https://www.hq.nasa.gov/alsj/alsj-reseau.html

[349] https://youtu.be/5KsH2M4m4zM?t=22m53s

[350] https://www.hq.nasa.gov/alsj/a15/AS15-88-11901.jpg

[351] https://www.hq.nasa.gov/alsj/a17/AS17-140-21354HR.jpg

[352] https://www.hq.nasa.gov/alsj/a17/AS17-135-20544HR.jpg

[353] http://lroc.sese.asu.edu/news/uploads/LROCiotw/LM3view.png

[354] http://sci.esa.int/hubble/31496-the-resolving-power-of-the-hubble-space-telescope/

[355] http://hyperphysics.phy-astr.gsu.edu/hbase/Solar/hst.html

[356] https://nssdc.gsfc.nasa.gov/nmc/spacecraftDisplay.do?id=2009-031A

[357] https://www.nasa.gov/mission_pages/LRO/multimedia/lroimages/lroc_200911109_apollo11.html

[358] https://www.nasa.gov/mission_pages/LRO/multimedia/lroimages/apollosites.html

[359] https://www.nasa.gov/mission_pages/LRO/multimedia/lroimages/lroc-20100708-apollo16.html

[360] https://www.youtube.com/watch?v=qr3YrmTOQaY

[361] https://content.satimagingcorp.com/static/galleryimages/geoeye-1-copacabana-beach-sm.jpg

[362] https://www.satimagingcorp.com/gallery/geoeye-1/

[363] https://www.youtube.com/watch?v=ndvmFlg1WmE

[364] https://www.youtube.com/watch?v=o6FX-Y6fIXw

[365] https://www.youtube.com/watch?v=5aDSYTMqyQw

[366] https://youtu.be/5aDSYTMqyQw?t=23m0s

[367] http://www.armaghplanet.com/blog/wp-content/uploads/2011/08/Image-of-LM-general-arrangement.gif

[368] http://heroicrelics.org/info/saturn-v/saturn-v-general.html

[369] https://www.youtube.com/watch?v=rHF5EcdLxQo

[370] https://nssdc.gsfc.nasa.gov/planetary/lunar/apollo_lrv.html

[371] https://www.youtube.com/watch?v=mGl0EoEo38U

[372] https://www.hq.nasa.gov/alsj/a17/a17.homeward.html

[373] https://www.youtube.com/watch?v=-fs8gkiap6U

[374] https://www.colorado.edu/today/2017/05/09/frequent-flyers-beware-radiation-risk-may-be-rise

[375] https://youtu.be/Qr6Vcvl0OeU?t=18m30s

[376] http://www.clavius.org/envrad.html

[377] https://www.youtube.com/watch?v=KyZqSWWKmHQ

[378] https://www.sciencedaily.com/releases/2008/10/081008231029.htm

[379] http://www.aulis.com/moonbase2014.htm

[380] http://www.aulis.com/moonbase2015.htm

[381] http://www.aulis.com/moonbase2016.htm

[382] https://www.youtube.com/watch?v=atIKUEfW-Vw

[383] http://www.nbcnews.com/id/3341669/ns/technology_and_science-space/t/nasas-vision-due-adjustment/

[384] https://web.archive.org/web/20040211125133/http:/www.moontomars.org/

[385] http://www.nasa.gov/pdf/60736main_M2M_report_small.pdf

[386] http://www.thespacereview.com/article/1706/1

[387] http://www.nasa.gov/news/media/trans/obama_ksc_trans.html

[388] https://www.youtube.com/watch?v=1w63sz3dBEA

[389] https://www.youtube.com/channel/UCmY9NKVfOdOooHvZMFJu7hA

[390] https://newswithviews.com/bill-clinton-and-the-deadly-arkansas-tainted-blood-scandal/

[391] http://news.bbc.co.uk/1/hi/special_report/1998/03/98/gagarin/72184.stm

[392] http://www.heraldscotland.com/news/11938446.Lord_Bruce-Gardyne_dies_after_a_long_illness/

[393] https://www.imdb.com/title/tt0205873/

[394] https://history.nasa.gov/apsr/Apollopt2-2.pdf

[395] https://nsarchive2.gwu.edu/NSAEBB/NSAEBB132/press20051201.htm

[396] http://history.nasa.gov/alsj/a17/a17schmitt.face.jpg

[397] https://history.nasa.gov/alsj/a17/a17schmitt.face.jpg

[398] https://bigelowaerospace.com/pages/whoweare/

[399] http://www.cohenufo.org/Nids-Investigates-The-Flying-Triangle-Enigma.htm

[400] https://www.cbsnews.com/news/bigelow-aerospace-founder-says-commercial-world-will-lead-in-space/

[401] https://www.google.co.uk/maps/@36.2110784,-115.1627636,3a,15y,277.78h,95.18t/data=!3m9!1e1!3m7!1siPl0IZzMT0R4v1_Nygf8qw!2e0!7i13312!8i6656!9m2!1b1!2i38

[402] https://www.space.com/18993-virgin-galactic.html

[403] https://www.space.com/11562-nasa-american-spaceflight-alan-shepard-spaceflight-faq.html

[404] http://uk.businessinsider.com/companies-elon-musk-invests-in-2016-4?op=1&r=US&IR=T/

[405] https://www.youtube.com/watch?v=PULkWGHeIQQ

[406] https://www.space.com/19341-jeff-bezos.html

[407] https://www.youtube.com/watch?v=ex-ui7ldnD0

[408] https://www.nasa.gov/mission_pages/shuttle/shuttlemissions/archives/sts-82.html

[409] http://www.spacepirations.com/2009/11/buran-russian-space-shuttle.html

[410] https://ralphmirebs.livejournal.com/219949.html

[411] https://commons.wikimedia.org/wiki/File:Vandenberg_AFB_Space_Launch_Complex_6_in_1985_(with_Enterprise).jpg

[412] https://www.airspacemag.com/space/secret-space-shuttles-35318554/

[413] http://www.pbs.org/wgbh/nova/military/astrospies.html

[414] https://www.youtube.com/watch?v=irQC5OCWGKU

[415] http://www.nro.gov/foia/declass/mol/6.pdf

[416] https://ipfs.io/ipfs/QmXoypizjW3WknFiJnKLwHCnL72vedxjQkDDP1mXWo6uco/wiki/Manned_Orbiting_Laboratory.html

[417] http://www.collectspace.com/ubb/Forum20/HTML/000857.html

[418] http://space.skyrocket.de/doc_sdat/almaz-1v.htm

[419] http://www.space.com/images/i/000/000/098/original/080212-almaz-02.jpg?interpolation=lanczos-none&fit=around|400:280&crop=400:280;*,*

[420] https://www.space.com/25275-x37b-space-plane.html

[421] https://www.airspacemag.com/space/spaceplane-x-37-180957777/

[422] https://www.bibliotecapleyades.net/luna/esp_luna_46.htm

[423] https://www.amazon.co.uk/Reagan-Diaries-Ronald/dp/006087600X/

[424] https://www.bibliotecapleyades.net/exopolitica/exopolitics_reagan05.htm

[425] https://www.reaganlibrary.gov/sites/default/files/digitallibrary/dailydiary/1985-06.pdf

[426] https://www.youtube.com/watch?v=LGvgNRbpLME

[427] http://www.ufo-disclosure.net/blog/ben-rich-esp-and-the-100th-monkey

[428] https://www.youtube.com/watch?v=OMYsFFtyKz4

[429] https://www.youtube.com/watch?v=8pa3ZT6QHjo

[430] https://www.youtube.com/watch?v=_4hycqDNnPE

[431] http://www.disclosureproject.org/access/docs/pdf/DisclosureProjectBriefingDocument.pdf

[432] http://freegary.org.uk/

[433] https://www.amazon.co.uk/Saving-Gary-McKinnon-Mothers-Story/dp/184954574X/ref=sr_1_1/279-0360622-4862253?s=books&ie=UTF8&qid=1378026088&sr=1-1

[434] https://www.youtube.com/watch?v=B4PkNPCEnJM

[435] http://news.bbc.co.uk/1/hi/programmes/click_online/4977134.stm

[436] http://projectcamelot.org/lang/en/gary_mckinnon_interview_transcript_en.html

[437] http://news.findlaw.com/hdocs/docs/cyberlaw/usmck1102vaind.pdf

[438] https://www.judiciary.senate.gov/imo/media/doc/mcnulty_testimony_10_21_03.pdf

[439] https://www.nytimes.com/2015/02/04/us/zacarias-moussaoui-calls-saudi-princes-patrons-of-al-qaeda.html

[440] https://www.youtube.com/watch?v=tEBLmWhx1K0

[441] http://www.theregister.co.uk/2012/11/07/uk_mulls_re_opening_mckinnon_investigation/

[442] http://www.richplanet.net/starship_main.php?ref=214&part=1

[443] http://www.themarsrecords.com/wp/the-mars-records/mars-records-book-1-preface/

[444] http://www.checktheevidence.com/cms/index.php?option=com_content&task=view&id=465&Itemid=59

[445] http://www.ufo-disclosure.net/blog/man-made-flying-disc-travels-faster-than-light

[446] https://www.youtube.com/watch?v=ua0MMXJl3FM

[447] https://www.counterpunch.org/2002/02/01/the-strange-career-of-frank-carlucci/

[448] https://www.spherebeingalliance.com/media/img/1600x0/2017-05/9_cutaway_view_of_interior_of_ARV.jpg

[449] https://web.archive.org/web/20091228154447/http:/ufocasebook.conforums.com/index.cgi?action=display&board=video&num=1035492108&start=60

[450] https://www.youtube.com/watch?v=lRxyp4F0jvo

[451] http://www.ufocasebook.com/maslinbeach1993.html

[452] https://www.spherebeingalliance.com/blog/transcript-cosmic-disclosure-zero-point-energy-and-advanced-propulsion-technology.html

[453] https://www.bbc.co.uk/news/uk-england-suffolk-29280298

[454] https://youtu.be/XoeeivrCK-Y

[455] http://freespiritproductions.com/site/?portfolio_item=david-adair-at-area-51

[456] http://digitalarchive.wilsoncenter.org/resource/cold-war-history/curtis-lemay

[457] https://www.history.com/this-day-in-history/u-s-air-force-reports-on-roswell

[458] http://www.greatdreams.com/david-adair.htm

[459] https://www.americasfallfromspace.com/

[460] https://allaboutdavidadair.wordpress.com/2014/01/05/jan-williams-daughter-of-colonel-bailey-arthur-williams-speaks-about-david-adair/

[461] https://www.youtube.com/watch?v=s2sLUyfQWs0

[462] https://www.youtube.com/watch?v=P0yIxZYrSxQ

[463] https://www.coasttocoastam.com/show/2018/02/20

[464] http://stantonfriedman.com/index.php?ptp=book_reviews&fdt=1997.12.10

[465] https://archive.org/details/PhilipJ.Corso-DawnOfANewAge

[466] https://fas.org/spp/military/program/track/overview.htm

[467] http://www.history.army.mil/faq/horizon/Horizon_V1.pdf

[468] https://www.defensemedianetwork.com/stories/an-army-base-on-the-moon/

[469] http://www.theufochronicles.com/2014/02/roswells-memory-metal-secret.html

[470] https://www.youtube.com/watch?v=DpffcjnXMx0

[471] http://www.stargate-chronicles.com/site/incidents/chasing-juno/

[472] http://www.angelfire.com/zine/UFORCE/page72.html

[473] https://archive.org/details/SaganIL

[474] https://youtu.be/Jy-3nSW9rbQ?t=27m50s

[475] https://www.youtube.com/watch?v=pKnDQ2qDuwk

[476] https://www.amazon.com/Alien-Rapture-Rothschild-Fouche-Steiger/

[477] https://www.lulu.com/shop/edgar-fouche-and-brad-steiger/alien-rapture-the-chosen/ebook/product-22921507.html

[478] http://www.checktheevidence.co.uk/cms/index.php?option=com_content&task=view&id=390&Itemid=51

[479] https://www.leagle.com/decision/19841037692p2d34511035

[480] https://www.youtube.com/watch?v=DlxrRAFaEvs

[481] https://www.britannica.com/science/quasicrystal

[482] https://www.theguardian.com/science/2013/jan/06/dan-shechtman-nobel-prize-chemistry-interview

[483] https://www.nobelprize.org/nobel_prizes/chemistry/laureates/2011/

[484] https://youtu.be/ogNKrQCH1Kk?t=34m55s

[485] http://large.stanford.edu/courses/2011/ph241/hamerly1/

[486] http://stantonfriedman.com/index.php?ptp=stans_bio

[487] https://www.holybooks.com/vaimanika-shastra/

[488] http://iopscience.iop.org/0295-5075/91/4/45002

[489] http://www-naweb.iaea.org/napc/physics/2ndgenconf/data/Proceedings 1958/papers Vol32/Paper20_Vol32.pdf

[490] http://www.richplanet.net/starship_main.php?ref=184&part=1

[491] http://articles.adsabs.harvard.edu/cgi-bin/nph-iarticle_query?1996Ap%26SS.242...93P&data_type=PDF_HIGH&whole_paper=YES&type=PRINTER&filetype=.pdf

[492] http://astrophysicsformulas.com/astronomy-formulas-astrophysics-formulas/temperature-in-kelvin-to-kev-conversion/

[493] https://www.translatorscafe.com/unit-converter/en/energy/11-65/electron-volt-kelvin/

[494] https://www.youtube.com/watch?v=Tx34w0Rtslk

[495] https://www.youtube.com/watch?v=0RrgrHAyRv8

[496] http://www.checktheevidence.com/pdf/Flying Triangle Mystery - Omar Fowler.pdf

[497] http://www.checktheevidence.co.uk/cms/index.php?option=com_content&task=view&id=152&Itemid=51

[498] http://www.openminds.tv/michael-schratt-are-black-triangle-ufos-secret-military-projects-april-17-2018/41684

[499] https://www.youtube.com/watch?v=pFWza6LTMrY

[500] http://www.richplanet.net/rp_genre.php?ref=158&part=1&gen=1

[501] http://www.richplanet.net/rp_genre.php?ref=115&part=1&gen=1

[502] https://www.youtube.com/watch?v=VxA-Y4enohQ

[503] https://www.independent.co.uk/news/secret-us-spyplane-crash-may-be-kept-under-wraps-1272714.html

[504] https://www.youtube.com/watch?v=mHA6Blu_dFY

[505] https://www.space.com/302-silent-running-black-triangle-sightings-rise.html

[506] https://www.youtube.com/channel/UCsJsNYEzl1WE42KVyZHYzgA

[507] http://www.youtube.com/watch?v=hXVi-_0YzV8

[508] https://www.youtube.com/watch?v=ff7SG-QN2hE

[509] https://www.coasttocoastam.com/guest/grimsley-ed/7185

[510] https://www.youtube.com/watch?v=bz7ENgrOqSw

[511] https://www.youtube.com/watch?v=Z4UxPF73cVk

[512] https://www.youtube.com/watch?v=pKJg8LeFqfI

[513] http://www.nicap.org/images/kasher_111.jpg

[514] http://www.nicap.org/muj_kasher_sts48.htm

[515] https://www.youtube.com/watch?v=XG0V_wNlcwo

[516] https://www.youtube.com/watch?v=0RrgrHAyRv8&t=2006

[517] http://www.youtube.com/watch?v=BCrOEmHHHAA

[518] http://postimage.org/image/ndey4jdq5/00316115/

[519] https://solarscience.msfc.nasa.gov/CMEs.shtml

[520] https://sohowww.nascom.nasa.gov/about/orbit.html

[521] https://map.gsfc.nasa.gov/mission/observatory_l2.html

[522] http://www.checktheevidence.com/cms/index.php?option=com_content&task=view&id=413&Itemid=63

[523] https://www.telegraph.co.uk/news/obituaries/12145101/Edgar-Mitchell-astronaut-obituary.html

[524] https://www.youtube.com/watch?v=cYPCKIL7oVw

[525]
https://web.archive.org/web/20061011075415/http:/pqasb.pqarchiver.com/sptimes/access/546961411.html?FMT=FT&FMTS=FT&date=Feb+18,+2004&desc=Former+astronaut:+We%27ve+had+otherworldly+visitors

[526] http://www.sptimes.com/2004/02/18/Neighborhoodtimes/Astronaut__We_ve_had_.shtml

[527] https://www.youtube.com/watch?v=8QKW6Gt3KKs

[528] https://www.youtube.com/watch?v=MiGsLA_sdmg

[529] https://www.youtube.com/watch?v=q6OKaMfPh4I

[530] https://www.youtube.com/watch?v=d3CxapWyA8g

[531] https://youtu.be/q6OKaMfPh4I?t=4m32s

[532] http://www.noufors.com/Documents/evidence.pdf

[533] http://www.noetic.org/directory/person/edgar-mitchell

[534] http://www.noetic.org/about/overview

[535] http://site.uri-geller.com/watch_the_secret_life_of_uri_gellr

[536] http://www.dailymotion.com/video/x2pmu72_ed-mitchell-and-cia_tv

[537]
http://www.checktheevidence.co.uk/cms/index.php?option=com_events&task=view_detail&agid=92&year=2015&month=5&day=26&Itemid=51

[538] https://www.youtube.com/watch?v=weFpGufxq0I

[539] https://youtu.be/uGRsQZx6zWA?t=4m20s

[540] http://www.richplanet.net/

[541] http://brucedepalma.com/

[542] http://www.enterprisemission.com/dtran5.html

[543] http://apollotruth.atspace.co.uk/

[544] mailto:nasascam@yahoo.com?subject=Apollo%20Hoax

[545] http://nsarchive.gwu.edu/radiation/dir/mstreet/commeet/meet4/brief4.gfr/tab_j/br4j1.txt

[546] http://www.drjudywood.com/pdf/080324_SAIC_AffdJW91_150.pdf

[547] http://files.shareholder.com/downloads/SAIC/0x0x208149/64117BC7-5895-497E-A8EB-158A6E57012C/AR_2004.pdf

[548] https://protected.networkshosting.com/depsor/DEPSpages/sponsors.html

[549] https://digwithin.net/2014/06/04/andrews-and-saic/

[550] http://aviationweek.com/awin/us-special-ops-command-psyops-related-contracts-go-saic-sycoleman-lincoln-group

[551] http://tinyurl.com/somaero

[552] http://web.archive.org/web/20080222044619/http:/www.aero2012.com:80/en/risk.mhtml

[553] http://www.disclosureproject.org/docs/pdf/disclosure911.pdf

[554] https://youtu.be/oHxGQjirV-c?t=3h45m32s

[555] http://www.disclosureproject.org/transcripts/JeanNoelBassior-Nov2005.htm

[556] http://www.bastison.net/RESSOURCES/Farce/57_Jones_Jesus.pdf

[557] https://archive.org/stream/HiddenTruthForbiddenKnowledgeStevenM.Greer/Hidden Truth Forbidden Knowledge Steven M. Greer_djvu.txt

[558] https://www.bibliotecapleyades.net/disclosure/briefing/disclosure13.htm

[559] https://www.biography.com/people/barry-goldwater-9314846

[560] https://www.youtube.com/watch?v=MtJo6vKnY54

[561] https://www.youtube.com/watch?v=CPEumptc82g

[562] http://www.presidentialufo.com/component/content/article/78-document-downloads/411-senator-barry-goldwater-ufo-files-now-posted

[563] https://siriusdisclosure.com/neil-armstrongs-ufo-secret/

[564] https://www.revolvy.com/main/index.php?s=L.%20Macon%20Epps

[565] https://www.youtube.com/watch?v=NlWFnBGvMPo

[566] http://exopolitics.blogs.com/files/steve.greer.dead.mans.trigger.pdf

[567] http://www.paradigmresearchgroup.org/stephenbassett.html

[568] https://youtu.be/MCSpgJNjop0?t=21m40s

[569] https://www.youtube.com/channel/UCA9iiA4pxNRYxZDW_6oWbKw/videos

[570] http://www.johnbalexander.com/biography

[571] http://www.checktheevidence.com/audio/911/Steve Bassett on 911 - UFO Truth Conf - 29 Apr 2017.mp3

[572] https://youtu.be/mjG7DchEe2E?t=29m27s

[573] http://www.checktheevidence.com/cms/index.php?option=com_content&task=view&id=460&Itemid=51http://www.checktheevidence.com/cms/index.php?option=com_content&task=view&id=460&Itemid=51

[574] https://www.dailystar.co.uk/news/weird-news/679045/Aliens-news-WW3-US-Russia-nuclear-weapons-ufo-video-Trump-Putin

[575] http://alienscientist.proboards.com/thread/1/edgar-fouche-author-rapture-welcome

[576] http://www.theoutpostforum.com/tof/showthread.php?1693-Edgar-Fouche-Career-Documents-Thoughts-on-Technology-and-Military&p=33521&viewfull=1

[577] http://military.wikia.com/wiki/San_Antonio_Air_Logistics_Center

[578] https://www.youtube.com/watch?&v=i-NvbLq94OE

[579] http://www.rense.com/general96/blck.html

[580] http://www.redicecreations.com/radio/2006/12dec/RICR-061214.html

[581] http://www.dfrc.nasa.gov/Gallery/Photo/STS/HTML/EC90-065-16.html

[582] http://eol.jsc.nasa.gov/scripts/sseop/photo.pl?mission=STS61C&roll=31&frame=2

[583] http://eol.jsc.nasa.gov/scripts/sseop/photo.pl?mission=STS61C&roll=31&frame=3

[584] https://www.dailymotion.com/video/x20qfdd

[585] http://www.drboylan.com/usspacefleet.html

[586] http://www.huffingtonpost.co.uk/darren-perks/solar-warden-the-secret-space-program_b_1659192.html

[587] http://fierycelt.tripod.com/xposeufotruth/perks_cosford_account_2008.html

[588] https://archive.org/details/B-001-014-055

[589] http://extraterrestrialcontact.com/tag/jo-ann-richards/

[590] http://tinyurl.com/911ftb

[591] http://edhca.org/about-us/

[592] http://www.starshipearththebigpicture.com/2017/05/01/capt-mark-richards-earth-defense-with-kerry-cassidy-in-total-recall-part-6-project-camelot-video/

[593] https://www.bibliotecapleyades.net/sociopolitica/serpo/index_serpo.htm

[594] https://projectcamelotportal.com/

[595] http://projectavalon.net/

[596] http://www.checktheevidence.com/cms/index.php?option=com_content&task=view&id=410&Itemid=60

[597] https://www.bibliotecapleyades.net/sociopolitica/serpo/operations_manual.htm

[598] http://www.projectpegasus.net/

[599] https://www.coasttocoastam.com/guest/basiago-andrew-d/42085

[600] https://www.exopolitics.org/jump-room-to-mars-did-cia-groom-obama-basiago-as-future-presidents/

[601] https://answers.yahoo.com/question/index?qid=20080730073058AA46WMR&guccounter=1

[602] http://www.exopolitics.org/tag/earth-defense-force/

[603] https://www.exopolitics.org/?s=Corey+Goode

[604] https://www.exopolitics.org/goode-secret-space-program-claims-go-viral-critics-attacks/

[605] https://www.youtube.com/watch?v=09eH4PdH69M&list=PL5EE2D04BD5A3BDCE

[606] https://www.express.co.uk/life-style/life/595391/A-liar-and-a-fantasist-James-Calbolt-sentenced-to-12-years

[607] https://www.thetruthseeker.co.uk/?p=8405

[608] http://department47.com/space-junk-t524-45.html

[609] mailto:webmaster@rense.com

[610] http://tinyurl.com/911htb

[611] http://alienscientist.com/

[612] https://www.youtube.com/user/AlienScientist

[613] http://tinyurl.com/911wrh

[614] http://www.checktheevidence.co.uk/cms/index.php?option=com_content&task=view&id=337&Itemid=60

[615] http://youtu.be/qK2NT-udlGc?t=4m21s

[616] http://www.alienscientist.com/judywood.html

[617] https://www.facebook.com/Alienscientist/info

[618] https://www.youtube.com/watch?v=a9QrcV9-kA8

[619] http://www.nbcnews.com/id/33330516/ns/us_news-life/t/feared-lost-balloon-boy-found-home/

[620] https://www.youtube.com/watch?v=YcYGcBYzvWs

[621] http://edition.cnn.com/2010/CRIME/01/11/heene.balloon.boy/index.html?iref=allsearch

[622] http://www.americanantigravity.com/link/alien-scientist

[623] http://www.checktheevidence.com/Misc/alienscientist-richard Heene-BalloonBoy.jpg

[624] http://www.r2controls.com/toc.htm

[625] http://www.invensys.com/en/AboutUs/

[626] https://www.youtube.com/watch?v=oTbSCUY1Z8o

[627] http://www.alienscientist.com/forum/forum.php

[628] https://www.youtube.com/user/checktheevidence

[629] https://web.archive.org/web/20140807121440/http:/patch.com/massachusetts/mansfield/police-log-road-rage-incident-involving-ax-reported-xfinity-center

[630] https://www.facebook.com/jeremy.rys.1

[631] https://www.youtube.com/playlist?list=PLduAC5d8yaEa6i3jx0jf_DxCvxu8Ax6Z0

[632] https://www.facebook.com/profile.php?id=890945646

[633] http://www.richplanet.net/rp_genre.php?ref=184&part=1&gen=1

[634] https://www.youtube.com/user/davidhilton1066/

[635] http://openseti.org/Hall.html

[636] http://www.richplanet.net/rp_genre.php?ref=52&part=1&gen=1

[637] https://www.youtube.com/watch?v=UZa4CKz7z_w

[638] http://www.theoutpostforum.com/tof/showthread.php?1693-Edgar-Fouche-Career-Documents-Thoughts-on-Technology-and-Military&p=33529&viewfull=1

[639] https://www.youtube.com/watch?v=KReD9nyPQpo

[640] http://www.theoutpostforum.com/tof/showthread.php?1495-Demise-of-alienscientist-forum&p=28102&viewfull=1

[641] http://www.theoutpostforum.com/tof/showthread.php?1495-Demise-of-alienscientist-forum&p=32674&viewfull=1

[642] https://www.youtube.com/watch?v=wg8WFmAJQhg

[643] https://www.youtube.com/watch?v=-qDAXk-3qYQ

[644] https://www.youtube.com/watch?v=UCZ5tfS7cAA

[645] http://www.checktheevidence.co.uk/cms/index.php?option=com_content&task=view&id=159&Itemid=59

[646] https://www.google.co.uk/search?q=invisible%20man&tbm=isch&ei=rq22U4OvKZKg7AaI1IGYBw

[647] https://web.archive.org/web/20140524035131/http:/www.thesleuthjournal.com/us-military-caught-manipulating-social-media-running-mass-propaganda-accounts/

[648] https://www.youtube.com/watch?v=L6ZAN15etZc

[649] https://www.youtube.com/watch?v=J2eyl_vUuSk

[650] http://www.alienscientist.com/fouche.html

[651] http://www.channel5.com/show/conspiracy/

[652] http://www.richplanet.net/starship_main.php?ref=170&part=1

[653] https://www.imdb.com/title/tt1256041/

[654] https://www.st-andrews.ac.uk/history/staff/jerrydegroot.html

[655] mailto:alison.higgins@channel5.com

[656] https://www.bbc.co.uk/programmes/b0752f85

[657] https://ichef.bbci.co.uk/images/ic/336xn/p03ntgb3.jpg

[658] http://tardis.wikia.com/wiki/Peter_Capaldi

[659] https://subsaga.com/bbc/documentaries/science/horizon/2015-2016/2-project-greenglow-the-quest-for-gravity-control.srt

[660] https://www.imperial.ac.uk/centenary/timeline/1960.shtml

[661] http://www.bbc.co.uk/news/magazine-35861334

[662] https://www.imdb.com/title/tt3776826/

[663] https://www.imdb.com/title/tt2718440/

[664] https://www.theverge.com/2016/1/28/10855450/operation-avalanche-review-moon-landing-hoax-sundance-2016

[665] https://www.imdb.com/title/tt0081505/

[666] https://web.archive.org/web/20101123205658/http:/www.jayweidner.com/ShiningSecrets.html

[667] https://www.youtube.com/watch?v=ipRkQFWEvSA

[668] https://www.youtube.com/watch?v=KHfFLB_Yb8g

[669] https://www.imdb.com/title/tt0057012/

[670] https://www.historyonthenet.com/dr-strangeloves-real-life-air-force-general-curtis-lemay/

[671] http://saturndeathcult.com/crimes-of-the-saturn-death-cult/stanley-kubrick-and-the-saturn-death-cult/

[672] https://jaysanalysis.com/2011/01/25/eyes-wide-shut-1999-esoteric-analysis/

[673] https://web.archive.org/web/20151214015936/https:/yournewswire.com/stanley-kubrick-confesses-to-faking-the-moon-landings/

[674] http://www.richplanet.net/starship_main.php?ref=210&part=1

[675] https://www.express.co.uk/news/science/626119/MOON-LANDINGS-FAKE-Shock-video-Stanley-Kubrick-admit-historic-event-HOAX-NASA

[676] https://youtu.be/MraNNzf31gk?t=8m0s

[677] https://www.imdb.com/title/tt0344160/

[678] https://www.lulu.com/shop/albino-galuppini/the-fastest-pen-of-the-west-part-one/paperback/product-23469250.html

[679] http://www.billkaysing.com/notes.php

[680] https://www.imdb.com/title/tt0816692/?ref_=nm_flmg_wr_3

[681] https://jaysanalysis.com/2014/11/12/interstellar-2014-the-secret-revelation/

[682] http://www.checktheevidence.com/cms/index.php?option=com_content&task=view&id=458&Itemid=50

[683] http://www.tellyads.com/play_advert/?filename=TA0828&type=recent

[684] http://www.lyricinterpretations.com/rem/Man-on-the-Moon

[685] http://www.suite101.com/content/the-legendary-life-and-mythical-death-of-andy-kaufman-a32321676

[686] http://www.imdb.com/title/tt0125664/

[687] http://thetruthseekersguide.blogspot.co.uk/2012/06/star-trek-conspiracy-part-one.html

[688] http://whale.to/b/inside_the_lc.html

[689] http://trineday.com/paypal_store/product_pages/sinister_forces_1.html

[690] https://youtu.be/9JLrwAB4DrY?t=40m39s

[691] https://www.youtube.com/watch?v=Wz6NkjxNWJA

[692] https://infogalactic.com/info/LeVar_Burton

[693] http://www.planetary.org/about/board-of-directors/robert-picardo.html

[694] http://memory-alpha.wikia.com/wiki/The_Voyager_Conspiracy_(episode)

[695] https://www.youtube.com/watch?v=xJM6qVlUNM8

[696] https://www.amazon.com/Cold-Fusion-Narrator-James-Doohan/dp/B0007TKNWY

[697] https://trekmovie.com/2009/07/14/jonathan-frakes-narrating-new-science-doc-on-history-channel-clip-from-show-on-cloaking-devices/

[698] https://www.youtube.com/watch?v=nguXc-NpyDI

[699] http://www.starchildproject.com/

[700] http://www.celesteyarnall.com/

[701] http://tinyurl.com/sitsbook

Printed in Great
Britain
by Amazon